Remote Sensing in Mangrove

Remote Sensing in Mangroves

Editor

Chandra Giri

MDPI • Basel • Beijing • Wuhan • Barcelona • Belgrade • Manchester • Tokyo • Cluj • Tianjin

Editor
Chandra Giri
United States Environmental
Protection Agency
USA

Editorial Office
MDPI
St. Alban-Anlage 66
4052 Basel, Switzerland

This is a reprint of articles from the Special Issue published online in the open access journal *Remote Sensing* (ISSN 2072-4292) (available at: https://www.mdpi.com/journal/remotesensing/special_issues/remote_sensing_mangroves).

For citation purposes, cite each article independently as indicated on the article page online and as indicated below:

LastName, A.A.; LastName, B.B.; LastName, C.C. Article Title. *Journal Name* **Year**, *Volume Number*, Page Range.

ISBN 978-3-0365-0850-4 (Hbk)
ISBN 978-3-0365-0851-1 (PDF)

Cover image courtesy of Chandra Giri.

Contents

About the Editor

Chandra Giri received his BS in forest conservation from Tribhuvan University, Nepal, his MS in interdisciplinary natural resources planning and management, and his PhD in Remote Sensing and Geographic Information Systems (GIS) from the Asian Institute of Technology (AIT), Bangkok, Thailand. He is currently a senior scientist and senior advisor at the Office of Research and Development, Environmental Protection Agency (EPA) of the United States. He is also an Adjunct Professor at Nicholas School of Environment, Duke University. Prior to this, he was a research physical scientist at the U.S. Geological Survey (USGS)/Earth Resources Observation and Science (EROS) Center and served as a guest/adjunct faculty at South Dakota State University. Earlier, he had worked for Columbia University's Center for International Earth Science Information Network (CIESIN), United Nations Environment Program (UNEP), AIT, and Department of Forests, Nepal. At EPA, Giri conducts research on the mapping and monitoring of mangrove forests of the world, to identify rates, patterns, causes, and consequences of mangrove change. At EROS, he led the International Land Cover and Biodiversity program. His work focuses on global and continental-scale land use/land cover and characterization and mapping using remote sensing and geographic information systems (GIS). His research in the past two decades has focused on global mangrove forest over mapping and monitoring using earth observation satellite data, and on studying the impact, vulnerability, and adaptation of sea-level rise to mangrove ecosystems, integrating both biophysical and socioeconomic data. He has also been researching the development of remote sensing-based state-of-the-art methodologies to monitor carbon stocks for the Reducing Emissions from Deforestation and Forest Degradation (REDD) initiative. He has experience working in the private sector, academia, government, and international organizations, at national, continental, and global levels, in different parts of the world. He serves as an expert in national and international working groups. He has more than 100 scientific publications to his credit and has received several awards from EPA, USGS, NASA, and other organizations.

Preface to "Remote Sensing in Mangroves"

Mangrove forest characterization, mapping, and monitoring are the most important and typical applications of remotely sensed data. The availability and accessibility of accurate and timely mangrove forest cover datasets play an important role in many global change studies and national applications. Several national and international programs have emphasized the increased need for better mangrove forest cover and mangrove forest cover change information at local, national, continental, and global scales. These programs, such as Group on Earth Observations (GEO), the Blue Carbon Scientific Working Group, the International Geosphere-Biosphere Program (IGBP), U.S. Climate Change Science Program, Land Cover and Land Use Change (LCLUC) program of the National Aeronautics and Space Administration (NASA), Global Land Project, Global Observation of Forest and Land Cover Dynamics (GOFC/GOLD), and Group on Earth Observations (GEO), have been in the forefront of framing scientific research questions on mangrove science.

Recent developments in earth-observing satellite technology, information technology, computer hardware and software, and infrastructure have helped produce mangrove cover datasets of better quality. As a result, such datasets are becoming increasingly available, the user base is ever-widening, application areas are expanding, and the potential for many other applications is increasing. Despite such progress, a comprehensive book, such as "Remote Sensing of Mangroves", on this topic has not been available so far. This book aims at providing a synopsis of basic mangrove cover research questions and highlights remote sensing applications. It also offers an overview of bi-physical parameters such as biomass that can be derived from remotely sensed data.

Examples of application at global, continental, and national scales from around the world have been provided. Overall, the book highlights new frontiers in the remote sensing of mangrove forest cover by integrating current knowledge and scientific understanding and provides an outlook for the future. Specific topics emphasize current and emerging concepts in mangrove forest cover mapping, an overview of advanced and automated mangrove forest cover interpretation methodologies in various parts of the world. The book offers a new perspective on the subject by integrating decades of research conducted by leading scientists in the field.

The book is expected to be a guide or handbook for resource planners, managers, researchers, and students at all levels and a valuable resource for those just starting out in this field or those with some experience in the area of mangrove forest cover characterization and mapping. The book also contains some advanced topics useful for seasoned professionals. It can also be used as a textbook or as reference material in universities and colleges.

Chandra Giri
Editor

Editorial

Recent Advancement in Mangrove Forests Mapping and Monitoring of the World Using Earth Observation Satellite Data

Chandra Giri

Office of Research and Development, United States Environmental Protection Agency, 109 T.W. Alexander Drive, Research Triangle Park, NC 27709, USA; giri.chandra@epa.gov; Tel.: +1-605-274-6655

Citation: Giri, C. Recent Advancement in Mangrove Forests Mapping and Monitoring of the World Using Earth Observation Satellite Data. *Remote Sens.* **2021**, *13*, 563. https://doi.org/10.3390/rs13040563

Received: 24 January 2021
Accepted: 28 January 2021
Published: 5 February 2021

Publisher's Note: MDPI stays neutral with regard to jurisdictional claims in published maps and institutional affiliations.

Mangrove forests are distributed in the inter-tidal region between the sea and the land in the tropical and subtropical regions of the world largely between 30° N and 30° S latitude. The total mangrove forest area of the world in the year 2000 was 137,760 km^2 in 118 countries and territories, accounting for less than 1% of total tropical forests of the world (Figure 1) [1]. Prior to this study, accurate, up-to-date, and reliable information on mangrove distribution was not available. The estimates of world mangroves varied from ~110,000 to 240,000 km^2 [1].

Mangrove forests provide important ecosystem goods and services for human well-being. They are one of the most productive and biologically complex ecosystems in the world. Recent findings suggest that mangroves annually sequester two to four times more carbon compared to mature tropical forests, and store three to four times more carbon per equivalent area than tropic forests.

The protective role of mangrove forests from natural disasters is well recognized. Mangrove forests received special attention after the Asian Tsunami of 2004 and recent natural disasters such as hurricanes and cyclones.

Mangroves are in a constant flux due to both natural and anthropogenic forces. The forests have been declining at a faster rate than inland tropical forests and coral reefs. Anthropogenic causes are responsible for mangrove destruction at present, but relative sea-level rise could be the greatest threat to mangroves in the future. The continued decline of the forests is caused by conversion to agriculture, aquaculture, tourism, urban development, and overexploitation. Predictions suggest that 30–40% of coastal wetlands and 100% of mangrove forest functionality could be lost in the next 100 years if the present rate of loss continues. Therefore, important ecosystem goods and services (e.g., natural barrier, carbon sequestration, biodiversity) provided by mangrove forests will be diminished or lost.

Despite the importance of mangrove forests, reliable, accurate, and timely information on mangrove forests of the world is not available. Mangroves possess a very distinct spectral signature in remotely sensed data, particularly in the spectral range corresponding visible red, near infrared, and mid infrared, thus making it easier to classify compared to other land cover types. Advancement in remote sensing with the availability of higher spatial, spectral, and temporal resolution and availability of historical remote sensing data provides an opportunity to better characterize, map, and monitor mangrove forests.

Recent advancement in remote sensing data availability, image-processing methodologies, computing and information technology, and human resources development have provided an opportunity to observe and monitor mangroves from local to global scales on a consistent and regular basis. Spectral and spatial resolution of remote sensing data and their availability has improved, making it possible to observe and monitor mangroves with unprecedented spatial and thematic detail. Novel remote sensing platforms, such as unmanned aerial vehicles, and emerging sensors, such as Fourier transform infrared spectroscopy and LiDAR, can now be used for mangrove monitoring. Furthermore, it is now possible to store and analyze large volumes of data using cloud computing.

Remote Sens. **2021**, *13*, 563. https://doi.org/10.3390/rs13040563   https://www.mdpi.com/journal/remotesensing

Figure 1. Distribution of the mangrove forests (green) of the world for the year 2000 at 30 m spatial resolution [1].

High quality contributions emphasizing (but not limited to) the topic areas listed below were solicited for the special issue:

- Application of aerial ground remote sensing, photography, multi-spectral, multi-temporal and multi-resolution, satellite data, synthetic aperture radar (SAR) data, hyperspectral data, and LiDAR data.
- Application of advanced image pre-processing for geometric, radiometric, and atmospheric correction, cloud removal, image mosaicking
- Application of advanced image classification and validation techniques including supervised and unsupervised classification
- Application of remote sensing to derive spatio-temporal information on mangrove forests distribution, species discrimination, forest density, forest health, mangrove expansion and contraction, and other ongoing changes in mangrove ecosystems

In the last decade or so, significant improvement has been achieved in terms of remote sensing data availability, classification methodologies, computing infrastructure, and availability of expertise. We now have a large amount of data in need of the integration to answer critical science questions. To accomplish this requires the implementation of automated image pre-processing and classification approaches (Figure 2). At present, not everything can be automated, but many steps including pre-processing that normally constitute 50–60% of project time can be automated.

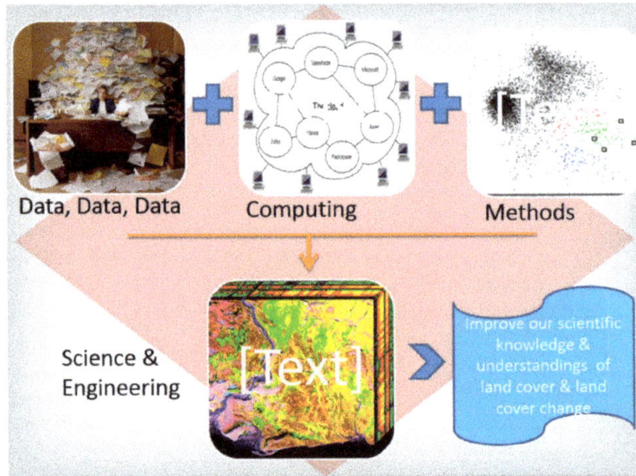

Figure 2. Conceptual diagram of the integration of data, computing, and methods using science and engineering to improve our scientific understanding of mangrove forest cover change.

Pre-processing of satellite data should be centralized, whereas classification and image interpretation can be decentralized (Figure 3). However, there should be an inflow of information from centralized to local levels and vice versa.

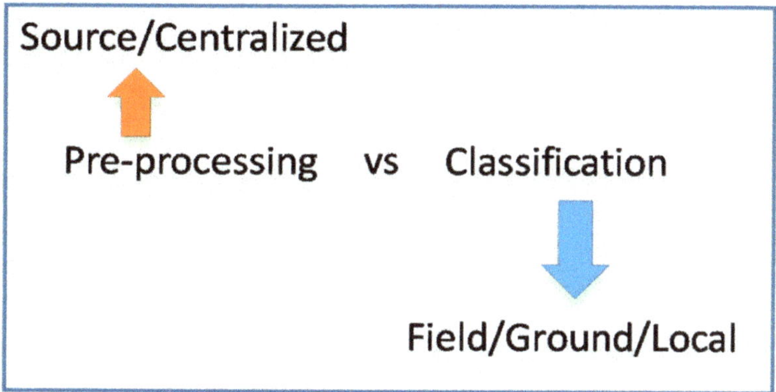

Figure 3. Conceptual framework of pre-processing and image classification showing centralized versus field/ground/local level processing.

The recent trend has been to perform image processing using cloud computing such as Google Earth Engine (GEE) and Amazon Web Services (AWS). Using parallel computing, users will have unlimited computer processing capabilities. Moreover, code and classification algorithms can be shared and discussed in the shared platform. The few disadvantages include a lack of full control of the cloud-computing platform, cost, and the fact that documents are not available or fully explained in some cases.

A brief summary of the twelve papers published in this special issue are presented below.

Younes [2] explored and developed a novel, data-driven approach to extract plant phenology of six different mangrove forests across Australia. They used Landsat imagery and Generalized Additive Models (GAMs) to derive phenology. They found that the Enhanced Vegetation Index (EVI) is directly related to leaf production rate, leaf gain, and net leaf production. Leaf production rate was verified using in-situ data, but the leaf gain and net leaf production was verified using published literature data. The authors also found that the EVI has a two-to-three-month lag time to respond to leaf gain in most cases. The paper concluded that satellite imagery can be useful to better understand mangrove phenology.

In a paper by Yancho et al. [3], a new tool was developed called the Google Earth Engine Mangrove Mapping Methodology (GEEMMM) to map and monitor mangrove forests of the world. The GEEMMM is an "intuitive, accessible and replicable approach" developed primarily for non-remote sensing users including coastal managers and decision makers. The tool was developed in a study conducted in the mangrove forests of Myanmar and is based on cloud computing capabilities GEE. Both qualitative and quantitative accuracy assessment were performed to test the tool. The accuracy assessment shows that the tool is suitable for mangrove mapping and monitoring worldwide. The tool may not be that effective to map large mangrove areas. In addition, internet connectivity may present a challenge in running GEEMMM.

Darmawan et al. [4] monitored the mangrove forests before and after the 1997 forest fire, identified the impact of forest and predicted for the future. The authors used Landsat satellite data acquired in 1989, 1998, 2002, and 2015 and integrated the Markov Chain and Cellular Automata model to compute mangrove forest cover change from 1989 to 2015 in Ambilang National Park Banyuasin Regency, South Sumatra, Indonesia. The change data was used to predict mangrove distribution in 2028. The study showed that approximately 9.6% of mangrove forest in the study area decreased from 1989 to 1998 primarily due to the 1997 forest fire. Mangrove forest has increased by 8.4% from 1998 to 2002, and 2.3% from 2002 to 2015. Future predictions showed continued increase of mangrove forests from 2015 to 2028 ranging from 27.4% to 31%.

Nabab et al. [5], studied the fifth largest mangrove forest of the world in Niger Delta, Nigeria. The forest is under immense pressure from overexploitation and degradation due to the oil and gas industries. The authors mapped the eight main land cover types using Landsat satellite data and L-band radar data of three epochs. They also examined the forest fragmentation of both healthy and degraded mangrove forests. The study concluded that mangrove forests decreased by 500 km^2 while built-up increased by 1740 km^2 from 1988 to 2013. The authors also concluded that the mangrove forests in the study was found to be more fragmented in 2013 compared to 1988. The major challenge in this area however, was the availability of cloud free images.

Toosi et al. [6] examined the applicability of multi-sensor remote sensing data to classify land cover classes in a mangrove ecosystem in Iran. They combined Sentinel-2 and WorldView-2 satellite data and classified eight land cover classes using an upscaling approach. The upscaling approach consists of "(i) extraction of reflectance values from Worldview-2 images, (ii) segmentation based on spectral and spatial features, and (iii) wall-to-wall prediction of the land cover based on Sentinel-2 images." They concluded that the information generated could be useful for the conservation and sustainable management of mangrove forests in Iran.

Quang et al. [7] examined the performance of four different image classification algorithms: Artificial Neural Network (ANN), Decision Tree (DT), Random Forest (RF), and Support Vector Machine (SVM). All four classification approaches are machine learning supervised classification approaches. They used Landsat, SPOT-7 and Sentinel-1 satellite data to classify mangrove forests of Red River estuary of northern, Vietnam. The authors mapped mangrove forest cover change, and age and species composition. The change analysis showed that the mangrove forest area increased from 1975 to 2019 due to successful plantation and forest protection efforts led by local community. The study concluded that SVM was the most accurate classifier out of four classifiers tested. This study concluded that SVM classifier will be valuable for monitoring mangrove plantation projects.

Biswas [8] developed a new method to delineate individual mangrove patches using Aerial Photography with a spatial resolution of 0.08 m, acquired in January 2017. The study was conducted in an area located adjacent the "Everglades National Park, in Florida, USA. This new method utilizes marker-based watershed segmentation. This segmentation methods detects markers using a "vegetation index and Otssu's automatic thresholding". The authors used fourteen vegetation indices. The Vegetation Index Excess Green (ExG) without shadow removal produced the most accurate results to detect individual mangrove patches and to detect individual trees.

Zhu et al. [9] estimated the Aboveground Biomass (AGB) of mangrove plantation forests in China. The authors used optical and radar a data obtained from Chinese satellite and Unmanned Aerial Vehicle (UAV) data. The optical data obtained from Geofen-2 (GF-2), SAR data obtained from Geofen-3 (GF-3), and UAV-based Digital Surface Model (DSM) data were used to estimate AGB of Qi'ao Island, China. Random forest classifier and collected field plot data were used for the classification and results validation. The study showed highest accuracy of AGB estimation when all three optical, SAR, and DSM were used. The lowest accuracy was achieved when only optical data was used, higher accuracy was achieved when both optical and SAR data were used. The paper highlighted the importance of combining multi-source data to improve the classification accuracy.

Hu et al. [10] used a combination of ground inventory data, spaceborne LiDAR, optical imagery, climate surfaces, and topographic maps to produce a global AGB map of the world for the year 2004 at 250-m resolution. Image classification was performed using random forest classification method. Training and validation data were obtained from published literature and free-access datasets. The study concluded that the average global mangrove "AGB density was 115.23 Mg/ha, with a standard deviation of 48.89 Mg/ha".

Total AGB storage of global mangrove forest was 1.52 Pg. This result was comparable to other AGB data of the mangrove forests of the world estimated using remotely sensed data. The new biomass map prepared during this study could help understand the global distribution of AGB at 250 m spatial resolution.

Pham et al. [11] investigated the usefulness of gradient boosting decision tree classification approach to estimate Above-Ground Biomass (AGB) of mangrove forests. This study was conducted in Can Gio Biosphere research in Vietnam. A synergistic use of optical and SAR data and a new gradient boosting regression technique called the extreme gradient boosting regression (XGBR) algorithm. The model results were verified using 121 sampling plot data collected during the dry season. Data fusion techniques were used to handle Sentinel-2 multispectral instrument (MSI) and the dual polarimetric (HH, HV) data of ALOS-2 PALSAR-2. Among all models, the XGBR model was the most accurate. The study demonstrated that the XGBR model and remotely sensed data such as Sentinel-1 and ALOS-2 PALSAR-2 data can accurately estimate the AGB of the study area.

Chamberlain et al. [12] combined remote sensing change analysis approach and conventional method of change detection to detect subtle transformations of land cover modification in a large estuarine region of Queensland, Australia. Landsat satellite data acquired in 2004, 2006, 2009, 2013, 2015 and 2017 were used for the classification and change analysis. Image classification was performed using supervised classification method and Maximum Likelihood clustering algorithm. Post classification change analysis was performed. Results from this study showed a steady decline (1146 ha), of mangrove from 2004 to 2017. They found a decreasing trend in the "vegetation extent of open forest, fringing mangroves, estuarine wetlands, saltmarsh grass, and grazing areas, but this was inconsistent across the study site". Results obtained from this study is expected to be useful to better understand the coastal ecosystem dynamics.

Hauser [13] used cloud computing capabilities of GEE and entire Landsat -7 and Landsat-8 archives to compute spatio-temporal dynamics of mangrove forests and land use changes. This study was conducted in Ngoc Hien District, Ca Mau province in the Mekong Delta of Vietnam. The Classification and Regression Trees (CART) classification method was used to classify (1) dense mangrove forest, (2) sparse mangroves, (3) aquaculture/waterbodies, and (4) built-up and barren lands, land cover classes. The study revealed that the annual rate of deforestation in the study area from 2001 to 2019 was 0.01%. This study contributes to the growing body of literature dealing with dense time series satellite data and cloud computing.

The twelve papers published in this special issue use a wide variety of satellite data and classification approaches to answer important mangrove conservation and management questions. The primary objective is to improve our scientific understanding on the distribution and dynamics of mangrove forests in different parts of the world. These studies help advance our scientific understanding of how various types of remotely sensed data can be utilized with different types of classification approaches to derive meaningful mangrove data and information in support of furthering the science needed to support a global monitoring effort.

Funding: Funding for this research was NASA Land-Cover/Land-Use Change Program, Grant number NNX17AI08G.

Conflicts of Interest: The author declares no conflict of interest.

References

1. Giri, C.; Ochieng, E.; Tieszen, L.L.; Zhu, Z.; Singh, A.; Loveland, T.; Masek, J.; Duke, N. Status and distribution of mangrove forests of the world using earth observation satellite data. *Glob. Ecol. Biogeogr.* **2011**, *20*, 154–159. [CrossRef]
2. Younes, N.; Northfield, T.D.; Joyce, K.E.; Maier, S.W.; Duke, N.C.; Lymburner, L. A Novel Approach to Modelling Mangrove Phenology from Satellite Images: A Case Study from Northern Australia. *Remote Sens.* **2020**, *12*, 4008. [CrossRef]
3. Yancho, J.M.M.; Jones, T.G.; Gandhi, S.R.; Ferster, C.; Lin, A.; Glass, L. The Google Earth Engine Mangrove Mapping Methodology (GEEMMM). *Remote Sens.* **2020**, *12*, 3758. [CrossRef]
4. Darmawan, S.; Sari, D.K.; Wikantika, K.; Tridawati, A.; Hernawati, R.; Sedu, M.K. Identification before-after Forest Fire and Prediction of Mangrove Forest Based on Markov-Cellular Automata in Part of Sembilang National Park, Banyuasin, South Sumatra, Indonesia. *Remote Sens.* **2020**, *12*, 3700. [CrossRef]
5. Nababa, I.I.; Symeonakis, E.; Koukoulas, S.; Higginbottom, T.P.; Cavan, G.; Marsden, S. Land Cover Dynamics and Mangrove Degradation in the Niger Delta Region. *Remote Sens.* **2020**, *12*, 3619. [CrossRef]
6. Bihamta Toosi, N.; Soffianian, A.R.; Fakheran, S.; Pourmanafi, S.; Ginzler, C.; Waser, L.T. Land Cover Classification in Mangrove Ecosystems Based on VHR Satellite Data and Machine Learning—An Upscaling Approach. *Remote Sens.* **2020**, *12*, 2684. [CrossRef]
7. Quang, N.H.; Quinn, C.H.; Stringer, L.C.; Carrie, R.; Hackney, C.R.; Van Hue, L.T.; Van Tan, D.; Nga, P.T.T. Multi-Decadal Changes in Mangrove Extent, Age and Species in the Red River Estuaries of Viet Nam. *Remote Sens.* **2020**, *12*, 2289. [CrossRef]
8. Biswas, H.; Zhang, K.; Ross, M.S.; Gann, D. Delineation of Tree Patches in a Mangrove-Marsh Transition Zone by Watershed Segmentation of Aerial Photographs. *Remote Sens.* **2020**, *12*, 2086. [CrossRef]
9. Zhu, Y.; Liu, K.; Myint, S.W.; Du, Z.; Li, Y.; Cao, J.; Liu, L.; Wu, Z. Integration of GF2 Optical, GF3 SAR, and UAV Data for Estimating Aboveground Biomass of China's Largest Artificially Planted Mangroves. *Remote Sens.* **2020**, *12*, 2039. [CrossRef]
10. Hu, T.; Zhang, Y.; Su, Y.; Zheng, Y.; Lin, G.; Guo, Q. Mapping the Global Mangrove Forest Aboveground Biomass Using Multisource Remote Sensing Data. *Remote Sens.* **2020**, *12*, 1690. [CrossRef]
11. Pham, T.D.; Le, N.N.; Ha, N.T.; Nguyen, L.V.; Xia, J.; Yokoya, N.; To, T.T.; Trinh, H.X.; Kieu, L.Q.; Takeuchi, W. Estimating Mangrove Above-Ground Biomass Using Extreme Gradient Boosting Decision Trees Algorithm with Fused Sentinel-2 and ALOS-2 PALSAR-2 Data in Can Gio Biosphere Reserve, Vietnam. *Remote Sens.* **2020**, *12*, 777. [CrossRef]
12. Chamberlain, D.; Phinn, S.; Possingham, H. Remote Sensing of Mangroves and Estuarine Communities in Central Queensland, Australia. *Remote Sens.* **2020**, *12*, 197. [CrossRef]
13. Hauser, L.T.; An Binh, N.; Viet Hoa, P.; Hong Quan, N.; Timmermans, J. Gap-Free Monitoring of Annual Mangrove Forest Dynamics in Ca Mau Province, Vietnamese Mekong Delta, Using the Landsat-7-8 Archives and Post-Classification Temporal Optimization. *Remote Sens.* **2020**, *12*, 3729. [CrossRef]

 *remote sensing* 

Article

Land Cover Classification in Mangrove Ecosystems Based on VHR Satellite Data and Machine Learning—An Upscaling Approach

Neda Bihamta Toosi [1,2,*], Ali Reza Soffianian [1], Sima Fakheran [1], Saeied Pourmanafi [1], Christian Ginzler [2] and Lars T. Waser [2]

1. Department of Natural Resources, Isfahan University of Technology, Isfahan 84156-83111, Iran; soffianian@cc.iut.ac.ir (A.R.S.); fakheran@cc.iut.ac.ir (S.F.); spourmanafi@cc.iut.ac.ir (S.P.)
2. Swiss Federal Institute for Forest, Snow, and Landscape Research WSL, CH-8903 Birmensdorf, Switzerland; christian.ginzler@wsl.ch (C.G.); waser@wsl.ch (L.T.W.)
* Correspondence: n.bihamtaitoosi@na.iut.ac.ir

Received: 9 July 2020; Accepted: 17 August 2020; Published: 19 August 2020

Abstract: Mangrove forests grow in the inter-tidal areas along coastlines, rivers, and tidal lands. They are highly productive ecosystems and provide numerous ecological and economic goods and services for humans. In order to develop programs for applying guided conservation and enhancing ecosystem management, accurate and regularly updated maps on their distribution, extent, and species composition are needed. Recent advances in remote sensing techniques have made it possible to gather the required information about mangrove ecosystems. Since costs are a limiting factor in generating land cover maps, the latest remote sensing techniques are advantageous. In this study, we investigated the potential of combining Sentinel-2 and Worldview-2 data to classify eight land cover classes in a mangrove ecosystem in Iran with an area of 768 km^2. The upscaling approach comprises (i) extraction of reflectance values from Worldview-2 images, (ii) segmentation based on spectral and spatial features, and (iii) wall-to-wall prediction of the land cover based on Sentinel-2 images. We used an upscaling approach to minimize the costs of commercial satellite images for collecting reference data and to focus on freely available satellite data for mapping land cover classes of mangrove ecosystems. The approach resulted in a 65.5% overall accuracy and a kappa coefficient of 0.63, and it produced the highest accuracies for deep water and closed mangrove canopy cover. Mapping accuracies improved with this approach, resulting in medium overall accuracy even though the user's accuracy of some classes, such as tidal zone and shallow water, was low. Conservation and sustainable management in these ecosystems can be improved in the future.

Keywords: ecosystem; mangrove; random forest; Sentinel-2; upscaling; Worldview-2

---

1. Introduction

Mangrove forests are considered one of the most important ecosystems on the earth. They occur in the inter-tidal zones along coasts in most tropical and semi-tropical areas [1,2]. Despite the large ecological benefits of mangrove forests, such as carbon sequestration, protection of land from erosion, purification of coastal water quality, and maintenance of ecological balance and biodiversity, mangroves have been destroyed worldwide as a result of climate change and human activities [3–6].

Qeshm Island, located off the southern coast of Iran in the Persian Gulf, is dominated by the cosmopolitan mangrove species *Avicennia marina*. Many studies have focused on the ecological and physiological characteristics of *A. marina* [7,8]. *Avicennia* species grow in oxygen-poor sediments that cannot supply the underground roots with sufficient oxygen. Consequently, their root system also includes vertically growing aerial roots (pneumatophores). These aerial roots also anchor the plants during the frequent inundation with seawater in the soft substrate of tidal systems, and they play a significant role in sustaining mangroves [9]. Sea-level rise, a main consequence of climate change, will have a significant influence on future growing conditions [10]. Recent estimates of the extent of mangrove forests indicate that their total area has already decreased substantially, by 50% during the last half-century [11–13].

Identification of the aerial root system at a high spatial resolution would enable efficient planning of reforestation in mangrove ecosystems, but this detailed information is currently missing. Image resolution is directly correlated with the ability to identify objects of the same type [14]. Despite the great value of Landsat images for numerous applications, the specifications are inappropriate for distinguishing mudflats with aerial roots from mudflats without aerial roots. This is also due to the spectral similarities of these classes and the influence of the soil in the tidal zone (dry and wet conditions). A more detailed mapping of the mangrove ecosystem, e.g., trees and aerial root systems, is required to improve assessments of their status and recommend appropriate protection measures.

In the last years, a range of low- to high-resolution aerial images [15–17], hyperspectral images [18], Synthetic Aperture Radar (SAR) data [19], and Light Detection and Ranging (LiDAR) data [20] has been used to map the extent and distribution of mangrove cover classes. In the past decade, data have become available from Very High-Resolution (VHR) satellites, such as Worldview-2 and Pléiades-1, leading to improved mapping of mangrove cover classes [21,22]. However, the main limiting factor is the high cost of data acquisition. Consequently, alternatives have been investigated, in particular combining satellite data of different spatial resolutions [23]. Only recently, studies focusing on the use of freely available VHR data have been completed [24–28]. For example, in the forestry sector, a combination of commercially available Worldview-2 (WV-2) images and Landsat time-series data has been used to map tree species [29]. Different classification techniques, such as traditional statistical regression [30], machine learning [31], artificial neural networks [32], and tree-based methods [33,34], have successfully been used with a large geographic extent and high level of detail.

Machine learning techniques such as Random Forest (RF), artificial neural networks (ANN) and Support Vector Machine (SVM) have gained exceptional attention to classify Land cover/Land use and identify mangrove forests because they perform better than traditional techniques [33,34]. These techniques use algorithms to learn the relationship between a response and its predictors and have been categorized into two sub-types: supervised and unsupervised techniques, respectively [35]. A main advantage is that they are all nonparametric classification techniques that require no assumptions about the distribution of the data and thus no prior knowledge about the characteristics of feature data is needed either [31]. Many studies in the field of Land cover/Land use classification have been carried out using different machine learning algorithms as well as comparing them among each other [35]. In the last decade, RF has recently become preferred for mapping land cover classes in several realms [36,37]. RF is a nonparametric technique based on a set of decision trees. Unlike parametric techniques, RF can be used to predict land cover classes even based on a small sample size and therefore reduces both cost and time [38]. Moreover, embedded feature selection in the model generation process makes it possible to obtain high mapping accuracy. Several studies have demonstrated that RF, in combination with satellite data (Landsat) [37] and a high spatial resolution [16], can be used to successfully map mangrove cover classes. Moreover, the latest advances in remote sensing data and techniques, i.e., increasing availability of datasets in combination with higher temporal, spatial and spectral resolutions (e.g., ESA Copernicus Program Sentinel-1/-2), enable improved characterization of mangrove ecosystems. They make it possible to derive leaf area index, height and biomass, map the mangrove forest extent, and monitor mangrove status over time [39]. Several studies have been carried out to explore satellite data of different spatial resolutions for improving land cover maps, i.e., in forestry that have combined data sets from Landsat and AVHRR [40] or Landsat and MODIS [41]. However, to the best of our knowledge, no study exist that combine Worldview-2 and Sentinel-2 images to classify mangrove ecosystems in greater detail which is a prerequisite for managing this ecosystem. Therefore, freely available Sentinel-2 data, in combination with commercially available high-spatial-resolution imagery, has great potential for mapping wall-to-wall mangrove cover at a high level of detail, i.e., distinguishing between land cover classes with similar spectral properties.

In the present study, we investigated whether the combination of Sentinel-2 and Worldview-2 imagery can be used to accurately map the most relevant land cover classes for mangrove ecosystem management. We developed a three-step approach: (i) extraction of reflectance values from high-resolution Worldview-2 imagery, (ii) segmentation based on spectral and spatial features, and (iii) wall-to-wall mapping of the eight land cover classes based on Sentinel-2 imagery.

The study aims at developing a cost-effective, accurate method that can be applied widely and in a standardized manner, particularly when field surveys are restricted.

2. Materials and Methods

In order to produce a wall-to-wall map of mangrove cover classes for Qeshm Island, a two-step method was applied: (i) Reference data were generated at a 0.5-m spatial resolution using an object-based method performed on Worldview-2 images. The Worldview-2 data were dispersed across the entire study area and covered 27% of the total land cover. (ii) Reference data based on Worldview-2 images were used for the upscaling.

2.1. Study Area

Qeshm Island is located a few kilometers off the southern coast of Iran, opposite the port cities of Bandar Abbas and Bandar Khamir. It is the largest island in the Persian Gulf and covers an area of 1491 km^2 (Figure 1). Most of the mangrove forests of Qeshm are located in the northern part of the island in the Hara Protected Area, a biosphere reserve that covers an area of approximately 20 by 20 km and is characterized by numerous tidal channels [42]. The mangroves are rooted in the saltwater of the Persian Gulf, but the special pores within their leaves extract the salt from the water. The whole forest area is affected by frequent boat trips, fishing and a small amount of leaf-cutting for livestock feed. The forests are the habitat for migratory birds, hooked turtles and venomous aquatic snakes, all of which are indigenous species.

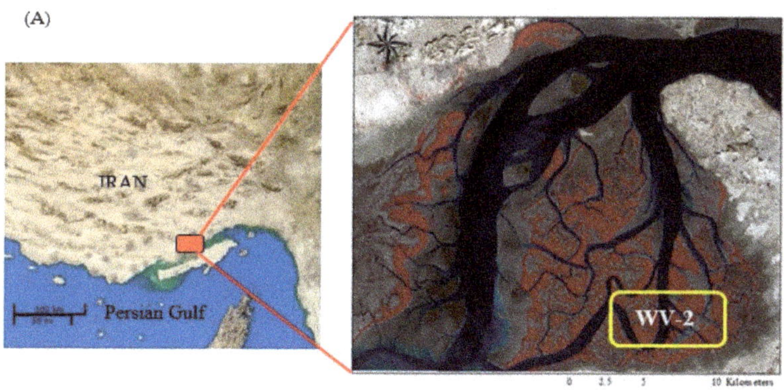

Figure 1. *Cont.*

Figure 1. (**A**) Left: location of Qeshm Island and the mangrove ecosystem, shown as a false-color Sentinel-2B image (2017, Combination of Bands 8-4-3); right: Worldview-2 image data used for the upscaling approach. (**B**) Aerial roots (pneumatophores) growing in a wide radius around the mangrove (*Avicenna marina*) are highlighted by the red polygon.

2.2. Field Data

The field survey revealed that *Avicenna marina* was the dominant mangrove species on Qeshm Island. Visual analysis of high-resolution images made it possible to distinguish between eight target classes of mangrove ecosystem, including three types of mangrove spatial pattern: closed canopy mangrove, open canopy mangrove, and individual mangrove trees (found in a small patch on the island). The remaining target classes in the study area were mudflat (either with or without aerial roots), tidal zone (sand, beaches or unvegetated area), shallow water (rivers or ponds), and deep (open) water.

During the field survey, a total of 170 GPS reference points (Garmin 629sc with spatial accuracy between 1 and 5 m) were collected and used for validation of the classification of the eight land cover classes. In order to minimize and avoid the negative impacts on the vulnerable ecosystem, the collection of field samples was restricted to easily accessible parts. In order to increase the number of samples for three types of mangrove and two types of mudflat, 53 points were additionally selected from Spot 6/7 data using image interpretation. Figure 2 shows the distribution of the samples for the eight land cover classes. The set of reference points collected from both GPS and from the Spot images are depicted for each class separately in Table 1.

Table 1. Overview of the two different sets of reference points collected from the GPS survey and the Spot 6/7 image interpretation.

	Land Cover Class								
Source of Reference Points	Closed Mangrove Cover	Open Mangrove Cover	Individual Mangrove Trees	Mudflats	Aerial Roots	Tidal Zone	Shallow Water	Deep Water	Total
GPS	5	12	0	27	7	6	7	28	92
Spot 6/7 images	12	15	11	15	25	0	0	0	78
Total	17	27	11	42	32	6	7	28	170

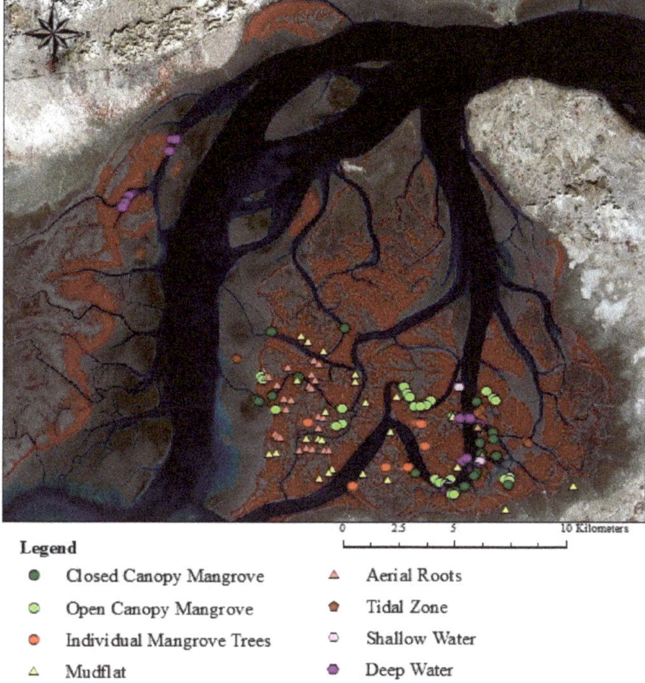

Figure 2. Distribution of the samples obtained from the field survey and from Spot 6/7 image interpretation of the whole study area.

2.3. Remote Sensing Data and Pre-Processing

Technical specifications of the Worldview-2 and Sentinel-2 imagery are given in Table 2. Images were cloud-free over coastal areas. The multispectral bands of Worldview-2 consist of four standard bands (red, green, blue and near-infrared 1) and four additional bands (coastal, yellow, red edge and near-infrared 2), which facilitated spatial and spectral analysis, mapping and monitoring of large areas at a more detailed level [43]. Sentinel-2 bands consist of four bands at a 10-m spatial resolution (blue, green, red and near-infrared), six bands at a 20-m spatial resolution (four narrow bands near the red edge and two wider SWIR), and three bands at a 60-m spatial resolution (aerosols, water vapor and cirrus) [44]. The obtained data were pre-georeferenced to the UTM zone 40 North projection using the WGS-84 datum. Sentinel-2 data were radiometrically calibrated to apparent surface reflectance by the FLAASH (Fast Line-of-sight Atmospheric Analysis of Hypercubes) atmospheric corrected algorithm [45] in ENVI 5.4 software. Fusion of panchromatic with multispectral images of Worldview-2 data resulted in an image with a 0.5-m spatial resolution. In the present study, the Gram Schmidt pan-sharpening algorithm was applied [46] because it preserves the primary spectral value of the objects and has successfully been applied to multispectral images. In this study, a Sentinel-2 level 1C product image was applied, acquired on a clear day and under the lowest tide condition over Qeshm Island.

2.4. Spectral Variability

VHR images show the required details of the mangrove ecosystem. Therefore, the Worldview-2 image was used to select the eight targeted land cover classes: (1) closed canopy mangrove, (2) open canopy mangrove, (3) individual mangrove trees, (4) mudflats, (5) aerial roots, (6) tidal zone, (7) shallow water, and (8) deep water. In order to better separate them and distinguish between the spectral signatures, based on the field survey and image interpretation, reflectance values of the target classes (by 100 points) were extracted from Worldview-2 image bands. The boxplots in Figure 3 show that the two classes closed canopy mangrove and open canopy mangrove are clearly distinguished by the blue band and the yellow band. Moreover, it shows that the aerial roots are clearly distinguished from the mudflats in the green, yellow and red bands. Figure 4 shows the reflectance values of the eight land cover classes for the Sentinel-2 bands.

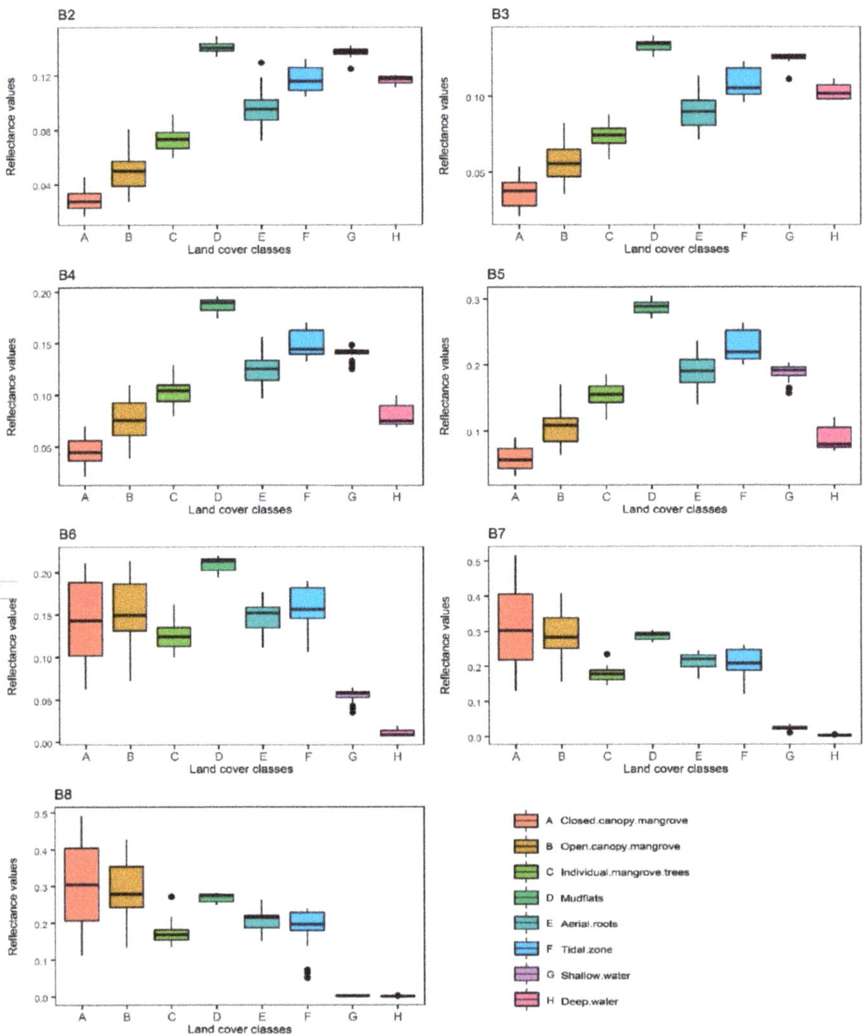

Figure 3. Reflectance values of the eight land cover classes for each Worldview-2 band: (**B2**) (Blue: 450–510 nm), (**B3**) (Green: 510–580 nm), (**B4**) (Yellow: 585–625 nm), (**B5**) (Red: 630–690 nm), (**B6**) (Red edge: 705–745 nm), (**B7**) (Near-infrared 1: 770–895 nm), and (**B8**) (Near-infrared 2: 860–1040 nm). The letters A to H show the land cover classes namely closed canopy mangrove, open canopy mangrove class, individual mangrove trees, mudflats, aerial roots, tidal zone, shallow water, and deep water, respectively.

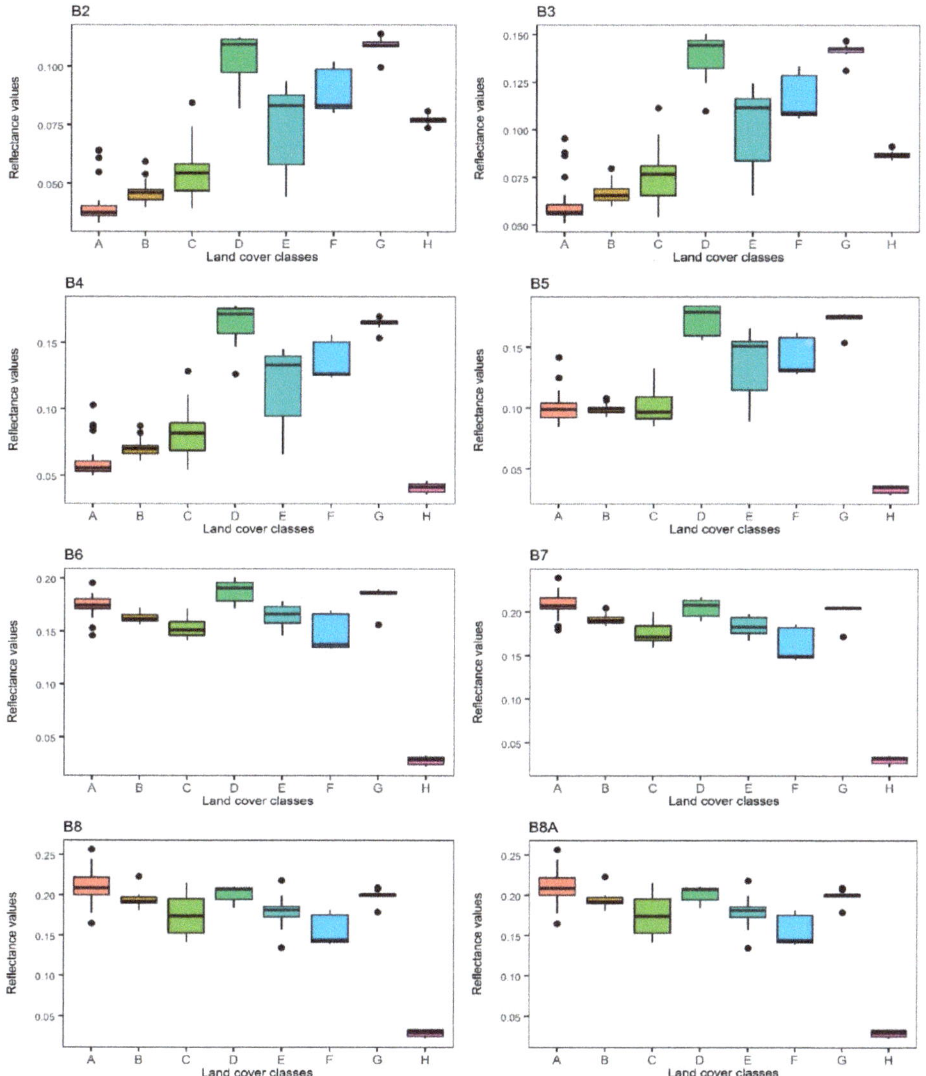

Figure 4. Reflectance values of the eight land cover classes for Sentinel-2: (**B2**) (Blue band 490 nm), (**B3**) (Green band 560 nm), (**B4**) (Red band 665 nm), (**B5**) (Vegetation Red Edge band 705 nm), (**B6**) (Vegetation Red Edge band 740 nm), (**B7**) (Vegetation Red Edge band 783 nm), (**B8**) (Near-infrared band 842 nm), and (**B8A**) (Vegetation Red Edge band 865 nm). The letters A to H show the land cover classes namely closed canopy mangrove, open canopy mangrove class, individual mangrove trees, mudflats, aerial roots, tidal zone, shallow water, and deep water, respectively.

Table 2. Sensor specifications of the Worldview-2 and Sentinel-2 imagery.

Sensor	Worldview-2	Sentinel-2
Acquisition date	26.12.2016	02.12.2017
Bands	8 multispectral 1 panchromatic	13 multispectral
Spatial resolution	2 m 0.5 m	10 m (bands: 2, 3, 4, 8) 20 m (bands: 5, 6, 7, 8A, 11, 12) 60 m (bands: 1, 9, 10)
Dynamic range	11 bits	12 bits
Swath width	16.4 km at nadir	290 km
Revisit time	1.1 day	10 days

2.5. Reference Data

The sampling of reference data used Object-Based Image Analysis (OBIA), which is based on segmentation [34,47]. The multi-resolution segmentation algorithm from eCognition 9.2 software (Trimble Inc., Munich, Germany) [48] was used, which classifies homogeneous image objects by using attributes of image objects rather than the attributes of individual pixels or a hierarchical object-oriented approach using a knowledge base. In the present study, a series of scale parameters, shape and compactness (from low to high) were tested to control the size of segmentation. In order to generate reliable reference samples, information from the Normalized Difference Vegetation Index (NDVI) layer and the Moran Index using the Worldview-2 bands was additionally included for image segmentation. In previous studies, NDVI has been successfully applied to display and quantify mangrove forest changes [12,49,50]. NDVI values were computed as:

$$NDVI = \frac{NIR - Red}{NIR + Red} \tag{1}$$

where NIR is band 8 and Red is band 5.

The Moran index provides the correlation between attributes at each location in a study area and the statistical mean of the values from neighboring locations. The Moran index has successfully been applied in almost all studies dealing with spatial autocorrelation (for a review see [51]). It evaluates the magnitude of homogeneity of a target image object to other objects surrounding it. If targets are attracted to (or repelled from) each other, the observations are dependent [52]. In addition, the Moran Index is similar to correlation coefficients and its value ranges from -1 to 1 [53]. Moreover, the Moran index provides quantitative clustering information that is used to select homogeneous regions. The Moran index measures the degree of spatial auto-correlation at each particular location [54]. Information and photos from the field observations, as well as a visual interpretation of Worldview-2 images, were used to develop the rule sets to select segmentations for each class as reference data (Ground Truth or OBIA training). In order to use spectral features (mean and standard deviation of blue, yellow, red edge bands and NDVI), additional geometric features such as shape and extent were used. The total number of variables selected was based on visual inspection of the reflectance values of the eight classes. The feature selection process was completed with the eCognition feature optimization tool using 100-point datasets.

2.6. Upscaling by Reference Data

After the generation of the reference data, RF was used to classify Worldview-2 and Sentinel-2 images. In this step, Sentinel-2 imagery was preliminarily mapped over the same extent as the Worldview-2 image with 70% of the reference data. The accuracy of the map of the RF algorithm was then checked, and the reference data were used for mapping mangrove classes to a larger extent. A layer stack was created from the NDVI, blue, green, red and near-infrared bands. Sentinel-2 data (10 m spatial resolution) served as input for the RF classification. RF was performed using the *Ranger*

Package in the R statistical software [55]. Figure 5 shows the main steps of the classification approach applied in this study.

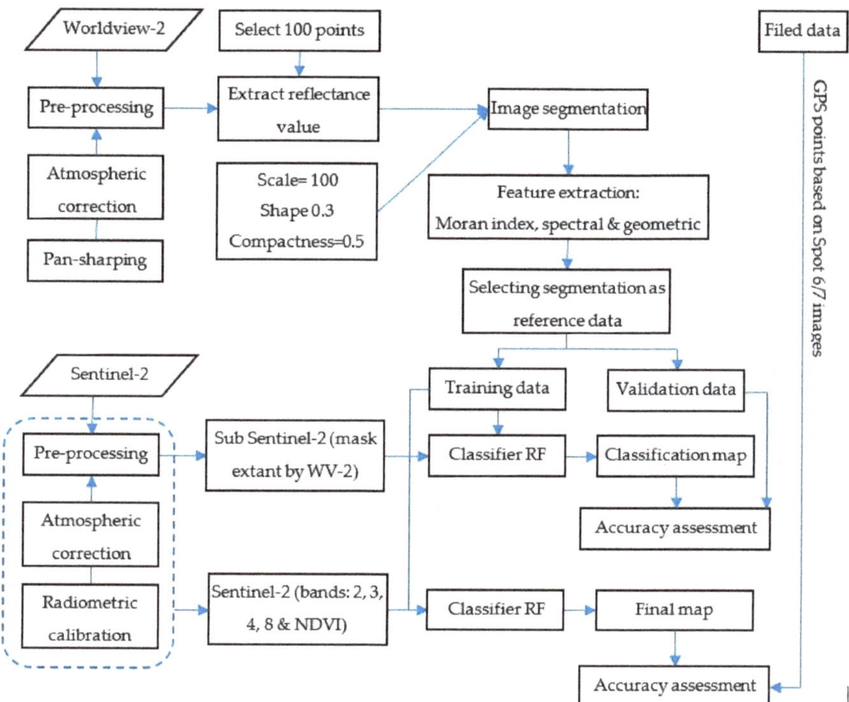

Figure 5. Flow chart of the upscaling approach for mapping land cover in mangrove ecosystems.

2.7. Accuracy Assessment

The land cover map based on the classification using Sentinel-2 images (same extent as Worldview-2) was assessed using 30% of the reference data, which was excluded from classification and from cross-validation. To assess the accuracy of the land cover map based on the upscaling approach, a confusion matrix was constructed, consisting of 167 validation points collected during the field survey, for image interpretation using the Spot 6/7 images. We used a leave-one-out cross-validation [56] because our sample was relatively small (30% of the training data did not cover the land cover map to a large extent).

This matrix provides the overall accuracy, the kappa coefficient, and the user's and producer's accuracies for each class. The producer's accuracy represents how well reference pixels of the ground cover type are classified. The validation points were rasterized to the 10-m resolution of the Sentinel-2 image. Furthermore, a Wilcoxon test (non-parametric statistical test that compares two paired groups) was applied in order to estimate the significance difference between the user's and producer's accuracies for the two classification maps [57].

3. Results

The mapped land cover classes of the Qeshm Island mangrove ecosystem are given in Figure 6.

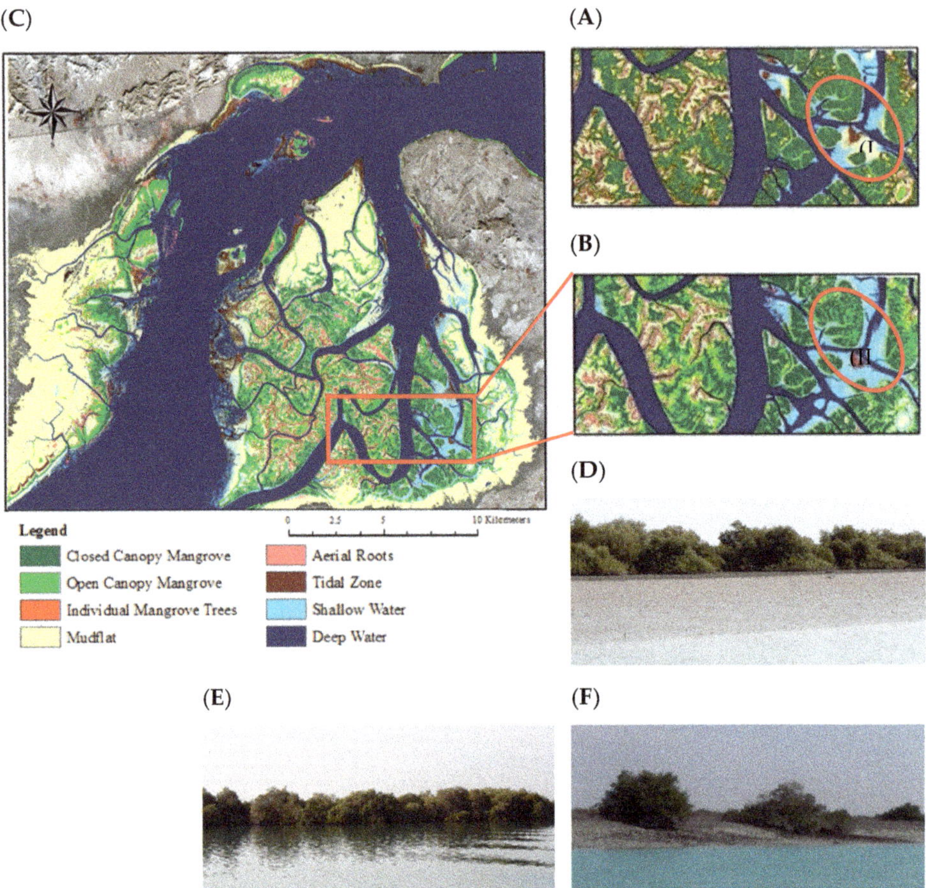

Figure 6. Classification map of the Worldview-2 image (**A**), classification map of the Sentinel-2 image with the same extent as the Worldview-2 image (**B**), and the final map based on the upscaling approach (**C**). The visible differences between (I) and (II) are related to misclassified shallow water. This error happened two reasons: First, the spectral profiles of the shallow water and tidal zone classes were similar in the Sentinel-2 image. Second, the date of the images differed, and the relative sea level rise had acted as an important factor in converting the tidal zone class (**D**) to shallow water (**E**). However, we were able to show more details of mangrove ecosystems with this approach, such as individual trees and aerial roots (**F**).

Model accuracies of the RF classification were assessed in two steps. In the first step, random reference data based on the segmentation of Worldview-2 images was used to validate the subset of Sentinel-2 imagery. An overall accuracy of 88% and a kappa coefficient of 0.85 were obtained. The validation revealed the producer's accuracy of the four classes shallow water (96.5%), deep water (94.8%), closed canopy mangrove (89.2%), and mudflat (83.1%) (Figure 7). In the second step of the validation, the overall accuracy of the upscaling approach was calculated at 65.5% and the kappa coefficient was 0.63. Whereas the user's accuracy for the two classes deep water (100%) and closed canopy mangrove (75.1%) was high, the producer's accuracy for the class mudflat with aerial roots (66.1%) and without aerial roots (73.3%) were lower (Figure 8). These two classes included a corollary omission error of 33.9% and 26.7%, respectively. The results of the confusion matrix of the different classification extents are given in Tables 3 and 4.

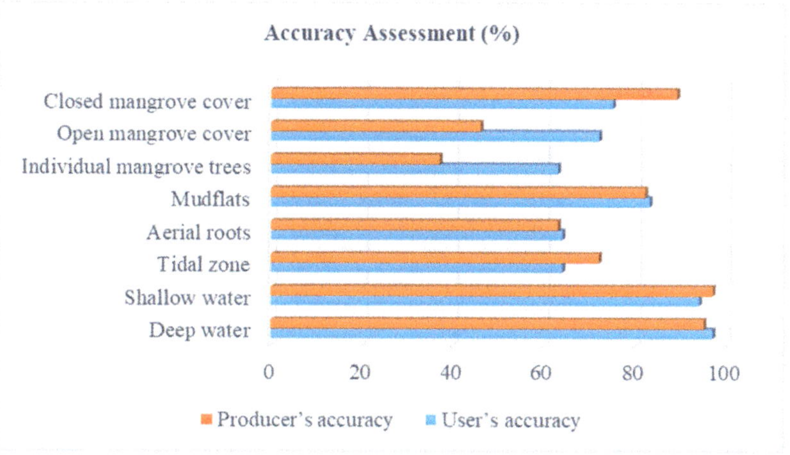

Figure 7. Accuracy statistics of the classification map of Sentinel-2 over the same extent as for Worldview-2.

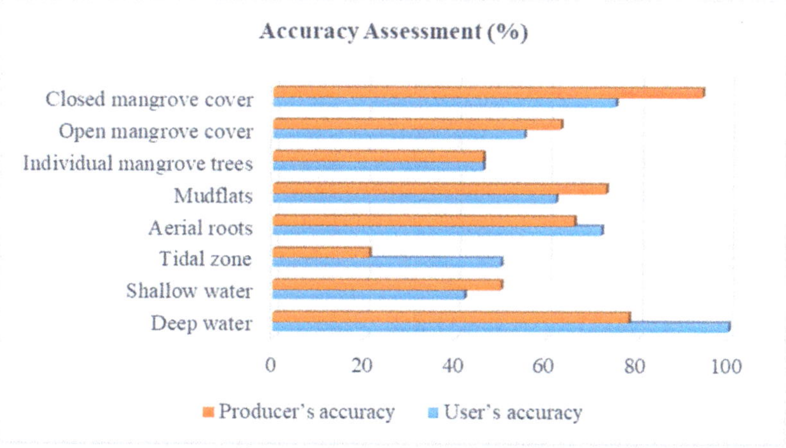

Figure 8. Accuracy statistics of the classification map of the upscaling approach.

Table 3. Confusion matrix for the classification map of Sentinel-2 over the same extent as for Worldview-2. Bold-faced numbers indicate the agreement between a class.

Classification	Reference Data							
	Closed Mangrove Cover	Open Mangrove Cover	Individual Mangrove Trees	Mudflats	Aerial Roots	Tidal Zone	Shallow Water	Deep Water
Closed mangrove cover	**1102**	119	235	1	3	0	0	2
Open mangrove cover	73	**696**	122	4	61	5	1	1
Individual mangrove trees	61	59	**23**	1	6	4	1	2
Mudflats	0	4	0	**991**	90	113	2	0
Aerial roots	0	160	19	38	**682**	163	2	0
Tidal zone	0	28	3	158	232	**738**	3	0
Shallow water	0	6	0	4	7	2	**1049**	51
Deep water	0	0	0	0	0	0	28	**1042**

Table 4. Confusion matrix for classification map of the upscaling approach. Bold-faced numbers indicate the agreement between a class.

Classification	Reference Data							
	Closed Mangrove Cover	Open Mangrove Cover	Individual Mangrove Trees	Mudflats	Aerial Roots	Tidal Zone	Shallow Water	Deep Water
Closed mangrove cover	**15**	2	0	0	0	0	0	0
Open mangrove cover	0	**15**	1	2	1	5	3	0
Individual mangrove trees	1	1	**5**	3	1	0	0	0
Mudflats	0	1	5	**22**	10	3	0	1
Aerial roots	0	2	0	2	**23**	4	0	1
Tidal zone	0	0	0	1	0	**3**	0	2
Shallow water	0	0	0	0	0	0	**3**	4
Deep water	0	0	0	0	0	0	0	**28**

The classification revealed that the largest area (27,678 ha) belongs to the class deep water and smallest (62 ha) to the class individual mangrove trees. The classes closed canopy mangrove, open canopy mangrove, mudflat, aerial root and tidal zone cover an area of 4857, 3474, 13,099, 2296, and 2026 ha, respectively. The p-value of the Wilcoxon test for differences in the user's and producer's accuracies between the two classification maps were 0.11 and 0.32, respectively, which is greater than the significance level alpha = 0.05. We can conclude that the accuracy assessments did not differ significantly between the two classification maps.

4. Discussion

4.1. General Comments

Mangrove forests typically grow in zones that are marshy and inaccessible [11]. Therefore, collecting GPS points as training data through field surveys is difficult [14]. Nowadays, new developments in remote sensing techniques have great potential to overcome the problem of acquiring field data in inaccessible areas of mangrove ecosystems [58]. Between 1970 and 2018, approximately 435 studies mapping the area of mangroves were conducted, and after the year 2000 the majority used Landsat images [14]. While Landsat imagery has the advantages of free availability, a large archive and extensive coverage, its relatively coarse spatial resolution of 30 m can be a major limitation. The potential of different datasets from Landsat, ALOS AVNIR-2, Worldview-2 and LIDAR to map a detailed land cover of mangrove ecosystems was recently evaluated [59]. The results clearly demonstrated the importance of a higher spatial resolution for mapping specific mangrove features, such as individual tree crowns and species communities.

With the present study, we contribute to this research with an efficient mapping of mangrove features using multi-resolution datasets. We add to existing knowledge gained in a previous study [37], which focused on comparing four classification algorithms based on Landsat images for predicting six land cover classes in the mangrove ecosystem: mangrove forest, mud flat, other land cover, tidal zone, water and settlement. The results of this earlier research demonstrated that using Landsat data enables to potentially distinguish between different mangrove forest stands and can be useful for detecting their changes over time. However, since mangrove forests usually consist of small patches, Landsat images are not suitable for extracting more details and are mainly only appropriate for detecting changes in mangrove forest canopies. This is in accordance with [14,59], in that only high-resolution images can be used to map more detailed land cover classes. By increasing the number of spectral bands and the spatial resolution, it is possible to discriminate between small objects and to detect small objects, such as individual trees and mudflats with aerial roots. Several studies have shown the potential of Worldview-2 data for detailed land cover mapping, including mangrove forest ecosystems [16,39,59–61]. However, the main reason for the limited use of such imagery is its high cost–in particular, for developing countries. Thus, in our study, an upscaling approach was applied that reduces costs while still enabling the generation of a more detailed map of land cover classes.

4.2. Modelling Approach

In the last decade, several studies have been carried out combining satellite data of different spatial resolutions to improve land cover maps in the forestry sector. Some investigations have considered the combination of Landsat data with datasets of higher spatial resolutions such as IKONOS [62], GeoEye-1 data [63] or Worldview-2 [29].

Comparison of the two confusion matrices clearly demonstrated that the accuracy and kappa of the upscale approach were lower than the accuracy and kappa of the map that had the same extent as the one based on the Worldview-2 imagery. The use of a large amount of reference data to predict the subset of Sentinel-2 data helped to reduce misclassification.

The confusion matrix of the upscaling approach (Table 4, Figure 8) indicates that the overall accuracy and kappa decreased with increasing map scale. There was a high incidence of misclassification of individual trees and tidal zone when Sentinel-2 data were used. Several possible reasons for this error exist. First, it might be due to the amount of reference data because the Worldview-2 data only cover about 27% of the Sentinel-2 image. On the other hand, in the Worldview-2 image, the area of these two classes is less than that of the other classes. It is well known that the number of reference samples from the Worldview-2 image affects classification accuracy. In a recent study, it was demonstrated that the large amount of reference data obtained from the Worldview-2 image was the main driving factor for the accuracy of the classification of two pine tree species by Landsat data [29]. Future work could include the collection of more training samples in order to further improve the distinction of these land cover classes. Second, the error could be a result of the similarity of the spectral profile of individual trees and open canopy mangrove forest. The use of fewer reference samples decreases the spectral separability of classes and potentially decreases the accuracy. Third, the decrease in accuracy could be related to the level pre-processing and viewing geometry of Sentinel-2 imagery.

Nevertheless, the present study demonstrates that areas with different canopy densities and mudflat areas (occurrence of aerial root systems) can be accurately classified using the upscaling approach with Sentinel-2 images. Overall, high accuracies were obtained for mapping closed canopy mangrove (75% user's accuracy, 94% producer's accuracy) and aerial roots (72%, 66%). Moreover, the combined use of Worldview-2 and Sentinel-2 images further increases map accuracies–in particular when the overall accuracy is not very high, and the user's accuracy is low in problematic classes.

4.3. Importantance of Mapping of Detailed Information on Mangrove Forests

Detailed maps of mangrove ecosystems are a prerequisite for successful protection and management. Since mangroves occur in areas with a high salt concentration in the soil, they have developed aerial roots for physiological functions and cover a large area within the Hara Protected Area [64]. This specialized root system reduces the power of sea waves and guarantees sustainable establishment of mangrove communities, as well as providing a protected place for aquatic animals [42]. In order to plan the development of mangrove forests, both naturally or artificially, the selection of potential suitable land is relevant. The land areas on the map that show the mangrove forests and mudflat with aerial roots are preferred to other areas that are not covered by vegetation. Moreover, the occurrence of mangrove is an indication that the land provides optimal conditions for the development of mangrove forests in terms of soil parameters such as salinity and pH. Mapping the details of mangrove ecosystems is an effective way to visualize, evaluate and better understand mangrove ecosystem development. Changes over a long period, as well as the recognition of unexpected changes due to natural or dramatic anthropogenic impacts, can be assessed at an early stage [65,66]. Moreover, assessing changes in the aerial root area can indicate the status of these forests because these roots are destroyed by an increase in water level or sediments.

5. Conclusions

In the present study, we demonstrate that field surveys in mangrove ecosystems are not always feasible, due to the high costs and inaccessibility of the area. Mangrove distribution mapping is a hot topic in the field of mangrove remote sensing [14]. Based on field observations, the mangrove forests in the present study have a uniform composition of the species *Avicenna marina* and the detectable differences are limited to canopy density, which consists of mangrove zonation patterns including forests of immature trees and of mature trees, and isolated trees. The use of VHR satellite imagery for sampling reference data in combination with freely available satellite data and machine learning is an effective and straightforward approach to further improve the details of land cover maps and to assess relevant forest parameters. Upscaling is a cost-efficient tool for producing accurate large-scale land cover maps in inaccessible ecosystems. The findings of the present study support the sustainable management of mangrove ecosystems and can be used to assess the efficiency of ecosystem services. Although the upscaling approach produced low user accuracies for the shallow water and tidal zone classes, overall accuracies were generally high.

With the proposed method, it is possible to distinguish between the two most relevant classes for management, i.e., canopy mangrove canopy and mudflat. Our findings confirm that advances in remote sensing data and techniques are favorable for developing novel methods to map mangrove ecosystems in greater detail. We conclude that the selection of appropriate images remains an important factor and that Sentinel-2 images have great potential for identifying different land cover types, thanks to their high spatial, temporal and spectral resolution. Continuity of the presented approach is guaranteed since Sentinel-2 data will be continuously acquired.

Author Contributions: N.B.T. is responsible for the study. N.B.T. developed the methodology, programming, statistical analysis and was the main writer of the manuscript. L.T.W. supported the development of the approach and methodology and writing the manuscript. C.G. supported the development of the approach and methodology. A.R.S., S.F. and S.P. conceived and provided useful suggestions to the manuscript. All authors have read and agreed to the published version of the manuscript.

Funding: This research received no external funding.

Acknowledgments: This study was carried out in the framework of the doctoral thesis and was supported by the Department of Natural Resources, Isfahan University of Technology, Isfahan, Iran. We thank Melissa Dawes for professional language editing.

Conflicts of Interest: The authors declare no conflict of interest.

References

1. Barbier, E.B.; Cox, M. Does economic development lead to mangrove loss? A cross-country analysis. *Contemp. Econ. Policy* **2003**, *21*, 418–432. [CrossRef]
2. Alongi, D.M. Present state and future of the world's mangrove forests. *Environ. Conserv.* **2002**, *29*, 331–349. [CrossRef]
3. Mazda, Y.; Magi, M.; Kogo, M. Phan Nguyen Hong Mangroves as a coastal protection from waves in the Tong King Delta, Vietnam. *Mangroves Salt Marshes* **1997**, *1*, 127–135. [CrossRef]
4. Donato, D.C.; Kauffman, J.B.; Mackenzie, R.A.; Ainsworth, A.; Pfleeger, A.Z. Whole-island carbon stocks in the tropical Pacific: Implications for mangrove conservation and upland restoration. *J. Environ. Manag.* **2012**, *97*, 89–96. [CrossRef] [PubMed]
5. Cornforth, W.A.; Fatoyinbo, T.E.; Freemantle, T.P.; Pettorelli, N. Advanced land observing satellite phased array type L-Band SAR (ALOS PALSAR) to inform the conservation of mangroves: Sundarbans as a case study. *Remote Sens.* **2013**, *5*, 224–237. [CrossRef]
6. Hamdan, O.; Khali Aziz, H.; Mohd Hasmadi, I. L-band ALOS PALSAR for biomass estimation of Matang Mangroves, Malaysia. *Remote Sens. Environ.* **2014**, *155*, 69–78. [CrossRef]
7. Jin-Eong, O. The ecology of mangrove conservation & management. *Hydrobiologia* **1995**, *295*, 343–351. [CrossRef]
8. Barlyn, G.P. The botany of mangroves. *Science* **1986**, *234*, 373–374. [CrossRef]

9. Vannucci, M. What is so special about mangroves? *Braz. J. Biol.* **2001**, *61*, 599–603. [CrossRef]

10. Lovelock, C.E.; Cahoon, D.R.; Friess, D.A.; Guntenspergen, G.R.; Krauss, K.W.; Reef, R.; Rogers, K.; Saunders, M.L.; Sidik, F.; Swales, A.; et al. The vulnerability of Indo-Pacific mangrove forests to sea-level rise. *Nature* **2015**, *526*, 559–563. [CrossRef]

11. Giri, C.; Ochieng, E.; Tieszen, L.L.; Zhu, Z.; Singh, A.; Loveland, T.; Masek, J.; Duke, N. Status and distribution of mangrove forests of the world using earth observation satellite data. *Glob. Ecol. Biogeogr.* **2011**, *20*, 154–159. [CrossRef]

12. Shi, T.; Liu, J.; Hu, Z.; Liu, H.; Wang, J.; Wu, G. New spectral metrics for mangrove forest identification. *Remote Sens. Lett.* **2016**, *7*, 885–894. [CrossRef]

13. Shi, C. An Analysis Comparing Mangrove Conditions under Different Management Scenarios in Southeast Asia. Master's Thesis, Duke University, Durham, NC, USA, 2017.

14. Wang, L.; Jia, M.; Yin, D.; Tian, J. A review of remote sensing for mangrove forests: 1956–2018. *Remote Sens. Environ.* **2019**, *231*, 111223. [CrossRef]

15. Koedsin, W.; Vaiphasa, C. Discrimination of tropical mangroves at the species level with EO-1 hyperion data. *Remote Sens.* **2013**, *5*, 3562–3582. [CrossRef]

16. Heenkenda, M.K.; Joyce, K.E.; Maier, S.W.; Bartolo, R. Mangrove species identification: Comparing WorldView-2 with aerial photographs. *Remote Sens.* **2014**, *6*, 6064–6088. [CrossRef]

17. Brown, M.I.; Pearce, T.; Leon, J.; Sidle, R.; Wilson, R. Using remote sensing and traditional ecological knowledge (TEK) to understand mangrove change on the Maroochy River, Queensland, Australia. *Appl. Geogr.* **2018**, *94*, 71–83. [CrossRef]

18. Kuenzer, C.; Bluemel, A.; Gebhardt, S.; Quoc, T.V.; Dech, S. Remote Sensing of Mangrove Ecosystems: A Review. *Remote Sens.* **2011**, *3*, 878–928. [CrossRef]

19. Zhang, H.; Wang, T.; Liu, M.; Jia, M.; Lin, H.; Chu, L.; Devlin, A. Potential of Combining Optical and Dual Polarimetric SAR Data for Improving Mangrove Species Discrimination Using Rotation Forest. *Remote Sens.* **2018**, *10*, 467. [CrossRef]

20. Olagoke, A.; Proisy, C.; Féret, J.-B.; Blanchard, E.; Fromard, F.; Mehlig, U.; de Menezes, M.M.; dos Santos, V.F.; Berger, U. Extended biomass allometric equations for large mangrove trees from terrestrial LiDAR data. *Trees* **2016**, *30*, 935–947. [CrossRef]

21. Proisy, C.; Viennois, G.; Sidik, F.; Andayani, A.; Enright, J.A.; Guitet, S.; Gusmawati, N.; Lemonnier, H.; Muthusankar, G.; Olagoke, A.; et al. Monitoring mangrove forests after aquaculture abandonment using time series of very high spatial resolution satellite images: A case study from the Perancak estuary, Bali, Indonesia. *Mar. Pollut. Bull.* **2018**, *131*, 61–71. [CrossRef]

22. Proisy, C.; Degenne, P.; Anthony, E.J.; Berger, U.; Blanchard, E.; Fromard, F.; Gardel, A.; Olagoke, A.; Santos, V.; Walcker, R.; et al. A Multiscale Simulation Approach for Linking Mangrove Dynamics to Coastal Processes using Remote Sensing Observations. *J. Coast. Res.* **2016**, *75*, 810–814. [CrossRef]

23. Whiteside, T.G.; Bartolo, R.E. Use of WorldView-2 time series to establish a wetland monitoring program for potential offsite impacts of mine site rehabilitation. *Int. J. Appl. Earth Obs. Geoinf.* **2015**, *42*, 24–37. [CrossRef]

24. Li, W.; El-Askary, H.; Qurban, M.A.; Li, J.; ManiKandan, K.P.; Piechota, T. Using multi-indices approach to quantify mangrove changes over the Western Arabian Gulf along Saudi Arabia coast. *Ecol. Indic.* **2019**, *102*, 734–745. [CrossRef]

25. Lucas, R.; Van De Kerchove, R.; Otero, V.; Lagomasino, D.; Fatoyinbo, L.; Omar, H.; Satyanarayana, B.; Dahdouh-Guebas, F. Structural characterisation of mangrove forests achieved through combining multiple sources of remote sensing data. *Remote Sens. Environ.* **2020**, *237*, 111543. [CrossRef]

26. Mondal, P.; Liu, X.; Fatoyinbo, T.E.; Lagomasino, D. Evaluating combinations of sentinel-2 data and machine-learning algorithms for mangrove mapping in West Africa. *Remote Sens.* **2019**, *11*, 2928. [CrossRef]

27. Qiu, P.; Wang, D.; Zou, X.; Yang, X.; Xie, G.; Xu, S.; Zhong, Z. Finer resolution estimation and mapping of mangrove biomass using UAV LiDAR and worldview-2 data. *Forests* **2019**, *10*, 871. [CrossRef]

28. Navarro, J.A.; Algeet, N.; Fernández-Landa, A.; Esteban, J.; Rodríguez-Noriega, P.; Guillén-Climent, M.L. Integration of UAV, Sentinel-1, and Sentinel-2 data for mangrove plantation aboveground biomass monitoring in Senegal. *Remote Sens.* **2019**, *11*, 77. [CrossRef]

29. Immitzer, M.; Böck, S.; Einzmann, K.; Vuolo, F.; Pinnel, N.; Wallner, A.; Atzberger, C. Remote Sensing of Environment Fractional cover mapping of spruce and pine at 1 ha resolution combining very high and medium spatial resolution satellite imagery. *Remote Sens. Environ.* **2018**, *204*, 690–703. [CrossRef]

30. Donmez, C.; Berberoglu, S.; Erdogan, M.A.; Tanriover, A.A.; Cilek, A. Response of the regression tree model to high resolution remote sensing data for predicting percent tree cover in a Mediterranean ecosystem. *Environ. Monit. Assess.* **2015**, *187*, 1–12. [CrossRef]

31. Cao, J.; Leng, W.; Liu, K.; Liu, L.; He, Z.; Zhu, Y. Object-Based mangrove species classification using unmanned aerial vehicle hyperspectral images and digital surface models. *Remote Sens.* **2018**, *10*, 89. [CrossRef]

32. Wang, L.; Silván-Cárdenas, J.L.; Sousa, W.P. Neural network classification of mangrove species from multi-seasonal Ikonos imagery. *Photogramm. Eng. Remote Sens.* **2008**, *74*, 921–927. [CrossRef]

33. Liu, K.; Li, X.; Shi, X.; Wang, S. Monitoring mangrove forest changes using remote sensing and GIS data With decision-tree learning. *Wetlands* **2008**, *28*, 336–346. [CrossRef]

34. Heumann, B.W. An object-based classification of mangroves using a hybrid decision tree-support vector machine approach. *Remote Sens.* **2011**, *3*, 2440–2460. [CrossRef]

35. Talukdar, S.; Singha, P.; Mahato, S.; Shahfahad; Pal, S.; Liou, Y.A.; Rahman, A. Land-use land-cover classification by machine learning classifiers for satellite observations-A review. *Remote Sens.* **2020**, *12*, 1135. [CrossRef]

36. White, J.C.; Stepper, C.; Tompalski, P.; Coops, N.C.; Wulder, M.A. Comparing ALS and image-based point cloud metrics and modelled forest inventory attributes in a complex coastal forest environment. *Forests* **2015**, *6*, 3704–3732. [CrossRef]

37. Toosi, N.B.; Soffianian, A.R.; Fakheran, S.; Pourmanafi, S.; Ginzler, C.; Waser, L.T. Comparing different classification algorithms for monitoring mangrove cover changes in southern Iran. *Glob. Ecol. Conserv.* **2019**, *19*, e00662. [CrossRef]

38. Colkesen, I.; Kavzoglu, T. The use of logistic model tree (LMT) for pixel- and object-based classifications using high-resolution WorldView-2 imagery. *Geocarto Int.* **2017**, *32*, 71–86. [CrossRef]

39. Immitzer, M.; Atzberger, C.; Koukal, T. Tree species classification with Random forest using very high spatial resolution 8-band worldView-2 satellite data. *Remote Sens.* **2012**, *4*, 2661–2693. [CrossRef]

40. De Fries, R.S.; Hansen, M.; Townshend, J.R.G.; Sohlberg, R. Global land cover classifications at 8 km spatial resolution: The use of training data derived from Landsat imagery in decision tree classifiers. *Int. J. Remote Sens.* **1998**, *19*, 3141–3168. [CrossRef]

41. Hansen, M.C.; Defries, R.S.; Townshend, J.R.G.; Sohlberg, R.; Defries, R.S.; Townshend, J.R.G.; Sohlberg, R. Global land cover classification at 1 km spatial resolution using a classification tree approach. *Int. J. Remote Sens.* **2010**, *21*, 1331–1346. [CrossRef]

42. Zahed, M.A.; Ruhani, F.; Mohajeri, S. An overview of Iranian mangrove ecosystem, northern part of the Persian Gulf and Oman Sea. *Electron. J. Environ. Agric. Food Chem.* **2010**, *9*, 411–417. [CrossRef]

43. DigitalGlobe Inc. Whitepaper: The benefits of the 8 spectral bands of worldview-2. *Retrieved* **2010**, *8*, 2011.

44. Drusch, M.; Del Bello, U.; Carlier, S.; Colin, O.; Fernandez, V.; Gascon, F.; Hoersch, B.; Isola, C.; Laberinti, P.; Martimort, P.; et al. Sentinel-2: ESA's optical high-resolution mission for GMES operational services. *Remote Sens. Environ.* **2012**, *120*, 25–36. [CrossRef]

45. Matthew, M.W.; Adler-Golden, S.M.; Berk, A.; Felde, G.W.; Anderson, G.P.; Gorodetzky, D.; Paswaters, S.E.; Shippert, M. Atmospheric correction of spectral imagery: Evaluation of the FLAASH algorithm with AVIRIS data, Algorithms and Technologies for Multispectral, Hyperspectral, and Ultra spectral Imagery IX, 23 September 2003. *Proc. SPIE* **2003**, *5093*, 157–163. [CrossRef]

46. Palubinskas, G. Fast, simple, and good pan-sharpening method. *J. Appl. Remote Sens.* **2013**, *7*, 073526. [CrossRef]

47. Kamal, M.; Phinn, S. Hyperspectral data for mangrove species mapping: A comparison of pixel-based and object-based approach. *Remote Sens.* **2011**, *3*, 2222–2242. [CrossRef]

48. Flanders, D.; Hall-Beyer, M.; Pereverzoff, J. Preliminary evaluation of eCognition object-based software for cut block delineation and feature extraction. *Can. J. Remote Sens.* **2003**, *29*, 441–452. [CrossRef]

49. Pavithra, B.; Kalaivani, K.; Ulagapriya, K. Remote sensing techniques for mangrove mapping. *Int. J. Eng. Adv. Technol.* **2019**, *8*, 27–30.

50. Xie, Y.; Sha, Z.; Yu, M. Remote sensing imagery in vegetation mapping: A review. *J. Plant. Ecol.* **2008**, *1*, 9–23. [CrossRef]

51. Cormack, R.M.; Upton, G.; Fingleton, B. *Spatial Data Analysis by Example. Volume 1: Point Pattern and Quantitative Data*; John Wiley & Sons Ltd.: Hoboken, NJ, USA, 1986; pp. 114–116. [CrossRef]

52. Sokal, R.R.; Oden, N.L.; Thomson, B.A. Local spatial autocorrelation in biological variables. *Biol. J. Linn. Soc.* **1998**, *65*, 41–62. [CrossRef]

53. Tu, J. Spatially varying relationships between land use and water quality across an urbanization gradient explored by geographically weighted regression. *Appl. Geogr.* **2011**, *31*, 376–392. [CrossRef]

54. Anselin, L. Local Indicators of Spatial Association—LISA. *Geogr. Anal.* **1995**, *27*, 93–115. [CrossRef]

55. R Development Core Team. *R: A Language and Environment for Statistical Computing*; R Development Core Team: Vienna, Austria, 2011; ISBN 3900051070.

56. Brovelli, M.A.; Crespi, M.; Fratarcangeli, F.; Giannone, F.; Realini, E. Accuracy assessment of high resolution satellite imagery orientation by leave-one-out method. *ISPRS J. Photogramm. Remote Sens.* **2008**, *63*, 427–440. [CrossRef]

57. Hogg, R. *Vinstructor's Solutions Manual Probability and Statistical Inference*; Prentice Hall: Upper Saddle River, NJ, USA, 2016.

58. Kumar, T.; Panigrahy, S.; Kumar, P.; Parihar, J.S. Classification of floristic composition of mangrove forests using hyperspectral data: Case study of Bhitarkanika National Park, India. *J. Coast. Conserv.* **2013**, *17*, 121–132. [CrossRef]

59. Kamal, M.; Phinn, S.; Johansen, K. Object-based approach for multi-scale mangrove composition mapping using multi-resolution image datasets. *Remote Sens.* **2015**, *7*, 4753–4783. [CrossRef]

60. Kux, H.J.H.; Souza, U.D.V. Object-Based Image Analysis of Worldview-2 Satellite Data for the Classification of Mangrove Areas in the City of São Luís, Maranhão State, Brazil. *ISPRS Ann. Photogramm. Remote Sens. Spat. Inf. Sci.* **2012**, *1*, 95–100. [CrossRef]

61. Nouri, H.; Beecham, S.; Anderson, S.; Nagler, P. High spatial resolution WorldView-2 imagery for mapping NDVI and its relationship to temporal urban landscape evapotranspiration factors. *Remote Sens.* **2013**, *6*, 580–602. [CrossRef]

62. Metzler, J.W.; Sader, S.A. Model development and comparison to predict softwood and hardwood per cent cover using high and medium spatial resolution imagery. *Int. J. Remote Sens.* **2005**, *26*, 3749–3761. [CrossRef]

63. Kumar, T.; Mandal, A.; Dutta, D.; Nagaraja, R.; Dadhwal, V.K. Discrimination and classification of mangrove forests using EO-1 Hyperion data: A case study of Indian Sundarbans. *Geocarto Int.* **2019**, *34*, 415–442. [CrossRef]

64. FAO. *The World's Mangroves 1980–2005. FAO Forestry Paper*; Food and Agriculture Organization of the United Nations: Rome, Italy, 2007; Volume 153, p. 77.

65. Jones, T.G.; Glass, L.; Gandhi, S.; Ravaoarinorotsihoarana, L.; Carro, A.; Benson, L.; Ratsimba, H.R.; Giri, C.; Randriamanatena, D.; Cripps, G. Madagascar's mangroves: Quantifying nation-wide and ecosystem specific dynamics, and detailed contemporary mapping of distinct ecosystems. *Remote Sens.* **2016**, *8*, 106. [CrossRef]

66. Song, X.F.; Cui, H.S.; Guo, Z.H. Remote sensing of mangrove wetlands identification. *Procedia Environ. Sci.* **2011**, *10*, 2287–2293. [CrossRef]

 *remote sensing*

Article

Delineation of Tree Patches in a Mangrove-Marsh Transition Zone by Watershed Segmentation of Aerial Photographs

Himadri Biswas [1,2,*], Keqi Zhang [1,3], Michael S. Ross [1,2] and Daniel Gann [2,4,5]

[1] Department of Earth and Environment, Florida International University, 11200 SW 8th Street, Miami, FL 33199, USA; zhangk@fiu.edu (K.Z.); rossm@fiu.edu (M.S.R.)

[2] Institute of Environment, Florida International University, 11200 SW 8th Street, OE 148, Miami, FL 33199, USA; gannd@fiu.edu

[3] Extreme Events Institute, Florida International University, 11200 SW 8th Street, AHC5 220, Miami, FL 33199, USA

[4] Department of Biological Sciences, Florida International University, 11200 SW 8th Street, Miami, FL 33199, USA

[5] GIS Center, Florida International University, 11200 SW 8th Street, GL 275, Miami, FL 33199, USA

* Correspondence: hbisw001@fiu.edu; Tel.: +1-305-348-0502

Received: 29 May 2020; Accepted: 23 June 2020; Published: 29 June 2020

Abstract: Mangrove migration, or transgression in response to global climatic changes or sea-level rise, is a slow process; to capture it, understanding both the present distribution of mangroves at individual patch (single- or clumped trees) scale, and their rates of change are essential. In this study, a new method was developed to delineate individual patches and to estimate mangrove cover from very high-resolution (0.08 m spatial resolution) true color (Red (R), Green (G), and Blue (B) spectral channels) aerial photography. The method utilizes marker-based watershed segmentation, where markers are detected using a vegetation index and Otsu's automatic thresholding. Fourteen commonly used vegetation indices were tested, and shadows were removed from the segmented images to determine their effect on the accuracy of tree detection, cover estimation, and patch delineation. According to point-based accuracy analysis, we obtained adjusted overall accuracies >90% in tree detection using seven vegetation indices. Likewise, using an object-based approach, the highest overlap accuracy between predicted and reference data was 95%. The vegetation index Excess Green (ExG) without shadow removal produced the most accurate mangrove maps by separating tree patches from shadows and background marsh vegetation and detecting more individual trees. The method provides high precision delineation of mangrove trees and patches, and the opportunity to analyze mangrove migration patterns at the scale of isolated individuals and patches.

Keywords: vegetation index; color; RGB; accuracy assessment; transgression

---

1. Introduction

Mangroves form an important coastal wetland ecosystem, dominating tropical and subtropical coastlines globally [1,2]. They are crucial not only for human economic activities, but also for a diverse group of terrestrial and marine species that are dependent on mangrove ecosystems for habitat [3,4]. Mangroves attenuate overland flow of water and therefore act as a shield that protects both natural and human infrastructure from storm surges [5]. Threatened by global climatic changes, sea-level rise, and human developments, mangrove response is variable, either retreating seaward or transgressing landward into other ecosystems [6–12]. To better comprehend these alternative trajectories, it is necessary to understand how mangroves are presently distributed and how their distributions have changed over time across a range of coastal environments. However, long-term monitoring of coastal and marine systems is rare [13], and, therefore, deciphering the changes in distribution of mangroves through time is a challenging task. In part, the reason is that mangroves typically occupy periodically inundated and remote regions where it is challenging, time-consuming, and cost-intensive to survey them through traditional field-based methods [2]. In contrast, information acquisition with greater coverage at lower cost is achievable through remote sensing methods. Remote sensing methods have been increasingly used in the past few decades to extract information for mapping and monitoring of forests [14].

Mangrove retreat or expansion is likely to be observed first in ecotones, the brackish transition zones between the coastal ecosystems and the interior freshwater ecosystems where mangrove trees mix with freshwater marsh vegetation. We expect that the leaves of evergreen mangrove trees will absorb more light in the blue and red spectra and reflect more light in the green spectrum, resulting in a large reflectance difference between green and red/blue bands. In contrast, partially senesced marsh vegetation, especially during the dry season, has a relatively small reflectance difference between green and red/blue bands. This distinct difference in spectral reflectance between mangroves and the graminoids that dominate in marshes, will allow the separation of these two vegetation growth forms using remote sensing imagery.

Vegetation mapping involving multispectral images are commonly applied in global studies [15,16]. Medium resolution multispectral images (e.g., Landsat, NASA, Greenbelt, MD, USA) are free of charge, have temporal coverage dating to the late 1970s, and spatial resolutions of 10s to 100s of meters that are adequate for detecting large-scale disturbances caused by episodic events such as hurricanes [17]. Though Giri et al. [15] mapped the global distribution of mangroves using medium resolution Landsat images and Global Land Survey data, and mangrove related vegetation mapping studies are becoming commonplace [18,19], we did not find any study that addressed tree crown detection, delineation, and cover estimation of mangroves at the individual patch level using true color or multi-spectral images. Detection of the early stages of mangrove invasion into freshwater marshes necessitates higher spatial resolution images (e.g., WorldView-2, DigitalGlobe, Westminster, CO, USA). Medium resolution imagery from satellites such as Landsat is too coarse to detect the subtle changes occurring at the patch or individual tree scale. However, acquisition of high-resolution images over large spatial extents with commercial satellites can be prohibitively expensive. In addition, mangrove transgression is inherently a slow process, and it takes multiple decades to detect mangroves as they mature starting from small seedlings. As such, the short temporal coverage of high-resolution multispectral images is insufficient to study mangrove transgression in much detail [20].

At the same time, a huge repository of high-resolution aerial photographs, some dating back as far as the early 1900s, are available for many parts of the world [21]. These aerial photographs are available as true color, infrared, or panchromatic photographs as hard or soft copies. The most commonly used method in mapping vegetation from aerial photography is manual digitization [22,23], which is not only time-consuming but also subject to the interpretation of the digitizing analyst, making repeatability and replication at the same accuracy and precision difficult.

Therefore, a desirable goal is to use automated detection, and delineation techniques to detect subtle changes in crown- and patch sizes at decadal time scales using high spatial resolution (sub-meter) true color (RGB), near-infrared and panchromatic aerial photographs that were acquired by conventional frame cameras. We present here an initial step toward that goal, an evaluation of the suitability of RGB aerial photography in a fully automated delineation process, differentiating tree patches against a graminoid marsh wetland matrix.

Researchers have successfully used true color (RGB) photographs in detection and delineation of tree crowns by various segmentation techniques [24–29]. Segmentation techniques separate an image into target plant and background components. Three widely used segmentation techniques are (i) color-index based segmentation, (ii) threshold-based segmentation, and (iii) learning-based segmentation [30]. Color-index or vegetation index is used to enhance the contrast between vegetation and non-vegetated classes. The rationale behind using color-based vegetation indices is to outline the vegetation region of interest, e.g., crops or trees, by combining information from several bands into a single grayscale image. Many color-based indices have been developed, among others Excess Green [31], Excess Red [32], Vegetative Index [33], Visible Atmospheric Resistance Index [34], Normalized Difference Index [35], Triangular Greenness Index [36], and Visible-band Difference Vegetation Index [28]. Other indices combine two or more vegetation indices such as Excess Green minus Excess Red [25], and the Combined index [27].

Despite promising outcomes, limitations of color-based indices to segment images have been reported when images are captured under variable light conditions [30]. Segmentation requires thresholding techniques which often depend on a user-selected threshold. Higher threshold selection may lead to under-segmentation, thereby merging plant pixels with background pixels, while lower threshold selection may lead to over-segmentation [30]. Among several thresholding techniques, Otsu's automatic thresholding method [37] is one of the most widely used. Because thresholds are determined automatically in Otsu's method, this approach is particularly applicable where several images must be processed, thereby reducing the time required to binarize the images.

Limitations of color-based vegetation indices and thresholding methods have prompted researchers to use machine learning approaches including both unsupervised [38] and supervised methods [39,40]. However, these approaches are complex and often require substantial user input and feedback at multiple stages of the process, making them labor intensive.

Wang et al. [41] categorized several other automatic recognition algorithms for individual tree delineation into four major types: contour-based, local maximum, template matching, and 3D-model. The contour-based method relies on intensity changes which in turn are scale dependent. Therefore, the biggest challenge with contour-based methods is to find a scale that is appropriate for all individual trees in the same image [41]. Local maximum methods underperform because of varied illumination conditions and irregular background phenomena in the image [41]. Model-based template matching requires detailed *a priori* knowledge about the object and is susceptible to varying illumination and noise in the image. Some researchers have applied 3D-based methods. One such method is the watershed segmentation algorithm, a region-based approach originally proposed by Digabel et al. [42] and revised by Beucher et al. [43]. Later, Meyer et al. [44] introduced marker-controlled watershed segmentation to overcome the problem of over-segmentation due to noise in the image [14]. The underlying principle stems from the geographical concept of watersheds and catchments.

Watershed segmentation requires a grayscale input image which is viewed as a topographic surface where the intensity (gray level) of each pixel represents elevation, and local maxima represent the tree crowns. To form catchment basins and delineate watersheds, the image is inverted so that local maxima become local minima, which form valleys [41,45]. As the surface is slowly flooded with water, water will start accumulating in the valleys (local minima) until it overflows into adjacent valleys. The idea is to prevent the water in neighboring catchments from merging by building dams on the watershed lines, thereby creating the boundary of each segment, or catchment basin [45]. Thus, a catchment basin becomes the tree crown or a contiguous patch region with several clumped trees, and the watershed lines become the edge of the crowns or patches.

There are two critical steps for accurate delineation of tree crowns by the watershed method:

1. Generating a binary grayscale image;
2. Delineating markers.

Various approaches have been used to implement these two steps [41,46–48]. Lamar et al. [48] developed an automated segmentation method to extract populations of hemlock trees for multi-temporal assessment from aerial images, using a spectrally classified binary image, and generated the markers by Euclidean distance map construction and Gaussian smoothing. Wang et al. [41] detected and delineated tree crowns from a high resolution multispectral aerial image. They identified and created two sets of treetops from the first component of a principal component analysis. The two sets were created using a local non-maximum suppression method, and a local maximum on morphologically transformed distance method, each producing a binary image of the treetops. The markers were generated by intersection of the two binary images based on well-defined criteria. Recently, Yin et al. [49] detected and delineated individual mangrove trees from light detection and ranging (LiDAR) data by seed region growing (SRG) and marker-controlled watershed segmentation (MCWS). The seeds/markers were assumed to be the treetops which were detected as local maxima from the canopy height model (CHM) using variable window filtering method. Although watershed segmentation holds the potential to use spectral imagery to differentiate and delineate tree crowns from a background matrix [48], this method has been evaluated mostly in non-mangrove forest settings.

Our objective was to fully automate an image segmentation technique to detect and delineate mangrove patches. By mangrove patches, we refer to mangroves that either occur as isolated individual trees that are large enough to be detected, or several trees that are clumped together. The mangroves were embedded in a graminoid dominated wetland landscape with a mixture of grasses, sedges, and rushes. Since true color aerial photographs have only three spectral bands (RGB), we evaluated which vegetation indices most effectively enhanced the contrast between target pixels (i.e., mangrove patches) and their background.

The application of a fully automated delineation of mangrove patches using the watershed algorithm to high-resolution true-color aerial photography was conducted in a two-step process: (1) Generation of a vegetation index and application of Otsu's thresholding method, followed by morphological operations to delineate markers; (2) Delineation of tree patches with marker-controlled watershed segmentation. In this paper we present the process that identified the vegetation indices and parameter settings that best delineate markers for watershed segmentation to detect mangrove patches. Assessment of the best method was evaluated on the basis of (1) agreement between algorithm-detected tree cover compared to actual cover, (2) overall and class-specific user's and producer's accuracies, and (3) object-based (patch) accuracy estimates.

The remaining sections of the paper are arranged as follows: Section 2 describes the study area, the components of the watershed algorithm and the metrics used to evaluate algorithm performance; Section 3 presents the results of the sensitivity analysis, and the success of individual tree detection and extraction of tree patches; Section 4 discusses the effects of parameter selection, vegetation indices, Otsu's thresholding method, and the presence of shadows on the detection and delineation of trees; and Section 5 presents the study's conclusions.

2. Materials and Methods

2.1. Study Area and Image Acquisition

The study area is located adjacent to Everglades National Park, in Florida, USA, approximately 300 m south of the C-111 Canal and 3.6 km west of South Dixie Highway (Figure 1).

The study area consists of heterogeneous freshwater herbaceous marsh vegetation with scattered occurrences of red mangroves (*Rhizophora mangle*). A georeferenced true color aerial photograph was used with a spatial resolution of 0.08 m (0.25 foot), acquired in the dry season on January 24, 2017 by Miami-Dade County [50]. The RGB image was acquired using Vexcel Ultracam Eagle (UCEagle) large format aerial sensor and was processed with Inpho (Trimble, Sunnyvale, CA, USA) Photogrammetry software. Each channel recorded 8-bit digital number (DN) brightness values ranging from 0 to 255. The methodology is presented in a flowchart in Figure 2 and the steps are described in detail in the following sections. Digitization and visual interpretation of reference samples was conducted in ArcGIS 10.5 [51]; index calculation, thresholding, and watershed processing were scripted in Python [52] using openCV [53] and scikit-learn [54]; and data analysis, and accuracy assessment were performed in R [55].

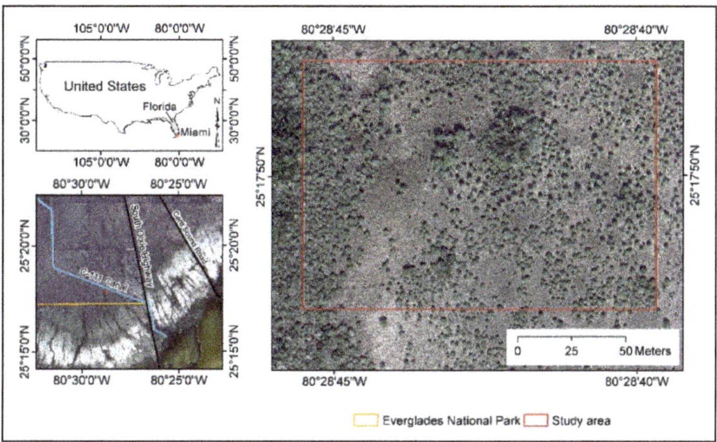

Figure 1. Study area in the southern Everglades, adjacent to Everglades National Park, Florida, USA.

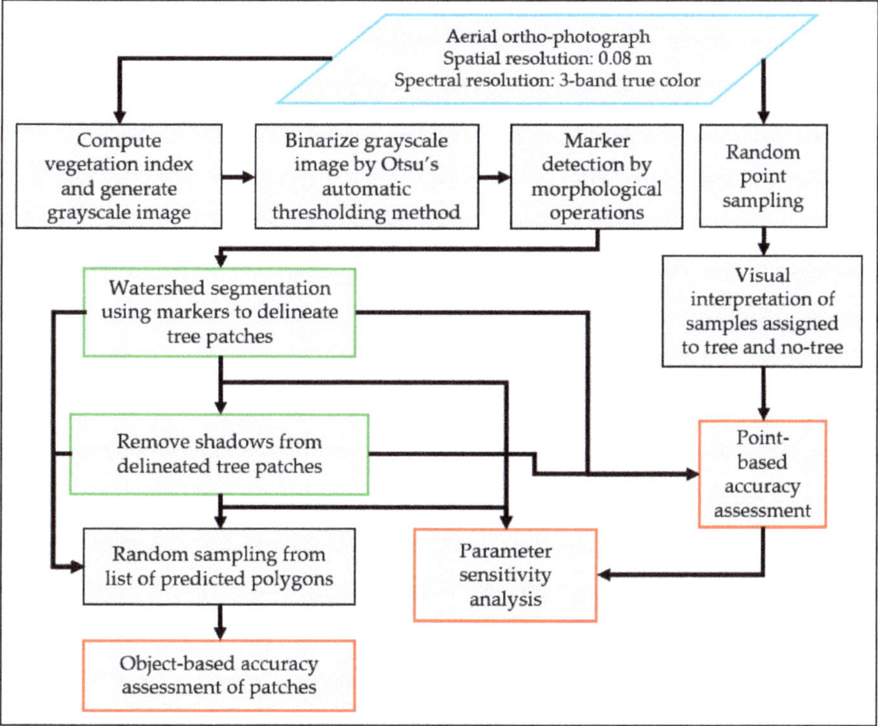

Figure 2. Flowchart of individual mangrove tree patches delineated from aerial photograph (blue box) using vegetation indices, Otsu's thresholding, and watershed segmentation. Parameter sensitivity analysis and accuracy assessments (red boxes) were performed on segmented images with and without shadow removal (green boxes).

2.2. Vegetation Indices

To determine the vegetation indices that delineated patches with highest accuracy, 14 commonly used indices were calculated (Table 1) from the RGB aerial image that covered the study area (Figure 1). The image contained individual and clumped mangrove trees within a graminoid marsh matrix. Through their interactions with incoming solar radiation, the two vegetation classes (marsh and mangrove) vary in absorption and reflection of electromagnetic radiation of different wavelengths, with trees also casting shadows onto other trees and marsh vegetation (Figure 3). As expected, mangroves reflected more light in the green spectrum than surrounding marsh vegetation or shadows (Figure 3). Suitable vegetation indices enhance the contrast between tree patches, marsh matrix, and shadows. The 14 vegetation indices (Table 1) were calculated with equations presented in Table 1.

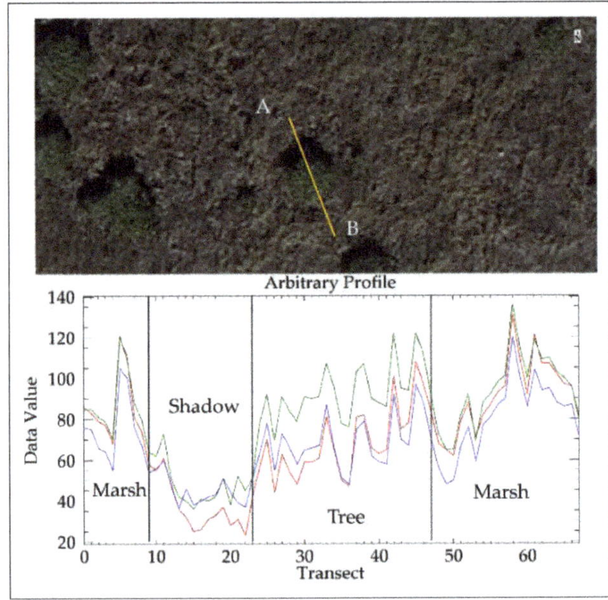

Figure 3. RGB spectral values shown across a selected linear profile AB extracted from the aerial photograph.

Table 1. Commonly used vegetation indices their equations and source references.

Vegetation Index	Equation	Reference
Excess Green (ExG)	$2g - r - b$	[31]
Excess Red (ExR)	$1.4 \times r - g$	[32]
Excess Green minus Excess Red (ExGR)	$ExG - ExR$	[25]
Vegetative Index (VEG)	$g/r^{0.667} \times b^{0.333}$	[33]
Color Index of Vegetation Extraction (CIVE)	$0.441 \times r - 0.881 \times g + 0.385 \times b + 18.78745$	[24]
Visible Atmospheric Resistant Index (VARI)	$(g - r)/(g + r - b)$	[34]
Combined Index (COM)	$0.25 \times ExG + 0.30 \times ExGR + 0.33 \times CIVE + 0.12 \times VEG$	[27]
Normalized Difference Index (NDI)	$(g - r)/(g + r)$	[35]
Triangular Greenness Index (TGI)	$g - 0.39 \times r - 0.61 \times b$	[36]

Table 1. *Cont.*

Vegetation Index	Equation	Reference
Visible-band Difference Vegetation Index (VDVI)	$(2g - b - r)/(2g + b + r)$	[28]
Red minus Green (R-G)	$r - g$	[31]
Green minus Blue (G-B)	$g - b$	[31]
Ratio (GB_RG)	$(g - b)/(r - g)$	[31]
GRB	$g \times r \times b$	[31]

No normalization was applied to the brightness values because the vegetation indices were mainly used to identify tree markers, and normalization does not necessarily enhance the contrast in index values between trees and marsh. The calculation of vegetation indices resulted in grayscale images as shown by ExG and ExR images (Figure 4). Gray index images were then binarized by Otsu's automatic thresholding method and used for delineation of markers.

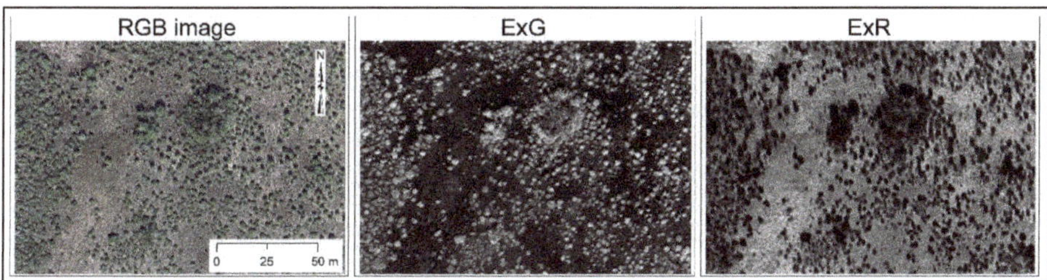

Figure 4. The original RGB aerial photograph and grayscale images of vegetation indices ExG and ExR.

2.3. Otsu's Thresholding Method

Otsu's automatic thresholding method [37] was used to generate the binary images for tree patches (with values of 1) and background marsh matrix (with values of 0). This thresholding method is a non-parametric approach which uses the histogram of the pixel brightness values derived from grayscale images representing two normal intensity distributions that show a bimodal distribution [25]. One distribution represents the target pixels (i.e., mangrove patches) and the other represents the background (i.e., marsh matrix). Figure 5a shows the histograms of five vegetation indices that display narrow to widely spread bimodal distributions. Otsu's method maximizes the between-class variance while minimizing the within-class variance of the intensity values in the image, thereby providing optimal thresholding for an index (Figure 5b).

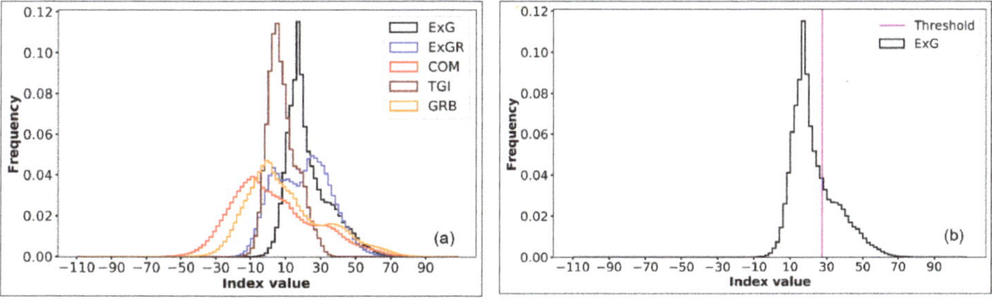

Figure 5. (a) Histogram of five vegetation grayscale indices (ExG, ExGR, COM, TGI, and GRB); (b) Otsu's automatic threshold for the ExG grayscale image.

2.4. Marker Detection and Watershed Segmentation

The markers for watershed segmentation were delineated from the binary image performing the following steps:

1. Opening morphological transformations that conduct erosions followed by dilations were applied to remove noise from patch boundaries and break up tree patches with thin connections [56]. Since we were interested in removing small and isolated noisy pixels from patch boundaries, we chose the frequently used square kernel shape because it is computationally efficient and effective. We used several opening iterations with varying square kernel sizes to determine their effect on marker delineation and watershed segmentation.

2. Unequivocal tree patch regions were identified by applying a distance transform followed by thresholding to the opening image generated in Step 1 (Figure 6b). The distance transform calculated the distances between the pixels inside a tree patch and the nearest background (marsh matrix) pixels. Euclidean distance was computed and the threshold was determined using the percentage of the maximum distance value in the image [57]. The optimum unequivocal tree patch image was generated by distance thresholding. This step ensured that the core portions of tree patches were identified.

3. A dilation was employed to expand tree patches in the opening image to include indeterminate regions (Figure 6c). Since tree patch boundaries were located in indeterminate regions between the outside boundaries of unequivocal patches and the outside boundaries of expanded tree patches (Figure 6c), consecutive dilations were conducted to ensure expanded tree patches were large enough to contain true patch boundaries. A number of consecutive dilations were tested to determine the effect of dilations on marker delineation and watershed segmentation.

4. Indeterminate regions were identified by subtracting the unequivocal patch image generated in Step 2 from the expanded patch image generated in Step 3 (Figure 6c).

5. Then, the marker image was generated by labeling connected regions in the unequivocal patch image with increasing integers from 1 to N and labeling the indeterminate regions as zero in the expanded patch image (Figure 6c).

6. Finally, watershed segmentation was executed on the color image utilizing the delineated markers to derive tree patch boundaries.

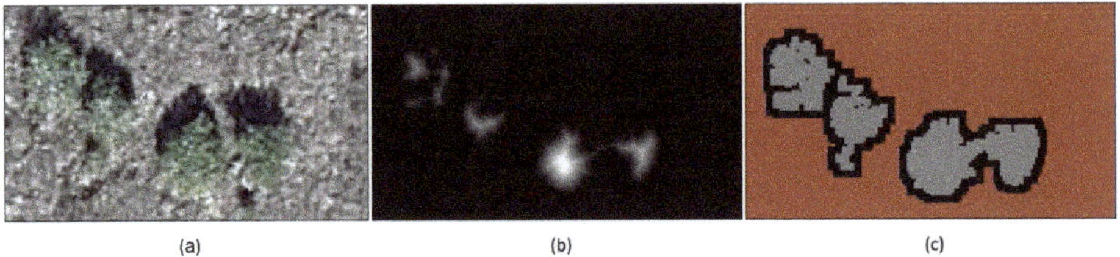

Figure 6. (**a**) Mangrove patches in the original RGB aerial photograph, (**b**) unequivocal tree patches derived from ExG index, and (**c**) marker image showing expanded tree patches (gray), background (brown), and indeterminate region (black) derived from ExG index.

2.5. Removal of Shadows

After applying the watershed segmentation algorithm, the following steps removed shadows:

1. A mean RGB (mRGB) index image was calculated by summing intensity values from all the bands and dividing by three. A mask image of same size as the mRGB index was created, where values from the mRGB index image was kept at 0 if mRGB values were less than the first percentile, and 1 if mRGB was equal to or greater than the first percentile.

2.   The shadows were removed by multiplying the patch mask and the original image.
3.   Isolated pixels in the shadow-removed image were eliminated by applying a morphological closing operation using a kernel of 4 pixels.
4.   The morphologically filtered image was labeled for connected components and small objects were removed by using a size threshold of 4 pixels to generate the final shadow-removed patches.

2.6. Parameter Sensitivity

The marker-detection process consisted of three morphological operations as described in Section 2.4. The parameters of these operations were values for the morphological kernel size (MKS) for opening and dilation, the opening and dilation iterations, and the distance transform coefficient (DTC). A sensitivity analysis was performed to determine the parameter values that enhanced segmentation, which was evaluated on the basis of overall accuracy of tree detection. The parameters and test values are provided in Table 2. The full-factorial design produced 90 model combinations per index image, resulting in a total of 2520 models. Point-based accuracy estimates as described in Section 2.7 for each of these 2520 models were used to determine optimal parameter combinations and indices.

Table 2. Parameters and their values used for sensitivity analysis.

Parameter	Values
Number of iterations for opening	{1, 2, 3}
Morphological kernel size	{3, 5}
Number of iterations for dilation	{1, 3, 5}
Distance transform coefficient	{0.01, 0.03, 0.05, 0.07, 0.1}

2.7. Tree-Cover Estimation from Random Samples and Tree Detection Accuracy

To evaluate the performance of each index, a simple random sample reference data set was generated. The first objective was to estimate the tree cover (area of patches) within the study area as a reference, and the second was to establish a reference for overall and class-specific omission and commission errors for each of the predicted tree cover maps. Since each map was to be evaluated with the same sample set, we chose a simple random sampling design [58]. The required minimum number of simple random sample points was calculated for a 2% precision (d = ±2%) estimate within a 95% confidence interval (z = 1.96) (Equation (1)) [58].

$$n = z^2 \times p \times (1 - p) \div d^2 \tag{1}$$

Considering the worst-case sampling scenario of p = 50% tree cover, a minimum of 2401 samples were required to estimate the tree-cover proportion within a 2% margin and a 95% confidence, and the sample points were randomly generated within the study area. Since resolution and contrast of the aerial photograph were high enough to visually distinguish trees from marsh and shadow, and because it is optimal to evaluate maps from their photo source data to avoid potential changes [59], we visually evaluated each random sample from the 2017 aerial photograph and assigned class labels (tree, marsh, or shadow). The visually interpreted random points were then used to estimate the tree cover within the study area. For this estimate the marsh and shadow classes were combined to a no-tree class.

To estimate overall and class-specific user's and producer's accuracy for each of the algorithm-predicted maps, the classified cover type was extracted for all random samples from each tree-cover map. The extracted values and the reference labels were then cross tabulated to generate confusion matrices. From the confusion matrices we estimated adjusted overall, and adjusted class-specific user's and producer's accuracies for both tree and no-tree classes, along with their standard errors [60], as well as adjusted tree cover proportions, factoring in the class proportion information of each map [58]. We used the terms overall, user's, and producer's accuracy in Section 3,

Section 4, Section 5 to refer to their adjusted values, respectively. Furthermore, we were interested in how the presence of shadows affected the accuracy of segmentation. For each index for which segmented images with and without shadow removal were generated, the differences in overall, user's, and producer's accuracy, and proportional area were calculated and compared.

2.8. Object-Based Overlap Accuracy Assessment

The performance of index images in delineating the patches was further evaluated by overlap analysis of automatically detected patches with a manually digitized reference dataset. We used an object-based approach with tree polygons as sampling units, and a post-classification simple random sampling design with equal probabilities for all polygons. Unlike point-based accuracy assessment where the same reference data can be used to evaluate the performance of all models, individual reference data have to be created for object-based evaluation of each model output, because each model generates a different number of polygons with different polygon sizes and, therefore, must be sampled individually. Consequently, it was not feasible to evaluate the performance of all 2520 models. Instead, we selected the two models with the highest point-based overall accuracy: one with shadows and the other with shadows removed. For both predicted tree cover maps, we selected 50 polygons using simple random sampling from a list frame. Random sampling from a list of all units within a population ensured equal selection probability for every polygon regardless of size. Point sampling would have increased the probability of including large polygons and over-representing large polygons at the cost of small polygons of individual trees [59]. We digitized tree patches manually from the original RGB aerial photograph. Since the patch polygons were of different sizes including either an individual tree or a group of trees (clumps), in addition to patch boundaries, when possible, we digitized individual tree crowns with their centers inside a predicted polygon. We then assigned the sample identifier of the predicted polygon to all digitized patches in order to evaluate the count of trees within each polygon that was delineated by the watershed segmentation. The spatial union of reference data and model-generated patches produced three types of areas: (1) Correctly predicted tree patches, i.e., areas where prediction and reference agreed; (2) areas of omission error, which included tree polygons in the reference data that were missed by the model; and (3) areas of commission error, i.e., algorithm-delineated portions of mangrove polygons that were not part of the reference data. We quantified the three area types using Equations (2)–(4).

$$Actual\ Tree\ Area = \frac{\Sigma Area\ of\ true\ overlap\ of\ patches\ from\ automatic\ segmentation}{\Sigma Area\ of\ patches\ from\ reference\ data} \tag{2}$$

$$Omission\ Error\ of\ predicted\ tree\ crowns = \frac{\Sigma Area\ of\ omitted\ patches\ from\ automatic\ segmentation}{\Sigma Area\ of\ patches\ from\ reference\ data} \tag{3}$$

$$Comission\ Error\ of\ predicted\ tree\ crowns = \frac{\Sigma Area\ of\ comitted\ patches\ from\ automatic\ segmentation}{\Sigma Area\ of\ true\ patches\ from\ automatic\ segmentation} \tag{4}$$

We also tabulated the total number of individual trees in each predicted polygon to determine if the detected tree was an individual or part of a clump of trees.

3. Results

3.1. Parameter Sensitivity Analysis

The sensitivity analysis to determine optimum values of three morphological operations used to delineate markers indicated that overall accuracy was maximized for all models, with and without shadow removed, when the MKS was three (Figure 7b, Table 3). Overall accuracy decreased for all models when opening iteration was increased beyond one, except for the two ExGR models, in which case an opening iteration of two maximized the overall accuracy (Figure 7a, Table 3). Higher overall accuracy was achieved for ExGR_s, ExGR_ns, COM_s, and GRB_s when a dilation iteration of one was used (Figure 7c, Table 3). For the remaining models, a dilation iteration of three increased the overall accuracy. Optimum values for the DTC were inconclusive (Figure 7d). The results indicate that, depending on the index image used for watershed segmentation, DTC can be selected accordingly to maximize accuracy (Figure 7d, Table 3). Across all models, overall accuracy ranged from $11.5 \pm 0.01\%$ for CIVE_s to $93.4 \pm 0.5\%$ for GRB_ns, the user's and producer's accuracies varied from 0% to $99.4 \pm 0.01\%$.

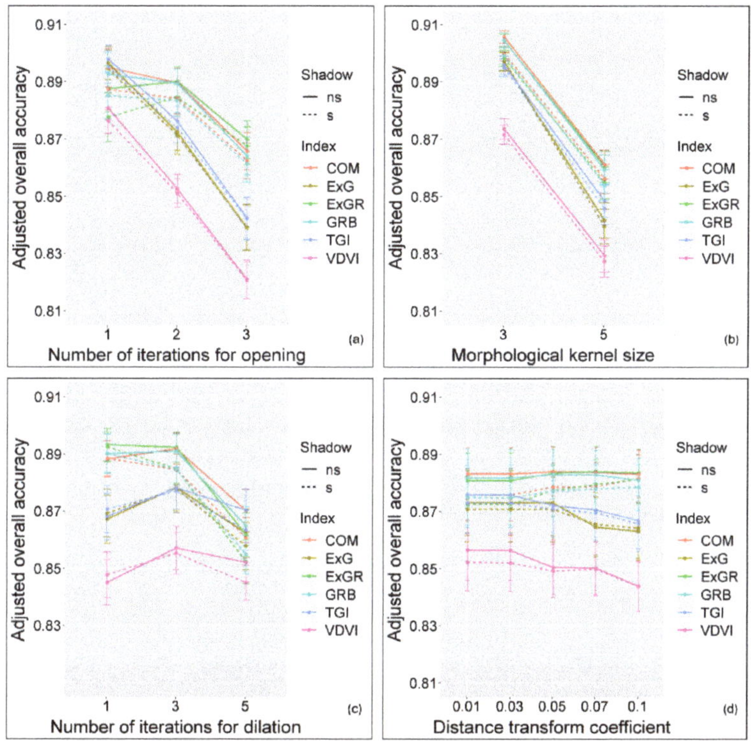

Figure 7. Sensitivity analysis results for optimum values of (**a**) number of iterations for opening, (**b**) morphological kernel size, (**c**) number of iterations for dilation, and (**d**) distance transform coefficient for several vegetation indices (Table 2). ns = no shadow, s = with shadow. Error bars indicate 95% confidence intervals of the mean computed from the standard error.

Table 3. Optimum parameter values for marker detection using vegetation indices. For description purpose, subscript "_ns" was added to the names of index images when shadows were removed after segmentation and "_s" was added when shadows were not removed, for example, shadow removed ExG index image were named ExG_ns and those with shadow present were named ExG_s. Index names as in Table 1.

Model	Morphological Kernel Size	Number of Iterations for Opening	Number of Iterations for Dilation	Distance Transform Coefficient
GRB_ns	3	1	3	0.05, 0.07
ExG_s	3	1	3	0.01, 0.03, 0.05
ExG_ns	3	1	3	0.01, 0.03, 0.05
GRB_s	3	1	1	0.05, 0.07, 0.1

Table 3. *Cont.*

Model	Morphological Kernel Size	Number of Iterations for Opening	Number of Iterations for Dilation	Distance Transform Coefficient
COM_ns	3	1	3	0.01, 0.03, 0.05, 0.07, 0.1
GRB_ns	3	1	3	0.05, 0.07
ExG_s	3	1	3	0.01, 0.03, 0.05
ExG_ns	3	1	3	0.01, 0.03, 0.05
GRB_s	3	1	1	0.05, 0.07, 0.1
COM_ns	3	1	3	0.01, 0.03, 0.05, 0.07, 0.1

3.2. Tree-Cover Area Estimation and Tree Detection Analysis

Further analysis only considered models that had an overall accuracy of tree detection greater than 90% and that fell inside the confidence interval of the reference area estimate. The reference area was estimated from the reference dataset consisting of 2401 random point samples. The number of tree and no-tree samples was 650 and 1751, respectively, thus, on the basis of the sampling design to provide a 2% precision with a 95% confidence level, the percent tree cover was 27.1 ± 2%. Six models with shadow (COM_s, ExG_s, ExGR_s, GRB_s, TGI_s, and VDVI_s) and seven models after shadow removal (COM_ns, ExG_ns, ExGR_ns, GRB_ns, R-G_ns, TGI_ns, and VDVI_ns) met both criteria (Table 4). Confusion matrix derived adjusted accuracy estimates for the selected shadow and shadow removed models are shown in Table 4. The overall accuracy for those 13 models ranged from 90.5 ± 0.6% to 93.4 ± 0.5% for VDVI_s and GRB_ns, respectively. User's accuracy was highest for GRB_ns (90.1 ± 1.2%) and lowest for VDVI_s (82.6 ± 1.5%), and producer's accuracy was highest for ExG_s (87.4 ± 1.2%) and lowest for VDVI_ns (81.1 ± 1.3%).

Table 4. Metrics derived from confusion matrix of segmented images by shadow removed and with shadow vegetation index models. Accuracies in percent ± standard errors. Index names as in Table 1.

Index Name	% Tree Cover	Adjusted User's Accuracy	Adjusted Producer's Accuracy	Adjusted Overall Accuracy
GRB_ns	25.8	90.1 ± 1.2	85.2 ± 1.3	93.4 ± 0.5
ExG_s	27.8	87.9 ± 1.3	87.4 ± 1.2	93.1 ± 0.5
ExG_ns	26.5	89.9 ± 1.2	84.6 ± 1.2	93.0 ± 0.5
GRB_s	26.5	87.6 ± 1.3	85.9 ± 1.3	92.9 ± 0.5
R-G_ns	25.8	89.3 ± 1.3	84.0 ± 1.3	92.9 ± 0.5
COM_ns	26.5	87.3 ± 1.3	85.1 ± 1.3	92.6 ± 0.5
TGI_ns	27.9	85.9 ± 1.4	87.2 ± 1.2	92.5 ± 0.5
TGI_s	26.0	88.9 ± 1.3	83.2 ± 1.3	92.4 ± 0.5
COM_s	27.3	84.8 ± 1.4	86.5 ± 1.2	92.3 ± 0.5
ExGR_ns	25.6	88.7 ± 1.3	82.2 ± 1.3	92.2 ± 0.5
ExGR_s	25.9	87.1 ± 1.4	82.6 ± 1.3	91.9 ± 0.6
VDVI_ns	25.8	86.9 ± 1.4	81.1 ± 1.3	91.4 ± 0.6
VDVI_s	27.5	82.6 ± 1.5	82.9 ± 1.3	90.5 ± 0.6

We found that GRB_ns model had highest overall accuracy of 93.4 ± 0.5%, closely followed by ExG_s (93.1 ± 0.5%). The GRB_ns also had the highest user's accuracy of 90.1 ± 1.2% followed by ExG_ns (89.9 ± 1.2%). Higher user's accuracy of trees implies that trees were detected with lower commission error. Although the user's accuracy of ExG_s was 87.9 ± 1.3%, this model had the highest producer's accuracy of 87.4 ± 1.2%. Higher producer's accuracy indicates better performance of the models in detection of actual trees with the lowest omission error. The VDVI models, VDVI_s and VDVI_ns, had the lowest user's and producer's accuracy respectively (Table 4). The commission error in the GRB model decreased after shadow removal but the omission error increased slightly. When GRB_ns was used, it attained the lowest commission error among all the index images (user's accuracy = 90.1 ± 1.2%), but had a higher omission error (producer's accuracy = 85.2 ± 1.3%) compared to ExG_s, TGI_ns, COM_s, and GRB_s. In contrast, ExG_s and TGI_ns had the lowest omission error (producer's accuracy = 87.4 ± 1.2% and 87.2 ± 1.2%, respectively). This indicated that the watershed segmentation using these two indices were able to detect trees with higher accuracy than other indices, but TGI_ns had higher commission error (user's accuracy = 85.9 ± 1.4%) than ExG_s (user's accuracy = 87.9 ± 1.3%).

We found that on an average overall accuracy and user's accuracy increased by 0.5% (standard deviation (SD) = 0.7%) and 2.9% (SD = 4%), respectively, when shadows were removed (Table 5). However, average producer's accuracy and the proportion of the area covered by trees decreased by 1.4% (SD = 3.2%) and 1.1% (SD = 2.1), respectively (Table 5). Although on an average accuracy increased or decreased only slightly, it must be noted that the overall accuracy and user's accuracy increased for six out of seven indices when shadows were removed, whereas producer's accuracy increased for only one index model (Table 5). The highest increase in user's accuracy of ~10% was observed when shadows were removed from the R-G derived segmented image followed by VDVI (~4%), although user's accuracy declined by 3% when TGI was used. The highest decrease in producer's accuracy after shadow removal was observed for the R-G index (6.7%), although producer's accuracy increased after shadow removal when TGI was used (4%). The estimated proportional area decreased in six index images when shadows were removed. The highest decrease in the proportion of tree-cover area was ~5% when the R-G index image was used (Table 5).

Table 5. Difference in proportional area, user's accuracy, producer's accuracy, and overall accuracy between shadow removed and with shadow vegetation index models. SD = Standard Deviation.

Model (ns-s)	Δ Percent Tree Cover	Δ Adjusted User's Accuracy	Δ Adjusted Producer's Accuracy	Δ Adjusted Overall Accuracy
COM	−0.76	2.53	−1.47	0.32
ExG	−1.32	2.07	−2.78	−0.10
ExGR	−0.32	1.52	−0.37	0.28
GRB	−0.70	2.45	−0.73	0.48
R-G	−5.14	10.35	−6.69	1.87
TGI	1.90	−3.02	4.06	0.12
VDVI	−1.65	4.26	−1.82	0.83
Mean ± SD	−1.14 ± 2.1	2.88 ± 3.99	−1.4 ± 3.2	0.54 ± 0.65

3.3. Object-Based Overlap Analysis

Two models (one with shadow (ExG_s) and the other without shadow (GRB_ns) that had the highest point-based overall accuracy were selected for overlap accuracy assessment. Using an object-based approach, 50 randomly sampled polygons for each of the two maps covered polygon size distributions including the 5th up to the 97th percentile for ExG_s, and from the smallest polygon up to the 99th percentiles for GRB_ns. The highest overlap accuracy between predicted and reference data was achieved by ExG_s (~95%) when compared to GRB_ns (88%) (Table 6). Although the GRB_ns model had the highest overall accuracy (93.4%) based on the point-based accuracy assessment, the ExG_s model performed better in delineation of actual crowns by as much as 7%. The omission area was very low when ExG_s model was used (~5%) compared to GRB_ns model (~12%), but the commission error was much higher with ExG_s model (21.4%). This is in line with the point-based accuracy assessment, where the GRB_ns model had higher user's accuracy and lower producer's accuracy compared to ExG_s.

Table 6. Percent (%) overlap, omission, and commission areas of predicted trees with reference trees by vegetation index models.

Index Image	Omitted Tree Area Sum (m^2)	Actual Tree Area Sum (m^2)	Committed Mean Tree Area (m^2)	Reference Tree Area (m^2)	Omitted Tree Area (%)	Actual Tree Area (%)	Committed Mean Tree Area (%)
GRB_ns	24.48	181.11	29.28	205.59	11.91	88.09	16.17
ExG_s	3.07	57.93	12.37	60.99	5.03	94.97	21.35

The mean patch sizes of tree clumps predicted by GRB_ns and ExG_s differed substantially. Patches predicted by GRB_ns were much larger than those predicted by ExG_s because ExG_s separated clumped trees better than GRB_ns. The mean size of patches delineated by GRB_ns was about 3.86 m^2 compared to roughly 1.16 m^2 by ExG_s. The data suggest that commission errors from GRB_ns- and ExG_s-predicted patches were similar, though GRB_ns had a higher mean commission error (0.58 m^2) compared to ExG_s (~0.25 m^2) (Figure 8). However, there was a significant difference in omission error between the two, in which GRB_ns had a higher mean omission error of 0.67 m^2 compared to 0.06 m^2 of ExG_s model (Figure 8). The total cover estimated by GRB_ns model was 0.49 hectare compared to ExG_s model which was estimated as 0.53 hectare.

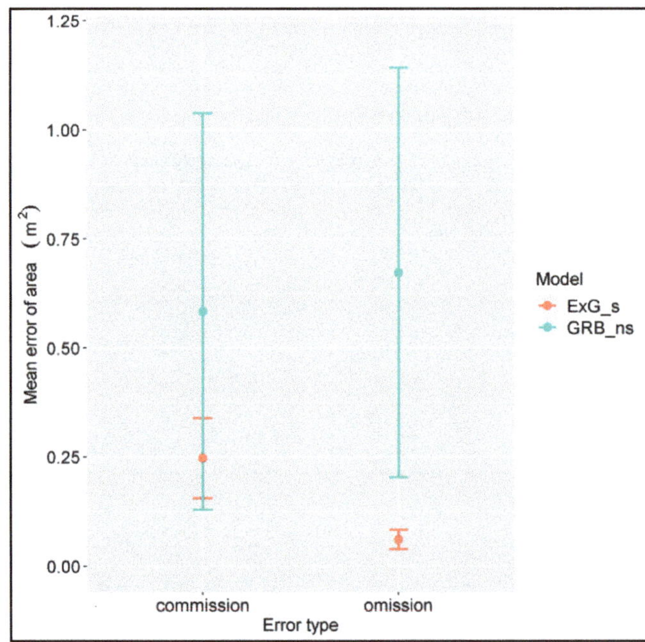

Figure 8. Comparison of omission and commission errors of patches predicted by GRB_ns and EXG_s models. Error bars are the 95% confidence intervals of the mean computed from the standard error.

Comparing the number of reference tree crowns that were fully within each of the predicted tree patches from the two models (GRB_ns and ExG_s), we found that ExG_s detected more trees as individuals compared to GRB_ns. The largest tree patch predicted by GRB_ns had eleven reference trees compared to only three in ExG_s. (Table 7). Individual predicted trees that coincided with one tree from the reference data were more common for ExG_s, whereas more tree clumps were delineated by GRB_ns (Table 7).

Table 7. Count of predicted tree crowns in patches versus number of reference tree crowns. 0 = tree not detected, 1 = individual isolated tree detected, >1 = number of tree crowns present in detected tree patch.

Number of Trees in a Polygon Patch	0	1	2	3	9	11
GRB_ns (count)	3	35	9	1	1	1
ExG_s (count)	0	49	0	1	0	0

4. Discussion

4.1. Feasibility of the Method

In this study, a new fully automated tree-patch-delineation method using vegetation indices derived from RGB aerial photography was developed. Markers were delineated from vegetation indices, using Otsu's automatic thresholding method, followed by the depiction of patch boundaries with watershed segmentation. The accurate delineation of markers is the key to the success of the method. The accuracy of tree patch delineation is dependent on two major factors: first, the contrast between the target tree pixels and the background in the index image, and second, Otsu's automatic thresholding to separate target tree pixels from the background matrix. The efficiency of vegetation indices in producing contrast between target pixels and background is dependent in part on the aerial image itself. Thus, the method will be ineffective when vegetation indices are derived from aerial photographs with homogenous brightness values, as Otsu's automatic thresholding will work well only when the brightness intensity values of a vegetation index produces a bimodal distribution. If the contrast is not high enough, or the bimodal distribution captures other properties of the landscape that are uncorrelated to the target pixel vs. background, then unreliable and imperfect threshold values are generated. Consequently, the delineation of markers will result in under- or over-detection of markers.

Correct marker delineation is also dependent on the parameter values of morphological operations. The parameters used for marker delineation were opening iterations, MKS, dilation iterations, and DTC. Morphological opening can be iterated and requires a kernel. It was found that increasing the number of iterations decreased the overall accuracy of tree detection as a direct consequence of incorrect marker delineation. This is to be expected because additional morphological openings not only remove noisy pixels from the tree patches but also valid tree pixels that resulted in a decline of marker delineation. Although dilation is the process to recover the objects of interest (i.e., mangrove trees) using the same kernel size, it cannot recover small objects that are completely removed by erosion [56]. In general, one iteration of morphological opening promoted higher overall accuracy (Figure 7a). The overall accuracy is high across all index images in watershed segmentation when MKS was 3 (Figure 7b), while an MKS of 5 removed actual tree pixels that led to significant decline in the overall accuracy (Figure 7b). The effect of the number of dilation iterations and DTC on marker delineation varies among index images. We therefore recommend using optimum values for these parameters based on the index image used for delineation of markers. Our results show that these indices obtained high overall accuracy and performed equally well when compared to each other except for VDVI (error bar overlap in Figure 7a–d). Although we found the best parameter values to use with these vegetation indices for marker detection, the values are specific to the geographical context of the acquired images. Therefore, the values should be used as a guide when this method is applied elsewhere, as optimal parameter values may change, because of lighting conditions that alter the contrast between foreground and background, or the heterogeneity of the vegetation matrix in which the trees are embedded.

Point-based accuracy showed that models of seven indices matched the proportional area estimate with overall accuracy estimates above 90%. The highest overall accuracy was obtained using GRB_ns model (93.4 ± 0.5%), and the highest producer's accuracy and user's accuracy were obtained using ExG_s (87.4 ± 1.2%) and GRB_ns (90.1 ± 1.2%) models, respectively. The object-based assessment indicated that the agreement between predicted and reference tree crowns was higher for the ExG_s (95%) when compared to the GRB_ns (88%) (Table 6), with a 7% lower omission error (Table 6, Figure 8).

Although there was little difference in the proportional area estimate between GRB_ns and ExG_s models, the average patch sizes of GRB_ns were three times larger than those of the ExG_s model, and in turn the GRB_ns model detected 1048 patches compared to 2600 detected by ExG_s. This indicates that GRB_ns grouped individual neighboring patches into a single larger patch (Figure 9), and therefore, we recommend the use of ExG_s when detection of individual trees is desired.

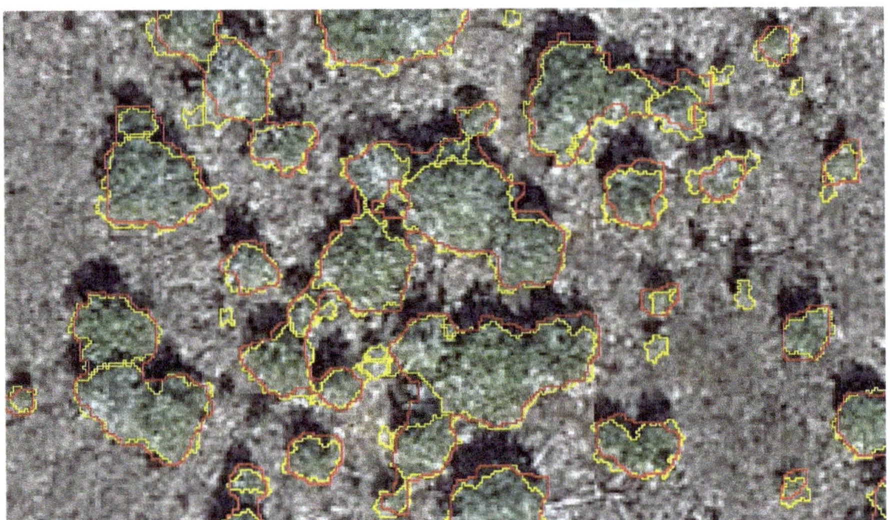

Figure 9. ExG_s (yellow polygons) obtained better separation between patches and shadows compared to GRB_ns (red polygons). ExG_s detected more trees inside clumped mangrove patches (number of yellow polygons inside a single red polygon). ExG_s also detected more isolated trees (yellow polygons without corresponding red polygons).

Shadow removal produced a mixed effect on delineation of tree patches. The commission errors in GRB_s were concentrated in transition areas that were shaded, removing shadows lead to a 2.5% increase in user's accuracy in GRB_ns (Table 5). However, removing shadows from ExG_s, though increasing user' accuracy by 2.1%, also eliminated many tree pixels along with shadows, thereby reducing producer's accuracy by 2.8% (Table 5). Since shadow removal is problematic with the current method, the vegetation index that performs best without shadow removal is preferred. An algorithm that corrects reflectance in shadow areas rather than removing shadow pixels should be developed to minimize the effect of shadows on tree detection and delineation. The ExG_s separated tree patches from shadows well without removing shadows (Figure 9). Most of the commission error for the ExG_s model occurred near the transition between the crown boundary and marsh matrix pixels (Figure 9). Unlike near the center of the tree crown, where the green intensity values are more homogenous, the separability in such transitional areas becomes more difficult because of the mixture of tree and marsh pixels. This uncertainty in boundary interpretation carries over to the manual digitization process, where some error is associated with imprecise digitization of tree boundaries. However, this source of error only marginally affected the proportional area estimation.

4.2. Comparison with Other Studies

In this study, we showed that our methodology is robust, efficiently achieving very high detection and delineation accuracies for mangrove patches in a graminoid background matrix. We identified individual patches of mangroves, consisting of either single or multiple crowns. Separation of each individual tree crown within a mangrove clump is not possible because of low contrast along neighboring tree boundaries in aerial photographs. For convex tree shapes or diverse tree heights in forests, height information may increase the performance of the watershed algorithm when delineating individual trees. However, using LiDAR derived height information, Yin et al. [49] achieved a detection accuracy of 76.9% for isolated trees but overall crown delineation accuracy was only 46%. Delineation of individual mangrove crowns with large branches can sometimes cause incorrect splits of single crowns into multiple trees. Applying size filters on optical data or height filters on LiDAR data may address some of these issues [61]. The combination of high-resolution spectral imagery and high-density LiDAR data may improve delineation of isolated and individual trees in patches, but this approach is limited for change detection because of the temporal coverage of LiDAR data.

Detection of encroachment or loss of woody vegetation in savannahs, prairies, other grasslands, and woodlands is of interest to many ecologists, and natural resource and protected area managers. Several studies have mapped the woody encroachment in grasslands such as savannah using multi-spectral imagery [62–64]. To understand the pattern of woody vegetation changes in grasslands, and graminoid wetlands requires detection of new emergence and growth of new trees at the individual tree level. Our method specifically aims at detecting these kinds of vegetation dynamics and can be applied to conduct studies that are interested in changes of woody vegetation in a graminoid dominated landscape.

4.3. Future Work and Challenges

In the small, homogenous, red mangrove-dominated wetland in which this pilot project was carried out, the method worked well for single isolated tree detection, tree patch delineation, and cover estimation, but not as well in delineation of individual tree crowns inside patches containing several clumped trees. With advances in technology to acquire very high resolution (sub-decimeter) images in the future, this method provides an opportunity to conduct baseline studies for long-term monitoring of woodlands. The results we achieved also provide a foundation for estimating and monitoring temporal changes in mangrove cover.

Some challenges may arise when applying this method to images acquired in different wetland settings or from aerial photography that has different spectral, radiometric, and spatial resolutions. First, not all wetlands exhibit as distinct a bimodal distribution as the area selected for this project. More sophisticated methods may need to be developed to threshold the multi-modal distribution of pixel brightness values from more heterogeneous landscapes. Second, images that enhance the contrast between grasses and trees are preferable. Because contrast is enhanced when trees are foliated and graminoid species are senescent, dried up, or dead, image acquisition time should be determined based on the phenological cycles of the dominant graminoids and the tree species of interest. Third, although older aerial photographs have high spatial resolution compared to medium-resolution multi-spectral images, their spatial resolution is low compared with the image used in this research. For meaningful comparison in change detection studies, more recent very-high resolution images may have to be downscaled to the resolution of older aerial photographs. Fourth, this method incorporates the usage of true color aerial image, therefore, use of infrared and panchromatic aerial images would require further modifications.

5. Conclusions

A new fully automated method was developed to successfully detect and delineate mangrove trees and patches in a coastal wetland environment from aerial photography. The introduced framework allows for the selection of the most suitable index images with and without shadow removal for detection and delineation of tree patches. High overall accuracy (>90%) with comparable user's and producer's accuracies in tree detection were obtained by using seven index images (COM, ExG, ExGR, GRB, R-G, TGI, and VDVI). The overlap accuracy of ExG_s (~95%) was better than GRB_ns (88%) in patch delineation. Despite having similar proportional area estimates, ExG_s performed better in separation of tree patches and shadows, and also delineated more trees than GRB_ns. The selection of optimum parameter values for morphological operations is crucial for the detection of markers for watershed segmentation. MKS of 3 produced the highest overall accuracy across all the index images. The parameter values of opening iterations, dilation iterations, and DTC affected marker delineation differently, and therefore, their values should be selected based on the index image used for watershed segmentation. The parameter values most effective in this study should be used only as an initial starting point when the method is applied in a different geographical setting, because optimal parameter values may change because of either lighting conditions or local contrast changes, which depends on the spatial distribution of trees within the surrounding vegetation matrix.

The shadow removal method had positive and negative effects; it increased the overall and the user's accuracy for the majority of models, but also reduced the producer's accuracy. Shadows in images are problematic and need to be dealt with carefully when applying automated delineation methods. To reduce shadow effects on the delineation of patches in the future, a more sophisticated algorithm to correct brightness values in shadows instead of removing shadowed areas deserves further study.

This method provides an opportunity to analyze mangrove migration patterns at the scale of isolated individuals and patches. It can be applied to reconstruction of change in mangrove distributions over time, and gain insight into the driving forces of their migration patterns. There is much potential in using widely available high-resolution aerial photography to understand not only mangrove transgression dynamics at the individual tree and patch levels, but also woody vegetation invasion in prairies and other grassland ecosystems.

Author Contributions: Conceptualization, K.Z. initiated the idea and H.B., K.Z., D.G., and M.S.R. developed the concept; methodology, K.Z. developed the watershed segmentation and shadow removal methods, D.G. and H.B. developed the methods to perform sensitivity and accuracy analysis; formal analysis, H.B. and D.G. processed images and conducted calculation and analysis; writing—original draft preparation, H.B. and D.G.; writing—review and editing, M.S.R., H.B., and D.G., and K.Z.; visualization, H.B. and D.G. All authors have read and agreed to the published version of the manuscript.

Funding: This research received no external funding.

Acknowledgments: This material is based upon work supported by the National Science Foundation under Grant No. HRD-1547798. This NSF Grant was awarded to Florida International University as part of the Centers for Research Excellence in Science and Technology (CREST) Program. This is contribution number 970 from the Southeast Environmental Research Center in the Institute of Environment at Florida International University.

Conflicts of Interest: The authors declare no conflict of interest.

References

1. Mohd, O.; Suryanna, N.; Sahib Sahibuddin, S.; Faizal Abdollah, M.; Rahayu Selamat, S. Thresholding and Fuzzy Rule-Based Classification Approaches in Handling Mangrove Forest Mixed Pixel Problems Associated with in QuickBird Remote Sensing Image Analysis. *Int. J. Agric. For.* **2012**, *2*, 300–306. [CrossRef]
2. Kuenzer, C.; Bluemel, A.; Gebhardt, S.; Quoc, T.V.; Dech, S. Remote Sensing of Mangrove Ecosystems: A Review. *Remote Sens.* **2011**, *3*, 878–928. [CrossRef]
3. Ewel, K.C.; Twilley, R.R.; Ong, J.E. Different Kinds of Mangrove Forests Provide Different Goods and Services. *Glob. Ecol. Biogeogr. Lett.* **1998**, *7*, 83–94. [CrossRef]

4. Costanza, R.; d'Arge, R.; de Groot, R.; Farber, S.; Grasso, M.; Hannon, B.; Limburg, K.; Naeem, S.; O'Neill, R.V.; Paruelo, J.; et al. The Total Value of the World's Ecosystem Services and Natural Capital. *Nature* **1996**, *387*, 253–260. [CrossRef]

5. Zhang, K.; Liu, H.; Li, Y.; Xu, H.; Shen, J.; Rhome, J.; Smith, T.J., III. The role of mangroves in attenuating storm surges. *Estuar. Coast. Shelf Sci.* **2012**, *102*, 11–23. [CrossRef]

6. Gilman, E.L.; Ellison, J.; Duke, N.C.; Field, C. Threats to mangroves from climate change and adaptation options: A review. *Aquat. Bot.* **2008**, *89*, 237–250. [CrossRef]

7. Alongi, D.M. The Impact of Climate Change on Mangrove Forests. *Curr. Clim. Chang. Rep.* **2015**, *1*, 30–39. [CrossRef]

8. Ross, M.S.; Meeder, J.F.; Sah, J.P.; Ruiz, P.L.; Telesnicki, G.J. The Southeast Saline Everglades revisited: 50 years of coastal vegetation change. *J. Veg. Sci.* **2000**, *11*, 101–112. [CrossRef]

9. Rogers, K.; Saintilan, N.; Heijnis, H. Mangrove encroachment of salt marsh in Western Port Bay, Victoria: The role of sedimentation, subsidence, and sea level rise. *Estuaries* **2005**, *28*, 551–559. [CrossRef]

10. Yao, Q.; Liu, K.B. Dynamics of marsh-mangrove ecotone since the mid-Holocene: A palynological study of mangrove encroachment and sea level rise in the Shark River Estuary, Florida. *PLoS ONE* **2017**, *12*, e0173670. [CrossRef]

11. Rodriguez, W.; Feller, I.C.; Cavanaugh, K.C. Spatio-temporal changes of a mangrove–saltmarsh ecotone in the northeastern coast of Florida, USA. *Glob. Ecol. Conserv.* **2016**, *7*, 245–261. [CrossRef]

12. Saintilan, N.; Williams, R.J. Mangrove transgression into saltmarsh environments in south-east Australia. *Glob. Ecol. Biogeogr.* **1999**, *8*, 117–124. [CrossRef]

13. Rosenzweig, C.; Karoly, D.; Vicarelli, M.; Neofotis, P.; Wu, Q.; Casassa, G.; Menzel, A.; Root, T.L.; Estrella, N.; Seguin, B.; et al. Attributing physical and biological impacts to anthropogenic climate change. *Nature* **2008**, *453*, 353. [CrossRef]

14. Ke, Y.; Quackenbush, L.J. A comparison of three methods for automatic tree crown detection and delineation from high spatial resolution imagery. *Int. J. Remote Sens.* **2011**, *32*, 3625–3647. [CrossRef]

15. Giri, C.; Ochieng, E.; Tieszen, L.L.; Zhu, Z.; Singh, A.; Loveland, T.; Masek, J.; Duke, N. Status and distribution of mangrove forests of the world using earth observation satellite data. *Glob. Ecol. Biogeogr.* **2011**, *20*, 154–159. [CrossRef]

16. Thomas, N.; Bunting, P.; Lucas, R.; Hardy, A.; Rosenqvist, A.; Fatoyinbo, T. Mapping mangrove extent and change: A globally applicable approach. *Remote Sens.* **2018**, *10*, 1466. [CrossRef]

17. Zhang, K.; Thapa, B.; Ross, M.; Gann, D. Remote sensing of seasonal changes and disturbances in mangrove forest: A case study from South Florida. *Ecosphere* **2016**, *7*, e01366. [CrossRef]

18. Simard, M.; Zhang, K.; Rivera-monroy, V.H.; Ross, M.S.; Ruiz, P.L.; Castañeda-moya, E.; Twilley, R.R.; Rodriguez, E. Mapping Height and Biomass of Mangrove Forests in Everglades National Park with SRTM Elevation Data. *Photogramm. Eng. Remote Sens.* **2006**, *72*, 299–311. [CrossRef]

19. Feliciano, E.A.; Wdowinski, S.; Potts, M.D.; Lee, S.K.; Fatoyinbo, T.E. Estimating mangrove canopy height and above-ground biomass in the Everglades National Park with airborne LiDAR and TanDEM-X data. *Remote Sens.* **2017**, *9*, 702. [CrossRef]

20. Giri, C.P.; Long, J. Mangrove reemergence in the northernmost range limit of eastern Florida. *Proc. Natl. Acad. Sci. USA* **2014**, *111*, E1447–E1448. [CrossRef]

21. USGS. *Looking for an Old Aerial Photograph*; Fact Sheet; Series Number 127-96; U.S. Geological Survey: Reston, VA, USA, 1997.

22. Freeman, M.; Stow, D.; Roberts, D. Object-based Image Mapping of Conifer Tree Mortality in San Diego County based on Multitemporal Aerial Ortho-imagery. *Photogramm. Eng. Remote Sens.* **2016**, *82*, 571–580. [CrossRef]

23. Rutchey, K.; Vilchek, L. Air photointerpretation and satellite imagery analysis techniques for mapping cattail coverage in a northern Everglades impoundment. *Photogramm. Eng. Remote Sens.* **1999**, *65*, 185–191.

24. Kataoka, T.; Kaneko, T.; Okamoto, H.; Hata, S. Crop growth estimation system using machine vision. In Proceedings of the 2003 IEEE/ASME International Conference on Advanced Intelligent Mechatronics (AIM 2003), Kobe, Japan, 20–24 July 2003; Volume 2, pp. 1079–1083.

25. Meyer, G.E.; Neto, J.C. Verification of color vegetation indices for automated crop imaging applications. *Comput. Electron. Agric.* **2008**, *63*, 282–293. [CrossRef]

26. Ponti, M.P. Segmentation of low-cost remote sensing images combining vegetation indices and mean shift. *IEEE Geosci. Remote Sens. Lett.* **2013**, *10*, 67–70. [CrossRef]

27. Yang, W.; Wang, S.; Zhao, X.; Zhang, J.; Feng, J. Greenness identification based on HSV decision tree. *Inf. Process. Agric.* **2015**, *2*, 149–160. [CrossRef]

28. Xiaoqin, W.; Miaomiao, W.; Shaoqiang, W.; Yundong, W. Extraction of vegetation information from visible unmanned aerial vehicle images. *Trans. Chin. Soc. Agric. Eng.* **2015**, *31*, 152–159.

29. Kazmi, W.; Garcia-Ruiz, F.J.; Nielsen, J.; Rasmussen, J.; Jørgen Andersen, H. Detecting creeping thistle in sugar beet fields using vegetation indices. *Comput. Electron. Agric.* **2015**, *112*, 10–19. [CrossRef]

30. Hamuda, E.; Glavin, M.; Jones, E. A survey of image processing techniques for plant extraction and segmentation in the field. *Comput. Electron. Agric.* **2016**, *125*, 184–199. [CrossRef]

31. Woebbecke, D.M.; Meyer, G.E.; Von Bargen, K.; Mortensen, D.A. Color Indices for Weed Identification Under Various Soil, Residue, and Lighting Conditions. *Trans. ASAE* **1995**, *38*, 259–269. [CrossRef]

32. Meyer, G.E.; Hindman, T.W.; Laksmi, K. Machine Vision Detection Parameters for Plant Species Identification. In *Precision Agriculture and Biological Quality*; SPIE Proceedings: Bellingham, WA, USA, 1999; Volume 3543, pp. 327–336.

33. Hague, T.; Tillett, N.D.; Wheeler, H. Automated Crop and Weed Monitoring in Widely Spaced Cereals. *Precis. Agric.* **2006**, *7*, 21–32. [CrossRef]

34. Gitelson, A.A.; Kaufman, Y.J.; Stark, R.; Rundquist, D. Novel algorithms for remote estimation of vegetation fraction. *Pap. Nat. Resour.* **2002**, *149*, 76–87. [CrossRef]

35. Woebbecke, D.M.; Meyer, G.E.; Von Bargen, K.; Mortensen, D.A. Plant Species Identification, Size, and Enumeration Using Machine Vision Techniques on Near-Binary Images. In *Optics in Agriculture and Forestry*; DeShazer, J.A., Meyer, G.E., Eds.; SPIE: Bellingham, WA, USA, 12 May 1993; Volume 1836, pp. 208–219.

36. Hunt, E.R.; Daughtry, C.S.T.; Eitel, J.U.H.; Long, D.S. Remote sensing leaf chlorophyll content using a visible band index. *Agron. J.* **2011**, *103*, 1090–1099. [CrossRef]

37. Otsu, N. A Threshold Selection Method from Gray-Level Histograms. *IEEE Trans. Syst. Man. Cybern.* **1979**, *9*, 62–66. [CrossRef]

38. Meyer, G.; Camargo Neto, J.; Jones, D.; Hindman, T. Intensified fuzzy clusters for classifying plant, soil, and residue regions of interest from color images. *Comput. Electron. Agric.* **2004**, *42*, 161–180. [CrossRef]

39. Tian, L.F.; Slaughter, D.C. Environmentally adaptive segmentation algorithm for outdoor image segmentation. *Comput. Electron. Agric.* **1998**, *21*, 153–168. [CrossRef]

40. Guo, W.; Rage, U.K.; Ninomiya, S. Illumination invariant segmentation of vegetation for time series wheat images based on decision tree model. *Comput. Electron. Agric.* **2013**, *96*, 58–66. [CrossRef]

41. Wang, L.; Gong, P.; Biging, G.S. Individual Tree-Crown Delineation and Treetop Detection in High-Spatial-Resolution Aerial Imagery. *Photogramm. Eng. Remote Sens.* **2004**, *70*, 351–357. [CrossRef]

42. Digabel, H.; Lantuejoul, C. Iterative Algorithms. In *Proceedings of the Actes du Second Symposium Europeen d'Analyse Quantitative des Microstructures en Sciences des Materiaux, Biologie et Medecine*; Chermant, J.-L., Ed.; Riederer: Stuttgart, Germany, 1978; pp. 85–99.

43. Beucher, S.; Lantuejoul, C. Use of Watersheds in Contour Detection. In Proceedings of the International Workshop on Image Processing CCETT, Rennes, France, 17 September 1979; pp. 2.1–2.12.

44. Meyer, F.; Beucher, S. Morphological segmentation. *J. Vis. Commun. Image Represent.* **1990**, *1*, 21–46. [CrossRef]

45. Ke, Y.; Quackenbush, L.J. A review of methods for automatic individual tree-crown detection and delineation from passive remote sensing. *Int. J. Remote Sens.* **2011**, *32*, 4725–4747. [CrossRef]

46. Huang, H.; Li, X.; Chen, C. Individual tree crown detection and delineation from very-high-resolution UAV images based on bias field and marker-controlled watershed segmentation algorithms. *IEEE J. Sel. Top. Appl. Earth Obs. Remote Sens.* **2018**, *11*, 2253–2262. [CrossRef]

47. Jing, L.; Hu, B.; Noland, T.; Li, J. An individual tree crown delineation method based on multi-scale segmentation of imagery. *ISPRS J. Photogramm. Remote Sens.* **2012**, *70*, 88–98. [CrossRef]

48. Lamar, W.R.; McGraw, J.B.; Warner, T.A. Multitemporal censusing of a population of eastern hemlock (Tsuga canadensis L.) from remotely sensed imagery using an automated segmentation and reconciliation procedure. *Remote Sens. Environ.* **2005**, *94*, 133–143. [CrossRef]

49. Yin, D.; Wang, L. Individual mangrove tree measurement using UAV-based LiDAR data: Possibilities and challenges. *Remote Sens. Environ.* **2019**, *223*, 34–49. [CrossRef]

50. MDC Miami-Dade County Aerial Photography Find and Download Application. 2017. Available online: https://gisweb.miamidade.gov/imagerydownload/ (accessed on 6 November 2018).
51. *ESRI ArcGIS Desktop: Release 10.5*; Environmental Systems Research Institute: Redlands, CA, USA, 2016.
52. Python Software Foundation Python Language Reference. 2016.
53. Bradski, G. The OpenCV Library. Dr. Dobb's J. Softw. Tools. 2000.
54. Pedregosa, F.; Varoquaux, G.; Gramfort, A.; Michel, V.; Thirion, B.; Grisel, O.; Blondel, M.; Prettenhofer, P.; Weiss, R.; Dubourg, V.; et al. Scikit-learn: Machine Learning in Python. *J. Mach. Learn. Res.* **2011**, *12*, 2825–2830.
55. R Core Team. *R: A language and Environment for Statistical Computing*; R Foundation for Statistical Computing: Vienna, Austria, 2013; ISBN 3-900051-07-0.
56. Soille, P. *Morphological Image Analysis: Principles and Applications*, 2nd ed.; Springer: Berlin/Heidelberg, Germany, 2004; ISBN 3540429883. (alk. paper).
57. Borgefors, G. Distance transformations in digital images. *Comput. Vision. Graph. Image Process.* **1986**, *34*, 344–371. [CrossRef]
58. Olofsson, P.; Foodu, G.M.; Martin, H.; Stehman Stephen, V.; Woodcock, C.E.; Wulder, M.A. Good practices for estimating area and assessing accuracy of land change. *Remote Sens. Environ.* **2014**, *148*, 42–57. [CrossRef]
59. Radoux, J.; Bogaert, P. Good Practices for Object-Based Accuracy Assessment. *Remote Sens.* **2017**, *9*, 646. [CrossRef]
60. Olofsson, P.; Foody, G.M.; Stehman, S.V.; Woodcock, C.E. Remote Sensing of Environment Making better use of accuracy data in land change studies: Estimating accuracy and area and quantifying uncertainty using strati fi ed estimation. *Remote Sens. Environ.* **2013**, *129*, 122–131. [CrossRef]
61. Leckie, D.; Gougeon, F.; Hill, D.; Quinn, R.; Armstrong, L.; Shreenan, R. Combined high-density lidar and multispectral imagery for individual tree crown analysis. *Can. J. Remote Sens.* **2003**, *29*, 633–649. [CrossRef]
62. Nagelkirk, R.L.; Dahlin, K.M. Woody cover fractions in African Savannas from landsat and high-resolution imagery. *Remote Sens.* **2020**, *12*, 813. [CrossRef]
63. Huang, C.Y.; Archer, S.R.; McClaran, M.P.; Marsh, S.E. Shrub encroachment into grasslands: End of an era? *PeerJ* **2018**, *2018*, 1–19. [CrossRef] [PubMed]
64. Mitchard, E.T.A.; Saatchi, S.S.; Gerard, F.F.; Lewis, S.L.; Meir, P. Measuring woody encroachment along a forest-savanna boundary in Central Africa. *Earth Interact.* **2009**, *13*, 1–29. [CrossRef]

 *remote sensing*

Article

The Google Earth Engine Mangrove Mapping Methodology (GEEMMM)

J. Maxwell M. Yancho [1], Trevor Gareth Jones [1,2,3,*], Samir R. Gandhi [1,4], Colin Ferster [5], Alice Lin [1] and Leah Glass [1]

[1] Blue Ventures Conservation—Mezzanine, The Old Library, Trinity Road, St Jude's, Bristol BS2 0NW, UK; yanchojo@gmail.com (J.M.M.Y.); samir@blueventures.org (S.R.G.); alin14@ucla.edu (A.L.); leah@blueventures.org (L.G.)
[2] Department of Forest Resources Management, University of British Columbia, Vancouver, BC V6T 1Z4, Canada
[3] Terra Spatialists, The Blue House, 660 West 13th Avenue, Vancouver, BC V5Z 1N9, Canada
[4] The Jolly Geographer, 7 Yorke Gate, Watford, Hertfordshire WD17 4NQ, UK
[5] Department of Geography, University of Victoria, P.O. Box 1700 STN CSC, Victoria, BC V8W 2Y2, Canada; cferster@uvic.ca
* Correspondence: trevor@blueventures.org

Received: 1 October 2020; Accepted: 9 November 2020; Published: 16 November 2020

Abstract: Mangroves are found globally throughout tropical and sub-tropical inter-tidal coastlines. These highly biodiverse and carbon-dense ecosystems have multi-faceted value, providing critical goods and services to millions living in coastal communities and making significant contributions to global climate change mitigation through carbon sequestration and storage. Despite their many values, mangrove loss continues to be widespread in many regions due primarily to anthropogenic activities. Accessible, intuitive tools that enable coastal managers to map and monitor mangrove cover are needed to stem this loss. Remotely sensed data have a proven record for successfully mapping and monitoring mangroves, but conventional methods are limited by imagery availability, computing resources and accessibility. In addition, the variable tidal levels in mangroves presents a unique mapping challenge, particularly over geographically large extents. Here we present a new tool—the Google Earth Engine Mangrove Mapping Methodology (GEEMMM)—an intuitive, accessible and replicable approach which caters to a wide audience of non-specialist coastal managers and decision makers. The GEEMMM was developed based on a thorough review and incorporation of relevant mangrove remote sensing literature and harnesses the power of cloud computing including a simplified image-based tidal calibration approach. We demonstrate the tool for all of coastal Myanmar (Burma)—a global mangrove loss hotspot—including an assessment of multi-date mapping and dynamics outputs and a comparison of GEEMMM results to existing studies. Results—including both quantitative and qualitative accuracy assessments and comparisons to existing studies—indicate that the GEEMMM provides an accessible approach to map and monitor mangrove ecosystems anywhere within their global distribution.

Keywords: GEEMMM; mangroves; remote sensing; google earth engine; Myanmar; cloud computing; digital earth

---

1. Introduction

Mangroves are a species of woody plants which comprise unique, halophytic communities in the tropical and sub-tropical inter-tidal coastlines of the world [1]. When meeting accepted definitions based on attributes including height, diameter and canopy closure, mangroves can qualify as forest [2]. Areas not qualifying as forest are peripheral parts of wider mangrove ecosystems, including expanses dominated by submerged, dwarf or scrub, and fringe plants [1,3–5]. Mangrove ecosystems —both forest and non-forest—are found in 102 countries and 21 territories [5]. The value of mangrove ecosystems is multifaceted, including the provisioning of critical goods (e.g., fuel wood, fish, shellfish, medicine, fiber, and timber) and services (e.g., shoreline stabilization, storm protection, and cultural, recreational and tourism opportunities) to millions of people residing in coastal communities [6–9]. In addition, mangrove ecosystems are incredibly biodiverse, providing habitat for numerous species, many of which are rare, at-risk, or endangered [10–12]. Mangrove forests are also incredibly carbon-dense and meet or exceed many of their terrestrial peers in sequestration and storage [13–15]. Increasingly, the conservation, restoration and managed-use of mangrove ecosystems is being pursued through payments for ecosystem services (PES) programs, including forest carbon initiatives (e.g., REDD+, Plan Vivo) [16,17].

Despite their multifaceted value, global mangrove loss is widespread. In the last two decades of the 20th century the world lost an estimated 35% of mangrove forest cover [18]. While globally the rate of loss has thus far slowed in the 21st century—an estimated 4% from 1996 to 2016—many parts of the world, notably SE Asia, remain loss hotspots [19–21]. The primary driver of mangrove loss is anthropogenic activities including aquaculture, agriculture, urban development, and unmanaged harvest [22]. Accurate, reliable, contemporary, and easily updated information representing the extent of mangrove ecosystems is required by decision makers and managers and to help countries pursue and meet environmental targets (e.g., Millennium Development Goals and Ramsar Convention on Wetlands of International Importance especially as Waterfowl Habitat) [23–25]. Remotely sensed data have a well-established utility for mapping and monitoring the multi-date distribution of mangrove ecosystems and quantifying change over time; however, the remote sensing of mangrove environments has its own unique set of challenges which must be overcome to produce accurate results, including the variable presence or absence of water associated with daily tidal fluctuations [23,26]. Fluctuating tides can drastically influence the spectral properties of mangrove ecosystems making information on tidal condition at time of image acquisition vital [27]. Many mangrove studies have ignored variable tidal conditions, combining images ranging from low to high tide [23]. Recently, studies have used image composites that include imagery acquired during selective tides (i.e., high and/or low); however, these have covered limited areas (e.g., a single bay within a single Landsat scene) where reliable local tidal stations or modeled tidal products are available, and have not evaluated dynamics [27–29]. Other studies demonstrated the potential to use remote sensing or models to calibrate tides across larger areas; however, these approaches depend on substantial expertise to run specialized or customized software and the models depend on high quality training data—which is not always available—making them too complex and inaccessible for most potential users [30–33].

Beyond tidal considerations, conventional mapping techniques—while successful and informative—remain limited by imagery availability, required computing resources, and necessary technical expertise [34]. A single uncompressed Landsat 8 scene is larger than 1.6 gigabytes, and applications using multiple scenes require computing resources that present a barrier to many practitioners [35]. Emerging tools and technologies are ushering in a new era for land-cover mapping and monitoring [26,36]. Cloud-based platforms, most notably Google Earth Engine (GEE), provide unprecedented volumes of ready-to-use geospatial data, including the entire Landsat archive (i.e., radiometrically and geometrically corrected), and tool and computing resources for rapid and seamless processing [34]. GEE stores data and completes processing on numerous remote servers (i.e., parallel processing), removing the need to download and process data on local stand-alone computers. This eliminates many barriers related to the hardware and technical expertise required for remote sensing. All that is required to use GEE is a computer capable of running a modern web browser and an internet connection—for development, research, or educational purposes, access is freely granted through Google, LLC (Limited Liability Corporation), by signing up through the GEE Homepage. These advancements allow for developing and carrying out mapping methodologies over unprecedented spatial extents with drastically increased speed (e.g., University of Maryland Global Forest Dynamics), making advanced remote sensing applications accessible to considerably broader audiences [34,37]. In addition, tools built for GEE and distributed over the Internet can facilitate methodological repeatability while providing opportunities for adaptability and customization [38].

To date, several studies have explored and demonstrated the utility of GEE for mapping mangroves yielding encouraging results and improvements over conventional methods [39–43]. While there is clear utility for mapping and monitoring mangrove ecosystems using GEE, published methodologies remain inaccessible to many would-be users. To replicate published methods requires an advanced level of specialized expertise with remote sensing, geospatial processing techniques, and/or coding. To date, no intuitive and accessible version of a mangrove mapping methodology within GEE has been proposed which caters to a wider audience of non-specialist conservation managers and decision makers. In addition, existing tools fail to fully capitalize on the wealth of local knowledge and understanding often held by coastal managers. Lastly, no single methodology comprehensively incorporates all of the best available options for mapping and monitoring mangrove ecosystems from across existing published studies and includes a widely applicable approach toward tidal calibration.

Herein we present a comprehensive, intuitive, accessible, and replicable methodology encapsulated in a new tool—the Google Earth Engine Mangrove Mapping Methodology (i.e., the GEEMMM). The GEEMMM was designed to provide a ready-to-go methodology for non-expert practitioners to map and monitor mangrove ecosystems, enabling them to combine their local knowledge with GEE's cloud computing capabilities. We developed the GEEMMM following a thorough review of mangrove remote sensing literature and incorporating the best available practices. In addition, our approach to tidal calibration operates completely within GEE based entirely on shoreline reflectance (i.e., image-based). To demonstrate the tool, we present an example of multi-date, desk-based (i.e., involving no field work) mapping and change assessment for Myanmar (Burma)—a global loss hotspot [19]. The GEEMMM—freely accessible to non-profit users—runs on detailed and well commented code within the GEE environment and is adaptable to any mangrove area of interest. GEEMMM outputs include multi-date classified maps, accuracies, and dynamic assessments. To set the stage for trailing the GEEMMM for Myanmar and contextualizing the outputs, and similar to methods detailed in Gandhi and Jones [19], all existing single- and multi-date mangrove maps for Myanmar were inventoried, described, and compared, with an emphasis on existing information on distribution and dynamics. We introduce the pilot area of interest (i.e., AOI), describe existing datasets, overview the GEEMMM tool, and compare the results to existing datasets.

2. Materials and Methods

2.1. Google Earth Engine Mangrove Mapping Methodology (GEEMMM) Pilot AOI

2.1.1. Regional Context

The region encompassing south (S) Asia, southeast (SE) Asia, and Asia-Pacific is home to approximately 46% of the world's mangroves [5,44]. This region includes some of the world's most productive, oldest, and biodiverse mangrove forests [45]. Regional loss—the highest in the world—is driven by conversion to aquaculture ponds (i.e., shrimp and fish farms), oil palm plantations and rice paddies, coastal development, and over-extraction for wood [11,46–57]. Natural processes and phenomena (e.g., rising ocean temperatures and sea-levels, severe tropical storms, and natural disasters) also contribute to regional dynamics [48,53,56,58–66]. Notably, SE Asia is exceptionally biodiverse containing 51 of the world's 73 documented mangrove species, compared to 10 in the Americas and Africa [5,67]. SE Asia alone contains an estimated 34% of the world's mangroves [5,68]. Recent studies show that mangrove areas in SE Asia are experiencing the highest prevalence of anthropogenic activity in the world [68,69].

2.1.2. Myanmar—A Regional and Global Loss Hotspot

Located within SE Asia, the preliminary AOI for this pilot study is all of coastal Myanmar (Figure 1). As confirmed by Gandhi and Jones [19], within SE Asia, mangrove loss is most notable in Myanmar, making the country both a regional and global loss hotspot. Giri et al. [70] reported a 35% decrease in mangrove extent from 1975–2005 whereas De Alban et al. [57] reported a 52% decrease from 1996–2016 [57,70]. According to De Alban et al. [57] and Estoque et al. [56], the primary anthropogenic drivers of this loss include conversion to rice paddies, oil palm and rubber plantations, and increasingly for aquaculture (e.g., shrimp, fish) [56,57]. Natural drivers include tsunamis triggered by seismic activity, and tropical storms [68,71,72]. Within Myanmar, according to Giri et al. [70], Saah et al. [73], Bunting et al. [74], De Alban et al. [57], and Clark Labs [75] sub-national loss hotspots include the northwestern (NW) coastline, much of the Ayeyarwady peninsula, and a smaller area slightly east of the Ayeyarwady peninsula (Figure 1).

2.1.3. Myanmar—Inventory, Summary and Acquisition of Existing Datasets

All national-level mangrove datasets providing single- or multi-date coverage for Myanmar up to July 2020 were inventoried through an exhaustive online search and literature review. When available, datasets were obtained from online repositories or through contacting authors. When not available, datasets were described based on associated literature. All datasets were summarized based on producer/organization/reference, single- vs. multi-date, temporal and spatial extent, availability, imagery source(s), mapping methods, and whether discrete or continuous (Table 1).

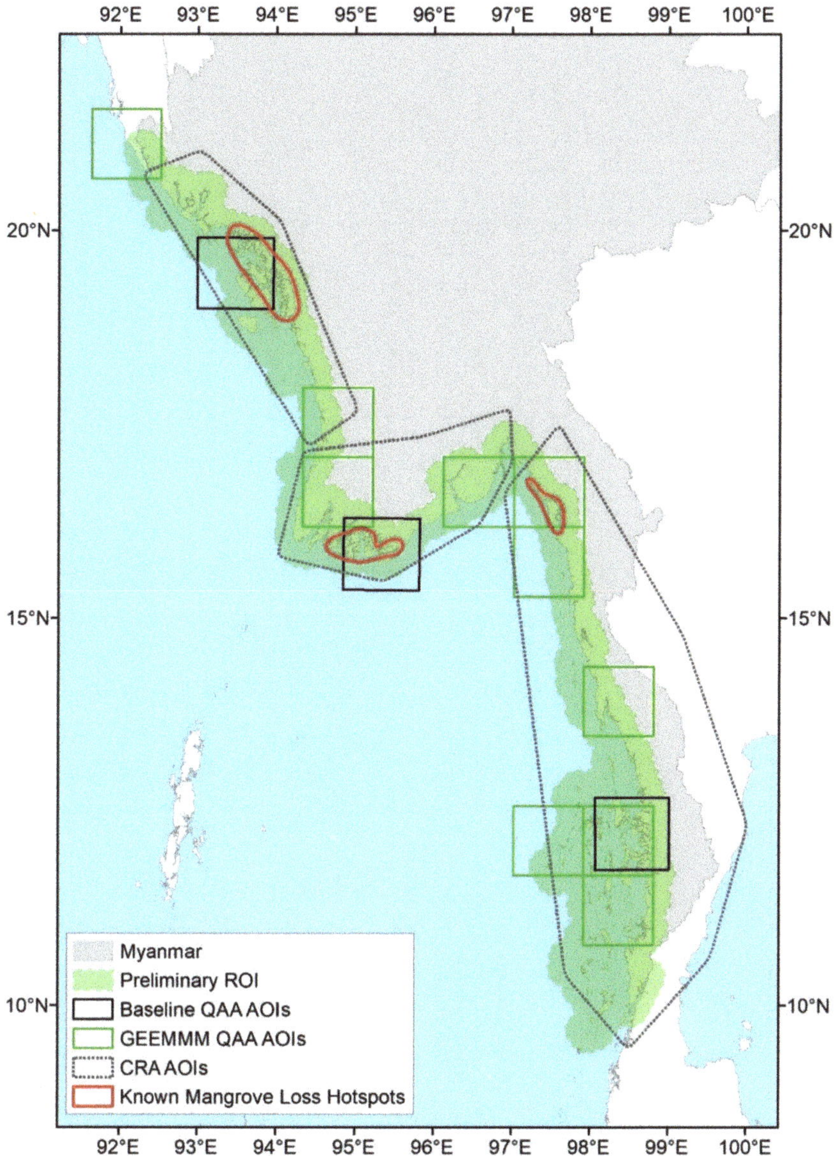

Figure 1. The preliminary region of interest (ROI) for the GEEMMM pilot representing coastal Myanmar; sub-national AOIs wherein qualitative accuracy assessments (QAAs) were untaken for existing maps (i.e., Baseline QAA AOIs); sub-national AOIs wherein GEEMMM QAAs were undertaken (i.e., GEEMMM QAA AOIs). Also shown are sub-national AOIs wherein classification reference areas (CRAs) were derived (i.e., CRA AOIs) and the location of known mangrove loss hotspots based on existing studies (i.e., Giri et al. [70], Saah et al. [73], GMW (Bunting et al. [74]), De Alban et al. [57], Clark Labs [75].

Table 1. Inventory and summary of existing national-level mangrove datasets for Myanmar—July 2020.

Author(s)	Year(s)	Spatial Extent/Resolution	Availability	Imagery Source(s)	Methods	Discrete/ Continuous
Giri et al. [70]	1975, 1990, 2000, 2005	6 tsunami-affected countries/30 m	Available from authors	Landsat	Hybrid supervised/unsupervised classification (ISODATA† clustering)	Discrete
SERVIR-Mekong Regional Land-Cover Monitoring System—Saah et al. [73]	1987–2018 (V3)	Greater Mekong region/30 m	Downloadable from SERVIR-Mekong website (at c. 120 m resolution; available from authors at 30 m)	Landsat, MODIS	Supervised classification (Support Vector Machine; Random Forest)	Discrete
Global Mangrove Watch—Bunting et al. [74]	1996, 2007–2010, 2015–2016	Global/25 m	Downloadable from Ocean Data Viewer	Jers-1, ALOS, ALOS-2, Landsat	Supervised classification (Random Forest); histogram thresholding [57]	Discrete
de Alban et al. [57]	1996, 2007, 2016	National/30 m	Available from authors	Landsat, JERS-1, ALOS, ALOS-2	Supervised classification (Random Forest)	Discrete
Stibig et al. [76]	1998–2000	S and SE Asia/1 km	Downloadable from JRC	SPOT-4	Unsupervised maximum likelihood classification	Discrete
Blasco et al. [77]	1999	Bangladesh and Myanmar/20 m	Not available	SPOT 1, 2, 3	Visual interpretation and supervised classification	Discrete
Clark Labs [75]	1999, 2014, 2018	Multi-national/30 m	Downloadable from Clark Labs website	Landsat	Mahalanobis classifier; hybrid supervised/ISOCLUST ‡ clustering	Discrete
World Atlas of Mangroves (WAM)— Spalding et al. * [5]	2000–2007	Global/30 m	Downloadable from Ocean Data Viewer	Landsat	Not disclosed	Discrete
Mangrove Forests of the World (MFW)— Giri et al. [44]	2000	Global/30 m	Downloadable from Ocean Data Viewer	Landsat	Hybrid supervised/unsupervised classification (ISODATA † clustering)	Discrete
CGMFC-21—Hamilton and Casey [78]	2000–2014	Global/30 m	2000–2012 data downloaded from CGMFC-21, 2013–2014 data available from authors	Landsat	Masked Global Forest Change (GFC) [47] maps using MFW [47]) to calculate dynamics	Continuous
Richards and Friess [47]	2000, 2012	SE Asia/30 m	Not available	Landsat	Masked GFC maps using MFW [56] to calculate loss	Continuous
Estoque et al. [56]	2000, 2014	National/30 m	Not available	Landsat	Unsupervised classification (ISODATA ‡ clustering)	Discrete

* WAM data over Myanmar from Ministry of Forestry's Remote Sensing and GIS Section, derived from Landsat imagery 2000–2007. † Iterative Self-Organizing Data Analysis Techniques. ‡ Iterative Self-Organizing Clustering.

2.1.4. Myanmar—Comparison of Existing Datasets and Baseline QAA

Once inventoried, all known datasets were compared based on mapped classes, mangrove distribution, accuracy, dynamics (when available), and known limitations. Where provided, mangrove distributions and dynamics were extracted from publications and supporting materials. If not readily apparent—and if the datasets were available—dynamics were calculated. Adding to the standard reported metrics, the accuracy was further qualitatively assessed for all available datasets through cross-checking in reference to high spatial resolution satellite imagery viewable through Google Earth Pro (GEP) [79]. This secondary qualitative accuracy assessment—or QAA—first reported in Gandhi and Jones [19], provides a more thorough understanding of existing mangrove datasets.

The QAA of existing maps (i.e., baseline QAA) was undertaken for the most recent entry in each discrete dataset, when available. Datasets were acquired in both raster and vector format, and in a range of coordinate systems, necessitating several pre-processing steps. For each baseline QAA, three 100 × 100 km sub-national AOIs were selected across Myanmar: in the north (Rakhine), in the center (Ayeyarwady Delta), and in the south (Tanintharyi) (Figure 1). Each baseline QAA AOI was divided into 10 × 10 km boxes, and working from NW to SE, every sixth box was selected for spot-checking, such that approximately 17% was systematically assessed. QAA of the Giri et al. [70] dataset was already conducted [19]. For the remaining datasets, each spot-check entailed comparing mangrove coverage to GEP imagery as close to the dataset's capture date as possible. In some instances, particularly in the southernmost Tanintharyi AOI, GEP imagery was partly/fully cloud-covered, limiting the ability to conduct QAA (limitations also noted by Estoque et al. [56]). A single mangrove class, representing the variability of canopy cover, height and stand structure in mangrove forests (as used in GEEMMM pilot classifications and defined below) was qualitatively assessed within each spot-check as either well-, under-, or over-represented. For each dataset, results help contextualize the representation of mangrove distribution and dynamics.

2.2. The Google Earth Engine Mangrove Mapping Methodology (GEEMMM)

The GEEMMM is intended to facilitate the mapping and monitoring of mangrove ecosystems anywhere in the world, without requiring a dedicated in-house geospatial expert. Intended users need basic computer skills and an understanding of the key steps required for mapping mangroves, but are not expected to hold advanced expertise in remote sensing, geospatial analysis, and/or coding. The interactive tool is broken into three modules—Module 1: defines customized region of interest (ROI) boundaries and generates multi-date imagery composites; Module 2: examines spectral separation between target map classes and undertakes multi-date classifications and accuracy assessments; Module 3: explores dynamics and offers an optional QAA. Each module is broken into thoroughly commented and referenced sections, bringing the user through all steps while making reference to this manuscript for full methodological details and context. Each module and the parameters used in this pilot study are described below. Table 2 provides a summary of all GEEMMM user inputs and variable selections for the Myanmar pilot study.

2.2.1. Module 1—Defining the ROI and Compositing Imagery

In the first step of Module 1: Section 1, the user must identify key datasets to be used in the GEEMMM. The first user-defined dataset is a preliminary ROI. This is generated using the 'drawing tools' function built into GEE and clips all subsequent user-defined datasets. The second user-defined dataset is the known extent of mangroves which is used to calculate elevation and slope thresholds and shoreline buffer distance. The user can select the baseline GEE data set representing global mangrove distribution circa 2000 (i.e., Giri et al. [44]) or upload their own. The third user-defined dataset is coastline, for which the user can select the baseline Large Scale International Boundary Polygons [80] or upload their own. The fourth user-defined dataset concerns elevation which is required to generate topographic masks (i.e., elevation and slope). The user can select the GEE JAXA-ALOS satellite radar

DSM (30 m) [81] or upload their own. For the Myanmar pilot, the preliminary ROI is shown in Figure 1, the GEE global mangrove distribution circa 2000 was used for known mangrove extent, the Global Administrative Boundaries database (GADM) Myanmar dataset (v3.6, www.gadm.org, an external source) for coastline, and the GEE JAXA-ALOS Global PALSAR-2/PALSAR Yearly Mosaic 25 m land-cover data for elevation [81,82].

Table 2. Summary of GEEMMM user inputs and selected variables used in the Myanmar pilot.

GEEMM User Inputs.			
Module	**Input**	**Type**	**Selected Variables**
Module 1	Preliminary ROI	Dataset (vector)	GUI Generated
	Known Mangrove Extent	Dataset (raster)	Giri et al. [44]
	Coast Line	Dataset (vector)	GADM—Myanmar [82]
	Digital Surface Model	Dataset (raster)	JAXA-ALOS DSM (30 m) [81]
	Contemporary Year(s)	Date range (YYYY)	2014–2018
	Historic Year(s)	Date range (YYYY)	2004–2008
	Month(s)	Date range (MM)	06–12
	Cloud Cover Limit	Integer (%)	30%
	Cloud Cover Mask	Variable	Aggressive
	Tidal Zone	Numeric (m)	1500 m
	Water Mask	Variable	Combined
	Fringe Mangroves	Boolean	False (not included)
	Topographic Mask	Variable	Uses Known Mangrove Extent [44]
Module 2	Classification Reference Areas (CRAs)	Dataset (vector)	See Table 4
	Class Names	Variable	See Table 4
	Class Numbers	Integer	Defined by Authors
	Classification Algorithm	Random Forest	Random Forest [82]
	Number of Trees	Integer	200
	Output Classification Maps	Variable	Hist. and Cont. Combined
Module 3	Classification Reference Areas (CRAs)	Dataset (vector)	See Table 4
	Class Names	Variable	See Table 4
	Class Numbers	Integer	Defined by Authors
	Classification View	Variable	See Figure 7
	Mangrove Class Number(s)	Integer	Defined by Authors

In the second step of Module 1: Section 1, the user defines input variables and sets how workflow thresholds are calculated. Table 2 lists all of the user variables and user inputs for the GEEMMM including those used in this pilot study. GEE provides unprecedented access to the Landsat catalog, offering approximately 1.3 M scenes from 1984 to present [34]. While it is certainly advantageous to have access to so many images, the choice of imagery based on parameters such as year(s) and time of year(s) must be considered carefully. Two variables define contemporary and historic year(s) of interest. There are two four-digit date inputs to bookend the historic and contemporary year windows. If the user wishes to isolate a single historic or contemporary year the same is selected for each book-end. Following the year(s) of interest, the month(s) of interest are selected. Seasonal variations can affect terrestrial vegetation adjacent to mangroves, and atmospheric conditions can change throughout the year, so the ability to target specific months is essential to generating optimal image composites [83–85]. The user identifies the month(s)-of-interest using two book-end numbers corresponding to the 12 months of the year; they may overlap the new year; e.g., "11" (Nov.) to "2" (Feb.). Next, the allowable cloud cover limit, an integer between 0 and 100, is used to filter the Landsat metadata [86]. Also related to cloud cover, the user decides whether to mask the imagery, and to what extent, i.e., setting a mild cloud mask using the USGS-provided (United States Geological Survey) quality band, or an aggressive cloud mask where pixels are excluded based on their 'whiteness' and a temperature band threshold [35]. For the sixth input, the approximate tidal zone—a numeric input (in m) that represents the tidally active zone buffered inland from the coastline—is entered. Approximate tidal zone helps isolate the portion of images subject to reflectance changes from tidal variation, while reducing influence from other non-tidal variability. The default value is 1000 m. Next, the user chooses how water is masked out of the imagery, either using a mask developed from the water present in the contemporary imagery alone, or a combination mask based on pixels determined to be water in both historic and contemporary imagery. A pixel is determined to be water if its value was greater than the 0.09 modified normalized difference water index (i.e., MNDWI) threshold established by Xu [87]. The modified normalized difference water index (MNDWI) was developed to detect water pixels by calculating the normalized difference between the green and short-wave infrared (e.g., Landsat 8 Operational Land Imager, 1.57–1.65 μm) bands, making it suitable for measuring the amount of water present in an acquisition. Topographic thresholds are set to generate masks based on elevation and slope. The user can either manually enter the elevation (m) and slope (%) thresholds, or have them automatically calculated based on the 99th percentile values extracted from within the known mangrove extent dataset. The user can further opt to search for inland-fringe mangroves, which have been documented as far as 85 km inland [75,88]. If inland-fringe mangroves are targets for the classification(s), the preliminary ROI is doubled for elevations lower than 5 m based on [89]. The last step in Module 1: Section 1 is the selection of spectral indices which the user would like to calculate for each image composite. After the workflow begins, the user chooses which indices they would like to calculate from a list of fourteen indices, including some which are mangrove-specific. The complete list of indices included in the GEEMMM can be found in Table 3. The contemporary and historic windows from which imagery was selected for the Myanmar pilot study were 2014–2018 and 2004–2008, respectively. The months of acquisition were limited to June through December, corresponding with the wet season and the months directly following that time [90]. The imagery was filtered using cloud cover information for each acquisition at a 30% threshold. All 14 spectral indices were selected for calculation.

Module 1: Section 2 determines the finalized ROI for processing. Numerous studies have demonstrated the utility of reducing the classification extent to the minimum required area—this approach helps reduce spectral confusion with unnecessary scene components [44,91]. The preliminary ROI is used to isolate a section of shoreline which is buffered at 5, 10, 15, 20, 25, 30, and 35 km intervals. 5 km intervals were used to ensure observable differences in buffer distances. 35 km was used as a maximum extent based on observations in several countries, including Myanmar. These buffer distances are used to calculate the area of known mangroves that falls within their respective bounds. The user either selects their buffer distance preference from a drop-down menu containing values in between, greater than, or less than the listed intervals.

In either case, the buffer distance is used to create the finalized ROI. This ROI is used to select Landsat path/row tiles and generate image composites, clip composite imagery and masks (i.e., elevation, slope, and water), define the classification and dynamics extent, and provide a visual aid for optional QAAs. The finalized ROI used in the pilot study was based on a 23 km buffered shoreline which represents the maximum observed distance between known mangrove extent (i.e., Giri et al. [44]) and Myanmar's coastline.

Module 1: Section 3 generates the imagery composites required for multi-date classifications. Given the daily dynamic nature of mangrove ecosystems—wherein tides inundate 2–3 times per day on average—tidal conditions and the associated presence (or lack thereof) of water must be considered—there are a growing number of mangrove detection indices which rely on the isolation of high and low tide imagery [29,92,93]. The GEEMMM uses an image-based approach to calibrate imagery based on high and low tide. For each available image, an MNDWI is generated and the land is masked out using JAXA-ALOS Global PALSAR-2/PALSAR Yearly Mosaic 25 m land-cover data. The MNDWI was selected as the key spectral index because it has been proven to be an improvement over the normalized difference water index (NDWI), and was developed explicitly for detecting water and non-water pixels [87]. The shoreline is buffered to the user-defined tidal zone value and mean

MNDWI is used to create a constant value band wherein the greater the MNDWI mean value, the more water present within the tidal zone, corresponding to higher-tidal conditions. A second value band is added to each available image by multiplying mean MNDWI by −1, isolating lower-tide conditions. Clouds, if present and opted to be, are masked prior to the calculation of mean MNDWI using only the pixel quality band or an aggressive approach where the three visible (red, greed, and blue) and thermal bands mask based on digital number reflectance thresholds. Under the aggressive filter, a pixel is considered to be a cloud if its visible spectrum bands digital number reflectance values are greater than or equal to 1850, and the thermal band (brightness temperature, Kelvin) digital number is less than or equal to 2955. For the Myanmar pilot, the aggressive cloud filter option was selected to filter the imagery in an effort to remove low-altitude clouds which were not correctly classified by the Landsat cloud detection algorithm. If/once clouds have been masked, all available images and their corresponding tidal value bands are used to create best available pixel-based highest observable tide (i.e., HOT) and lowest observable tide (i.e., LOT) composites. Composite generation works as if all available images were stacked and organized by desired tidal condition. For example, as the LOT composite is being generated, the imagery with the lowest observed tidal condition is placed on top, and any missing pixels in that image, e.g., clouds masked, would be filled by the next best tidal observation and so on until all the gaps are filled. This process takes place for both the contemporary and historic data sets, resulting in a maximum of four composites (i.e., HOT and LOT contemporary, HOT and LOT historic). Because tides are determined using value bands, it is possible that all of the pixels for HOT and/or LOT composites within a particular area may be from one image (e.g., if no clouds were present and that image represented best available tidal conditions). The GEEMMM employs USGS surface reflectance Landsat products, which are readily available within GEE [35,94].

Module 1: Section 4 calculates the user selected indices from Section 1 of Module 1 (Table 3). There are a growing number of Landsat-related spectral indices available, many of which relate directly to mangroves such as the submerged mangrove recognition index (SMRI) and the modular mangrove recognition index (MMRI) [29,43]. The GEEMMM provides the user with the option to select from 14 spectral indices, of which four are mangrove-specific. The selected indices are calculated for both contemporary and historic HOT and LOT composites and added as potential classification inputs. Figure 2 compares the appearance of a typical mangrove-dominated area in Myanmar across all of the available mangrove-specific spectral indices (i.e., combined mangrove recognition index (CMRI), MMRI, SMRI, MRI) in the GEEMMM [29,43,93,98].

In Module 1: Section 5 the classification extent is further reduced through masking. In accordance with numerous mangrove mapping studies (e.g., Jones et al. [91], Thomas et al. [68], and Weber et al. [105]), the GEEMMM incorporates cloud, water, slope, and elevation masks to produce a finalized AOI. The cloud mask is generated and applied before composites are produced. The water mask is calculated for each composite using the methodology established in Xu [87], where the MNDWI layer for historic and contemporary LOT composites are generated and then a threshold is applied. Pixels with a value greater than 0.09 are considered to be water and a binary mask is produced. Depending on user selection, the water mask is finalized by either using just contemporary or combining the historic and contemporary and selecting only pixels determined to be water in both composites. This pilot study used the combined water mask. The two topographic masks are generated through user-defined thresholds or automatically determined using the 99th percentile of elevation and slope for known mangroves. The Myanmar pilot study used the known mangrove extent to generate topographic masks based on elevation values > 39 m and slope values > 16%. Noting how minor elevation is within mangrove ecosystems, the elevation threshold actually represents an approximate combined elevation + canopy height past which mangroves are not found. The generated masks are combined to create a binary, single unified final mask which is applied to all composites within the finalized ROI.

Module outputs include: (1) HOT contemporary composite, (2) LOT contemporary composite, (3) HOT historic composite, (4) LOT historic composite, (5) finalized ROI, and (6) Finalized Mask.

Table 3. List of all spectral indices available in the GEEMMM including mangrove-specific.

Index	Abbreviation	Calculation	Citation
Simple Ratio	SR	NIR/Red	Jordan [95]
Normalized Difference Vegetation Index	NDVI	(NIR − Red)/(NIR + Red)	Tarpley et al. [96]
Normalized Difference Water Index	NDWI	(Green − NIR)/(Green + NIR)	Gao [97]
Modified Normalized Difference Water Index	MNDWI	(Green − SWIR1)/(Green + SWIR1)	Xu [87]
Combined Mangrove Recognition Index	CMRI *	NDVI − NDWI	Gupta et al. [98]
Modular Mangrove Recognition Index	MMRI *	(|MNDWI| − |NDVI|)/(|MNDWI| + |NDVI|)	Diniz et al. [43]
Soil-Adjusted Vegetation Index	SAVI	1.5*(NIR − Red)/(NIR + Red + 0.5)	Huete [99]
Optimized Soil-Adjusted Vegetation Index	OSAVI	(NIR − Red)/(NIR + Red + 0.16)	Rondeaux et al. [100]
Enhanced Vegetation Index	EVI	2.5*((NIR − red)/NIR + 6*Red − 7.5*Blue + 1))	Huete et al. [101]
Mangrove Recognition Index	MRI *	|GVI(l) − GVI(h)|*GVI(l)* (WI(l) + WI(h))	Zhang and Tian [93]
Submerged Mangrove Recognition Index	SMRI *	(NDVI(l) − NDVI(h))* ((NIR(l) − NIR(h))/(NIR(h))	Xia et al. [29]
Land Surface Water Index	LSWI	(NIR − SWIR1)/(NIR + SWIR1)	Chandrasekar et al. [102]
Normalized Difference Tillage Index	NDTI	(MIR − SWIR2)/(MIR + SWIR2)	Van Deventer et al. [103]
Enhanced Built-up and Bareness Index	EBBI	(SWIR1 − NIR)/(10*√(SWIR1 + LWIR))	As-syakur et al. [104]

* denotes mangrove-specific spectral index.

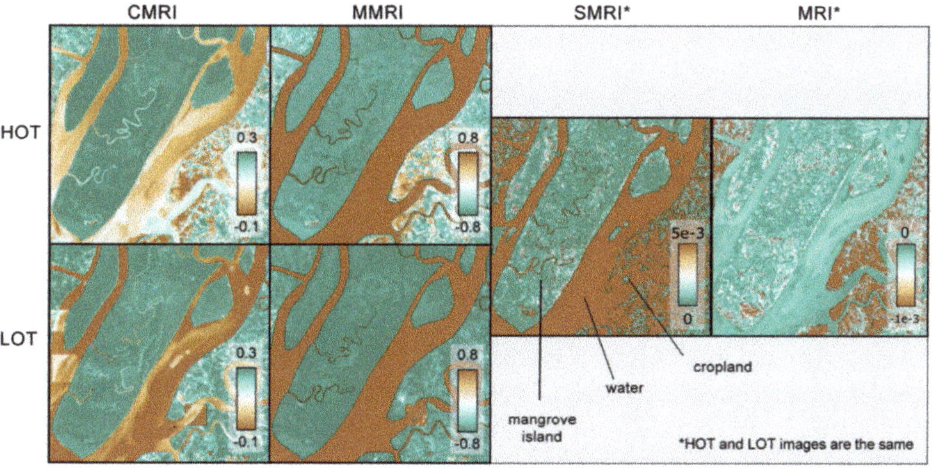

Figure 2. The appearance of a typical mangrove-dominated area in Myanmar across all of the available mangrove-specific spectral indices (i.e., CMRI, MMRI, SMRI, MRI) in the GEEMMM [91].

2.2.2. Module 2—Spectral Separability, Classifications and Accuracy Assessment

For Module 2: Section 1, user inputs address classification variables and settings. The user enters the asset path for historic and contemporary classification reference areas (CRAs) (i.e., the user-defined examples of target map classes) and identifies the unique column labels for class names and numeric codes. Next, the user identifies whether CRAs are spatio-temporally invariant (i.e., each CRA represents a class example in both contemporary and historic imagery). If the CRAs are not spatio-temporally invariant, the spectral properties of the contemporary CRAs are extracted and used to define class boundaries in the historic classification(s). For classification algorithm the single option is currently random forest [106]. The user determines how many trees are employed. The final input determines classification outputs. Users have the option to select outputs from either HOT or LOT composites for contemporary and historic inputs (i.e., four possible outputs), and/or a combined classification where HOT and LOT composites are merged to create single outputs (i.e., two more possible outputs), totaling six possible classification outputs. Zhang and Tian [93] demonstrated the utility of using combined HOT and LOT image composites as classification inputs. For the Myanmar pilot, 200 trees were selected with outputs based on combined (i.e., HOT and LOT) historic and contemporary classifications (i.e., two classifications).

In Module 2: Section 2, the user can examine correlation between potential spectral indices and the spectral separability of CRAs across all potential classification inputs. The Pearson's correlation is calculated for each selected index to all others and these values are used to generate a correlation matrix with values ranging from −1 to 1 [107]. A value of 1 means that the potential inputs have a perfect, positive, linear correlation, and a value of −1 indicates that the indices have a perfect, negative, linear correlation. Users are encouraged to select indices that are not highly correlated indicating that they provide unique information. As a general rule, correlation coefficients with absolute values greater than 0.7 are considered moderately to strongly correlated and thus present similar information [107]. Users are advised to consider that correlation coefficients are also impacted by the amount of variability in the data, the shapes of distributions, and the presence of outliers among other factors [108].

The spectral separability between target map classes as represented by CRAs is explored through the generation of three types of graphs. First, the user can view the spectral separability between each target class and each Landsat band—the user has the option to view this output for each of the four imagery composites. Box-and-whisker plots show the min, max, and inter-quartile range for each band and each map class. The second set of graphs is similar to the first, except that spectral separability is shown for individual indices across all of the target classes, showing only one index at a time. The final graph shows spectral feature space, where the x and y axes are user selected bands or indices. For the pilot study, and based on previously established precedents in Jones et al. [91], we included the visible, NIR and SWIR Landsat bands. Based on the correlation matrices and further the spectral separability they provided, the MNDWI, CMRI, MMRI, enhanced vegetation index (EVI), and Land Surface Water Index (LSWI) indices were selected as additional classification inputs.

For piloting the GEEMMM in Myanmar, six classes were initially targeted, including, (1) closed-canopy mangrove, (2) open-canopy mangrove, (3) terrestrial forest, (4) non-forest vegetation, (5) exposed/barren, and (6) residual water. Table 4 provides class descriptions and an overview of how many CRAs were digitized per class. CRAs can be derived within the GEE environment or externally. For this pilot, 90×90 m (i.e., 3×3 Landsat pixels) CRAs were derived externally. To ensure that internal class variability was captured for each class and across the AOI, three sub-national AOIs were used to define CRAs (Figure 1). CRAs were derived referring to finer spatial resolution satellite imagery viewable in Google Earth Pro (Google, Mountain View, CA, USA), existing contemporary land-cover maps for Myanmar (i.e., Giri et al. [44], Saah et al. [73], and De Alban et al. [57]), and expert interpretive knowledge gained with mapping mangroves in other regions of the world. Two mangrove classes were defined to ensure that the internal variability of mangrove forests based on stature, canopy cover and density was captured. Figure 3 shows examples of all targeted classes in HOT, LOT, a key spectral index, and finer spatial resolution imagery viewable in Google Earth Pro (Google, Mountain View, CA, USA) [79].

Table 4. Names and description of classes and numbers of classification reference areas (CRAs). Also shown is how many CRAs were derived within each sub-national CRA AOI (Figure 1).

Class	Class Description	Contemporary				Historic			
		AOI 1	AOI 2	AOI 3	Total	AOI 1	AOI 2	AOI 3	Total
Non-Forest Vegetation	Grass and/or shrubs dominate; some exposed soil + scattered trees; canopy < 30% closed; active cropland, vegetation appears green	10	8	7	25	3	7	0	10
Terrestrial Forest	Forested areas; canopy > 30% closed (includes plantations (e.g., palm))	10	8	7	25	1	9	0	10
Closed-Canopy Mangrove	Tall, mature stands; canopy > 60% closed	12	16	9	37	9	1	0	10
Open-Canopy Mangrove	Short-medium stands; canopy 30–60% closed	6	3	2	11	0	10	0	10
Exposed/Barren	Soil/sediment/sand dominates; includes senesced/unhealthy (i.e., inactive) crops, mudflats, recently deforested areas	4	4	4	12	2	4	4	10
Residual Water	Water areas missed from masking	4	3	3	10	3	4	4	11
					120				61

In Module 2: Section 3: once the user confirms their final choice of classification inputs and target classes, classification—the process by which remotely sensed data is assigned land-cover classes—can occur [109,110]. There are many established algorithms for classifying Landsat data to produce maps of mangrove distribution, including classification and regression trees (CART), support vector machines (SVM), unsupervised k-means, decision trees, and maximum likelihood (ML) [28,29,42,111,112]. Many of these algorithms are available to use within the GEE environment; however, random forests—also available in GEE—is well established and used to map mangroves across the world, with distinct success within the GEE environment [27,41,43,106,113]. The inputs for random forest include an imagery data set (i.e., selected Landsat bands and spectral indices), training data (i.e., randomly selected 70% of CRAs), and a numeric parameter determining the number of 'trees' to be employed. For each classification the output is a single band raster with the same spatial resolution as the input data (30 m), with each pixel assigned a map value based on target classes. Following classification, the user can choose to merge map classes—this is particularly advantageous in scenarios where initial map classes were used to capture variability, but for which confidence in class boundaries may be lacking. For example, in the Myanmar pilot, we merged the two mangrove classes (i.e., closed- and open-canopy) post-classification. This ensured capturing mangrove variability while not having to draw a distinct boundary between these potentially overlapping classes in the final map.

Classification accuracy—defined as "a comparison of the derived product to ground condition"—is not reported in numerous studies involving mangrove mapping [114,115]. Following classification and optional class merging, in Module 2: Section 3, the GEEMMM automatically produces resubstitution and error matrices for all output classifications [116]. The resubstitution matrices determine end land-cover class for the CRAs used for training the classifier. The error matrices use 30% of CRAs held back from classification to independently evaluate map accuracies. The overall accuracy is reported using the error matrix 'accuracy' tool, found within the GEE library. Overall accuracy is printed below both the error and resubstitution matrices. By reviewing the error matrices and visually inspecting the output maps the user may wish to collapse/further collapse classes (e.g., if two classes are very confused). If the user combines classes, they can opt to re-calculate accuracy, re-generating resubstitution and error matrices. The final step for all users to exporting the classification maps to their assets. Module 2 outputs include, (1) correlation and spectral separability graphs, (2) classified maps, and (3) accuracy assessments.

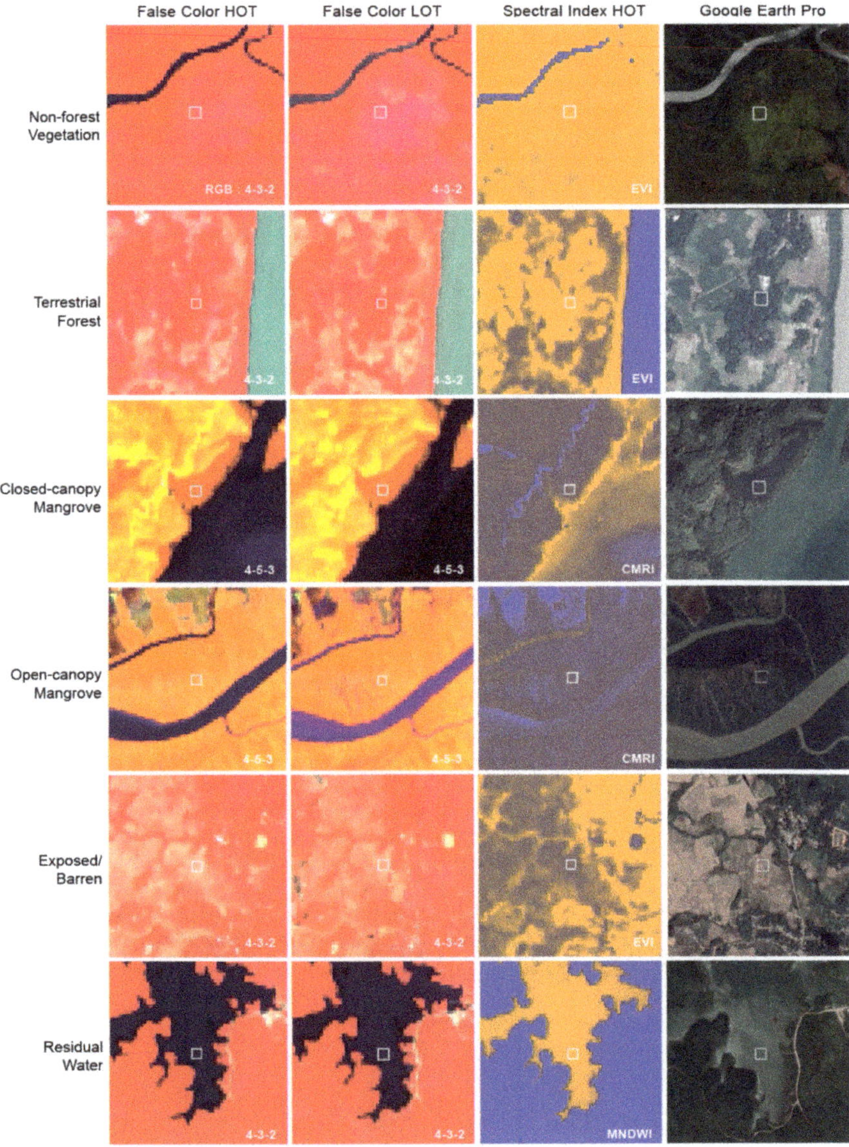

Figure 3. The appearance of all targeted classes in highest observable tide (HOT), lowest observable tide (LOT), key spectral indices, and fine spatial resolution satellite imagery viewable in Google Earth Pro (Google, Mountain View, CA, USA) [79]. The HOT and LOT composites represent 432 (R: NIR, G: red, B: green) or 453 (R: NIR, G: SWIR, B: red) false color. The spectral indices include enhanced vegetation index (EVI—[101]), combined mangrove recognition index (CMRI—[98]) and modified normalized difference water index (MNDWI—[87]).

2.2.3. Module 3—Dynamics and QAA

In Module 3: Section 1, the user indicates which classification(s) will be used to calculate dynamics and/or assess optional QAA. If desired, the user can further clip classifications to a country's boundary— if pertinent—using the GEE Large Country Boundary Polygons, or by uploading an external dataset. For the

Myanmar pilot we further clipped using a uniquely uploaded boundary (GADM) and exclusive economic zone (EEZ) from Marine Regions (v10 World EEZ,) [117]. For the QAA, the user enters CRA information (e.g., asset path, class names, and unique class numbers).

In Module 3: Section 2, multi-date outputs are used to quantify dynamics. This is foundational to understanding long-term trends and the effectiveness of conservation efforts. The user selects which map class they would like to view, and loss, persistence, and gain (i.e., LPG) are calculated. The automatically produced, self-masked layers are added to the GEE-GUI map interface. The resulting area for each dynamic assessment is printed to the console, expressed in hectares. Building on the inventory, description, acquisition and comparison of existing datasets, the dynamics resulting from this GEEMMM pilot were also compared to published values.

Module 3: Section 3—building on the previously referenced methods detailed in Gandhi and Jones [19]—facilitates an optional QAA. For this GEEMMM QAA, an interactive map is divided into three linked maps (Figure 1). In each map, two sets of grids are automatically generated, (1) 100 km by 100 km grids, and (2) within each of those cells, sub-divided 10 km by 10 km grids. The 100 km × 100 km grid cells are randomly selected, retaining 50% of the grid cells that intersect the ROI. In slight contrast to the baseline QAA described in Section 2.1.4, for the QAA tool in the GEEMMM, within each selected grid cell, 20% of the sub-grid cells are selected. The tool works by cycling through the sub grid cells, and giving the user the option to view simultaneously on linked maps showing Landsat composites where the date can be changed at the user's preference, the classifications produced in Module 2, and the imagery used for the classifications generated in Module 1. The user then has the ability to record in the GUI whether each map class is under-, well-, or over-represented, and record 'free comments' for each sub-cell.

Module 3 outputs include: automatically generated LPG as raster and—if performed—QAA grid (for viewing outside of GEE). The user also has the option to export the QAA table (containing the under, over, and well representation statistics, and the free comments) as a CSV (i.e., comma separated values) file at any point during the QAA.

3. Results and Discussion

3.1. Myanmar—Comparison of Existing Datasets

Table 5 provides a comparison of all single- and multi-date datasets based on dataset/authors, year, extent (ha), dynamics (ha and %), whether discrete or continuous, mapped classes, accuracy, and known limitations. Figure 4 provides a comparison of all distributions across time across all datasets. Results show that Myanmar's mangrove distribution ranged from 851,452–1,323,300 ha circa 1975–1987 (i.e., historic) to 475,637–1,002,098 ha circa 2014–2018 (i.e., contemporary). Of the 11 existing studies, only five provided quantitative accuracy assessments, with overall accuracies ranging from 76% (i.e., Saah et al. [73]) to 97% (i.e., Estoque et al. [56]), mangrove producer's accuracies ranging from 75% (i.e., De Alban et al. [57]) to 93.1% (i.e., also De Alban et al. [57]), and mangrove user's accuracies ranging from 92.3% (i.e., De Alban et al. [57]) to 98.1% (i.e., Clark Labs [75]). Of the existing studies, eight provided dynamics, including a loss of 300,091 ha/35.2% from 1975–2005 (Giri et al. [70]), 195,227 ha/16.3% from 1987–2018 (Saah et al. [73]), 43,208 ha/8.0% from 1996–2016 (Bunting et al. [74]), 694,600 ha/52.5% from 1996–2016 (De Alban et al. [57]), 76,465 ha/10.9% from 1999–2018 (Clark Labs [75]), 27,064 ha/9.7% from 2000–2014 (Hamilton and Casey [78]), 27,770 ha/5.5% from 2000–2012 (Richards and Friess [47]), and 191,122 ha/28.7% from 2000–2014 (Estoque et al. [56]). Two reported specifically on sub-national loss hotspots (i.e., De Alban et al. [57] and Estoque et al. [56]). According to De Alban et al. [57], Bago, Mon, Yangon—the three states immediately to the east of the Ayeyarwady delta—suffered greatest proportionate loss from 1996–2016 totaling more than 80% of their extents. In terms of absolute loss, from 2000–2014, Estoque et al. [56] reported Rakhine as the state with the greatest loss (75,494 ha/39.5% of Myanmar's total loss), followed by Ayeyarwady experiencing 69,431 ha/36.3% of Myanmar's total loss.

Table 5. Comparison of single- and multi-date datasets based on mapped classes, accuracy, mangrove distribution (ha), dynamics, and known limitations. Accuracy: OA = overall accuracy; UA = user's accuracy; PA = producer's accuracy.

Dataset/ Author(s)	Year	Extent (ha)	Dynamics (ha, %)	Discrete/Continuous	Mapped Classes	Accuracy	Known Limitations
Giri et al. [70]	1975	851,452	−300,091 −35.2%	Discrete	4 classes including Mangrove	Positional root mean square error of ± 1/2 pixel	Semantic differences in class definitions, positional, and classification errors. Mangrove patches smaller than 1 ha not mapped likely reducing distribution figures.
	2005	551,361					
SERVIR-Mekong (Saah et al. [73])	1987	1,197,325	−195,227 −16.3%	Discrete	21 classes including Mangrove	OA 76% (2016 map)	Gap in 2012 data due to removal of ETM+ imagery following Landsat 7 Scan Line Corrector failure. 2012 primitives interpolated using Whittaker smoothing algorithm. Bias in reference data toward more recent past, due to availability of high-resolution imagery.
	2018	1,002,098					
GMW (Bunting et al. [74])	1996	537,428	−43,208 −8.0%	Discrete	Mangrove presence vs no presence	OA 95.3% (2010 baseline map)	Fine-scale features commonly misclassified, e.g., aquaculture features, riverine environments, and coastal fringes. Minimum mapping unit of 1 ha suggested for end user mapping.
	2016	494,220					
De Alban et al. [57]	1996	1,323,300	−694,600 −52.5%	Discrete	10 classes including Mangrove	1996: OA 85.6% Mangrove UA 92.3% Mangrove PA 93.1% 2016: OA 89.2% Mangrove UA 97.5% Mangrove PA 75.0%	No significant limitations disclosed.
	2016	628,700					
Clark Labs [75]	1999	703,945	−76,465 −10.9%	Discrete	7 classes including Mangrove	OA 96.9% Mangrove UA 98.11% Mangrove PA 93.04%	No significant limitations disclosed.
	2018	627,480		Discrete	33 classes including Mangrove	2014: OA 93.7% Mangrove UA 94% Mangrove PA 92%	
Blasco et al. [77]	1999	690,000	n/a	Discrete	8, including 6 mangrove classes	Not disclosed	Limitations with use of 'quick look' data due to modest technical performance. The authors state that classification accuracy could be improved by 10% if NDVI and empirical thresholds were included.
MFW (Giri et al. [44])	2000	494,584	n/a	Discrete	Mangrove presence vs no presence	Positional root mean square error of ±1/2 pixel	Small patches of mangrove (<0.09–0.27 ha) not well captured.

Table 5. *Cont.*

Dataset/ Author(s)	Year	Extent (ha)	Dynamics (ha, %)	Discrete/Continuous	Mapped Classes	Accuracy	Known Limitations
CGMFC-21 (Hamilton and Casey [78])	2000	279,260	−27,064 −9.7%	Continuous	Mangrove canopy cover	Positional root mean square error of ±1/2 pixel	Pixels containing just 0.01% forest canopy cover are included as mangrove falling well below commonly used minimum canopy cover definitions (e.g., [78,118,119]).
	2014	252,196					
Richards and Friess [47]	2000	502,466	−27,770 −5.5%	Continuous	Mangrove deforestation	Positional root mean square error of ±1/2 pixel	Reported figures reflect rates of mangrove loss rather than net mangrove change, likely reducing areal figures.
	2012	474,696					
Estoque et al. [56]	2000	666,759	−191,122 −28.7%	Discrete	Mangrove presence vs no presence	2000: OA 91% 2014: OA 97%	No significant limitations disclosed.
	2014	475,637					
WAM (Spalding et al. [5] *)	2004	502,911	n/a	Discrete	Not disclosed	Not disclosed	No significant limitations disclosed.
GEEMMM (Yancho et al., 2020)	2004–2008	995,411	−352,752 −35.4%	Discrete	6 classes including combined Mangrove.	2004–2008: OA 97.01% 2014–2018: OA 96.08%	Refer to Results and Discussion; Conclusion.
	2014–2018	642,659					

* WAM data over Myanmar from Ministry of Forestry's Remote Sensing and GIS Section, derived from Landsat imagery 2000–2007.

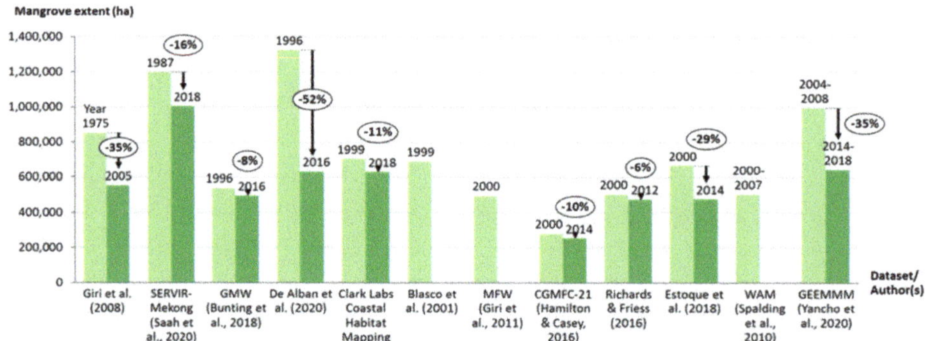

Figure 4. Comparison of distribution for all existing single- and multi-date mangrove distribution maps for Myanmar, including results of GEEEMMM pilot.

Direct comparisons of existing datasets are challenging due to differences in temporal coverage, methodologies, and imagery sources. Although most studies use optical imagery (typically medium-resolution Landsat), some of the more recent studies combine optical with radar imagery (e.g., Bunting et al. [74]; De Alban et al. [57]). Several different mapping techniques are employed, while two of the datasets (Hamilton and Casey [78]; Richards and Friess [47]) calculate and present continuous measures of mangrove canopy cover, rather than discrete (i.e., presence vs no presence). Interpreting continuous datasets for areal mangrove extent is problematic as pixels containing just 0.01% canopy cover are included as mangrove falling well below commonly used minimum definitions mangrove forest (e.g., 30%) [18,78,91].

Of the five datasets reporting, all achieve overall accuracies of >75%, with four >85% [56,57,74,75]. QAAs further identified Clark Labs [75] as mapping mangroves in Myanmar most consistently. Mangroves were under-represented in the remaining five datasets assessed, particularly in Giri et al. [70] and Saah et al. [73], but also in Bunting et al. [74] (Table 6).

Table 6. Results of QAA for available/acquired datasets (1 = best).

Rank	Dataset	AOI 1—Rakhine	AOI 2—Ayeyarwady	AOI 3—Tanintharyi	Overall Representation	Comments
1	Clark Labs [75]	Well-represented	Well-represented	Well-represented	Well-represented	Mangrove very well-represented
2	De Alban et al. [57]	Under-represented	Under-represented	Well-represented	Under-represented	Mangrove slightly under-represented; some confusion between cropland and mangrove
3	MFW (Giri et al. [44])	Well-represented	Under-represented	Under-represented	Under-represented	Mangrove under-represented
4	GMW (Bunting et al. [74])	Under-represented	Under-represented	Well-represented	Under-represented	Mangrove under-represented
5	Giri et al. [70]	Under-represented	Under-represented	Under-represented	Under-represented	Mangrove under-represented, considerably in places
6	SERVIR-Mekong (Saah et al. [73])	Under-represented	Under-represented	Under-represented	Under-represented	Mangrove under-represented, considerably in places

Remote Sens. **2020**, *12*, 3758

Existing studies clearly establish that Myanmar has experienced consequential mangrove loss; however, baseline distributions and dynamics (when available) are highly variable. These discrepancies are likely attributed to the differences highlighted in Table 5. In addition, the definitions for mangroves and surrounding land-cover classes and the actual examples used for classification (i.e., CRAs) likely further account for differences. Only with agreed upon conventions for defining mangroves and providing examples as CRAs can cross-study comparisons become standardized and optimized. Falling short of this, discrepancies will remain common.

3.2. Results of the Google Earth Engine Mangrove Mapping Methodology (GEEMMM)

3.2.1. Module 1—Defining AOI and Compositing Imagery

As confirmed through qualitative yet systematic spot-checks, the imagery generated from the Myanmar pilot reflects the selected inputs well—both historic and contemporary composites are mostly cloud- and artifact-free, and clearly represent distinct HOT and LOT conditions (Figure 5). Figure 6 shows a national overview of the AOI including contemporary and historic HOT and LOT composites. The challenges that have been identified can be attributed to the extent of the study area and trying to capture a long, complex coastline in a series of contiguous composites. The most notable challenge relates to seasonal variability observed primarily in areas where large clouds were masked out of one image and the pixels selected to fill captured seasonally different land-cover conditions. Notably, this issue was almost entirely associated with areas which undergo significant changes throughout the year, i.e., agricultural mosaics and non-forest vegetation. Even within the defined seasonal window, variability was observed. Users are advised to select meaningful seasonal windows that restrict such variability while still offering enough imagery to make optimal composites—this is constrained by the size of the AOI.

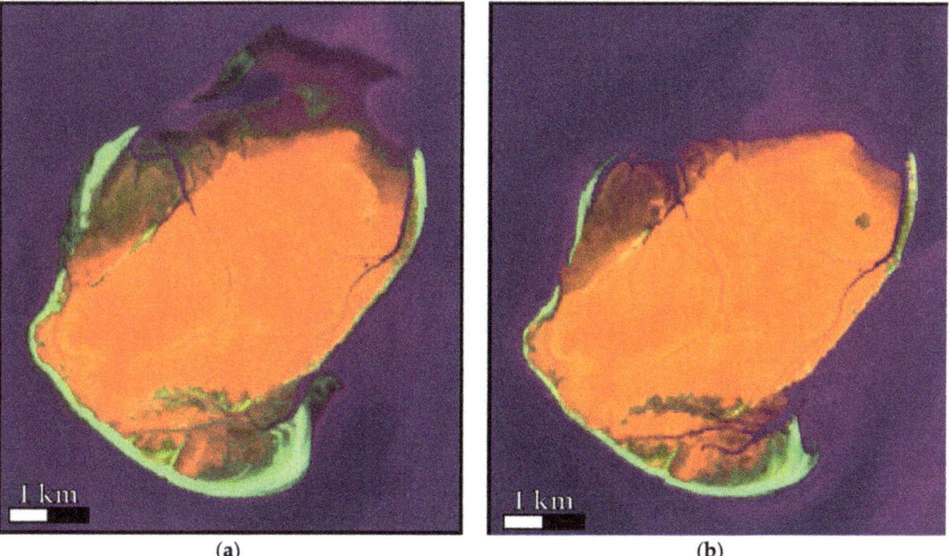

Figure 5. Examples of image composite outputs from the GEEMMM showing lowest observable tide (LOT), panel (**a**) and highest observable tide (HOT), panel (**b**). The north oriented, false colour (R: NIR G: SWIR B: Red) Landsat image is over Kaingthaung Island, Ayeyarwady Region, Myanmar.

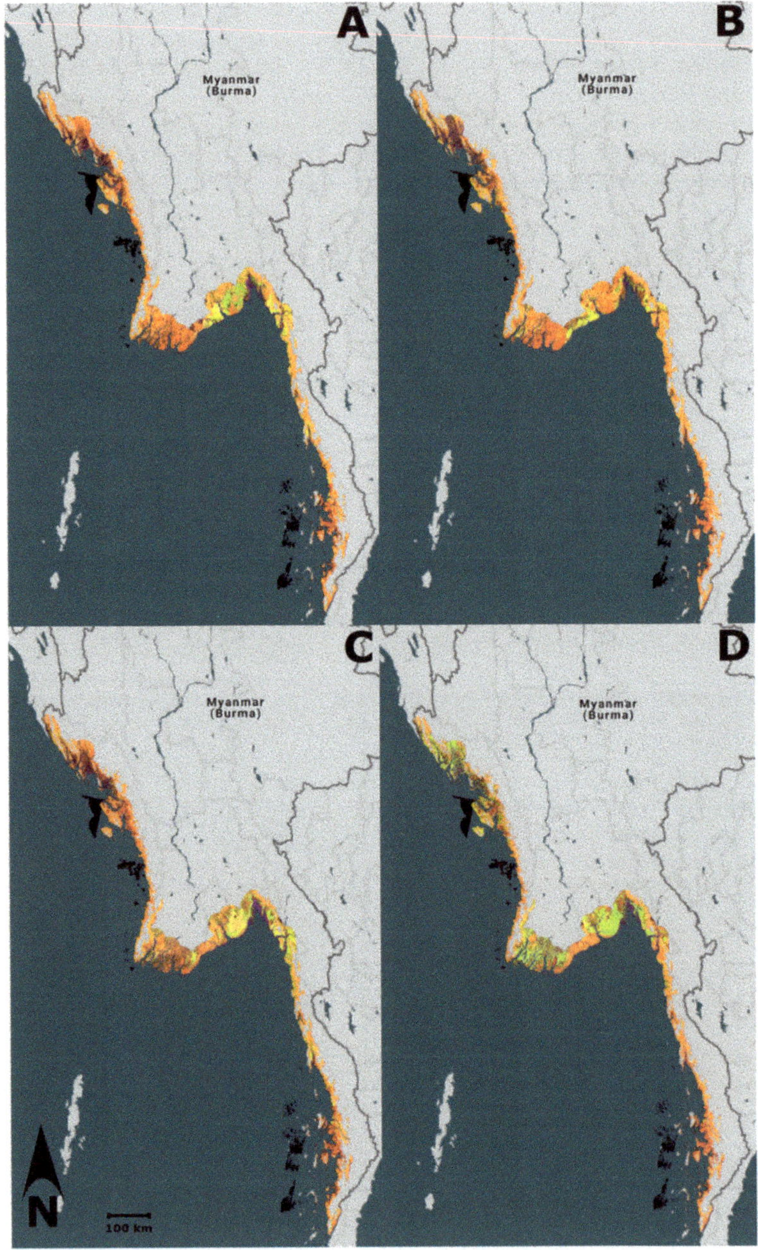

Figure 6. National overview of image composite outputs from the GEEMMM showing highest observable tide (HOT) and lowest observable tide (LOT) (false color composites, R: NIR, G: SWIR, B: Red). The composites were further reduced in area using topographic and combined water masks. (**A**) Contemporary HOT; (**B**) Contemporary LOT; (**C**) Historic HOT; (**D**) Historic LOT.

3.2.2. Module 2—Spectral Separability, Classifications and Accuracy Assessment

Based on correlation analysis of all available spectral indices, five (i.e., MNDWI, CMRI, MMRI, EVI, and LSWI) stood out as not correlated (Appendix A) and were selected as classification inputs. Using the spectral separability tools, all target classes as represented by CRAs were assessed across all non-thermal (red, green, blue, NIR, SWIR1, and SWIR2) Landsat bands (Figure 7), and each selected index was further evaluated to confirm that it provided additional separation for one or more classes (e.g., MMRI: Figure 8). Results indicate that bands SWIR1 and SWIR2 were particularly helpful in separating non-forest vegetation. Non-forest vegetation was most confused with other vegetation classes in the visible spectrum and MNDWI. For terrestrial forest, NIR, MNDWI, and MMRI provided separability. In particular, MNDWI provided good separation from mangroves; whereas within the visible spectrum and CMRI the most confusion was noted, particularly with other vegetation classes. Closed-canopy mangrove was best distinguished by LSWI, MMRI, and to a limited extent bands SWIR1 and SWIR2. In contrast, open-canopy mangrove was best distinguished by CMRI, MNDWI, NIR, SWIR1, and SWIR2. While there are meaningful and distinct differences between the two canopy-based mangrove classes, there is spectral overlap—this speaks to the advantage of capturing the variability within mangrove forests while subsequently merging into a single class post-classification. Field work is required to confidently define the boundaries between these sub-mangrove types to make them final map classes—following classification and prior to validation, mangroves were merged into a single class (i.e., mangrove). Taken together, the combined mangrove class exhibited some confusion with terrestrial forest and non-forest vegetation classes in EVI, the visible bands, and SWIR1 and SWIR2. The exposed/barren class had the most separability in indices CMRI, MMRI, and LSWI, and the most confusion with non-forest vegetation and terrestrial forest notably in MNDWI and residual water in the visible bands. Residual water was easily distinguished with MNDWI, and the non-visible bands, but was confused with exposed/barren in the visible bands, non-forest vegetation within CMRI, and all classes within EVI.

For both historic and contemporary classifications, resubstitution accuracies were 100%, indicating all training data was assigned to the correct land-cover class. Based on accuracy assessments using independent validation data, overall accuracies for historic and contemporary classifications were 97.0 and 98.5%, respectively (Table 7). For the contemporary classification, there was slight confusion between terrestrial forest and mangroves. Additionally, there was a small amount of two-way confusion between non-forest vegetation and terrestrial forest. The greatest source of error for the historic classification was the non-forest vegetation class, which was at-times confused with mangroves and the exposed/barren class.

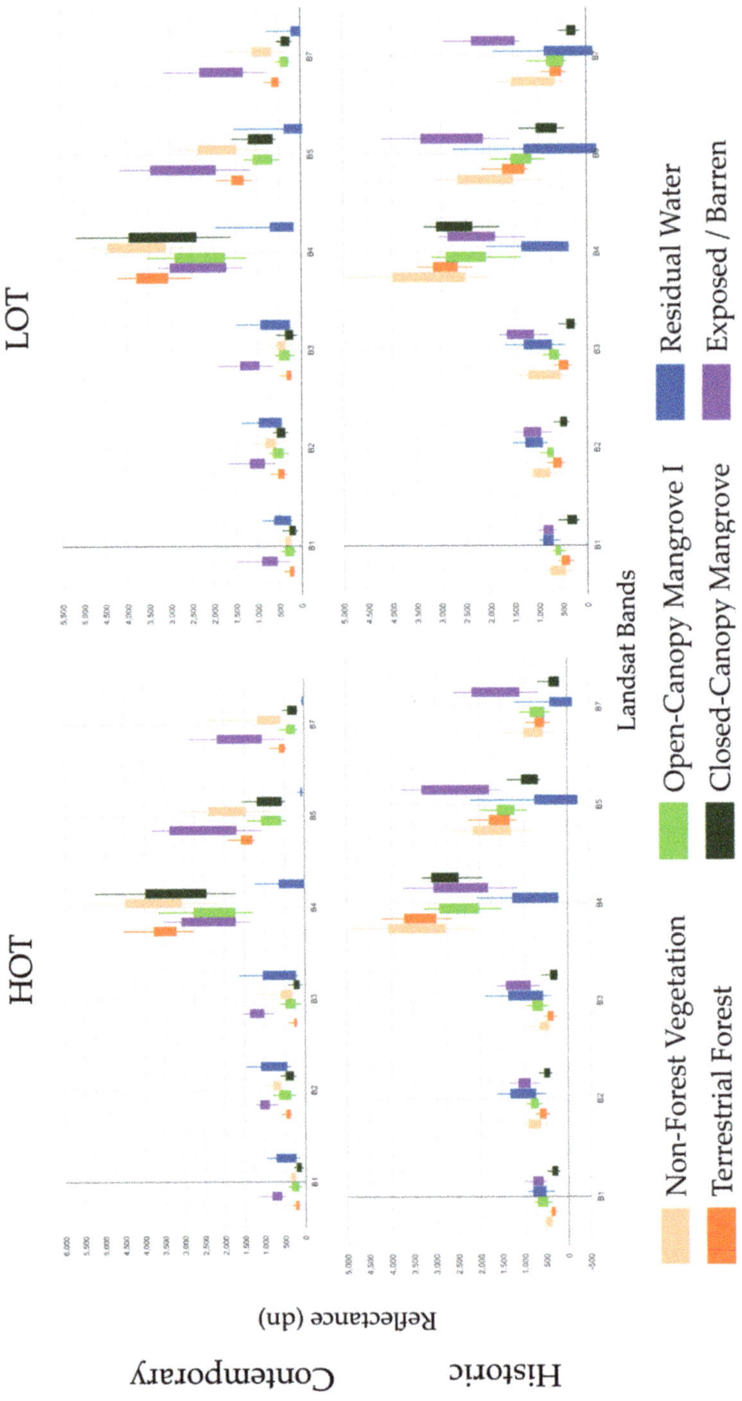

Figure 7. The spectral separability of all target classes as represented by CRAs across Landsat red (B1), green (B2), blue (B3), NIR (B4), SWIR1 (B5), and SWIR2 (B7) bands. The set of bar and whisker plots shows the min, max, and interquartile range.

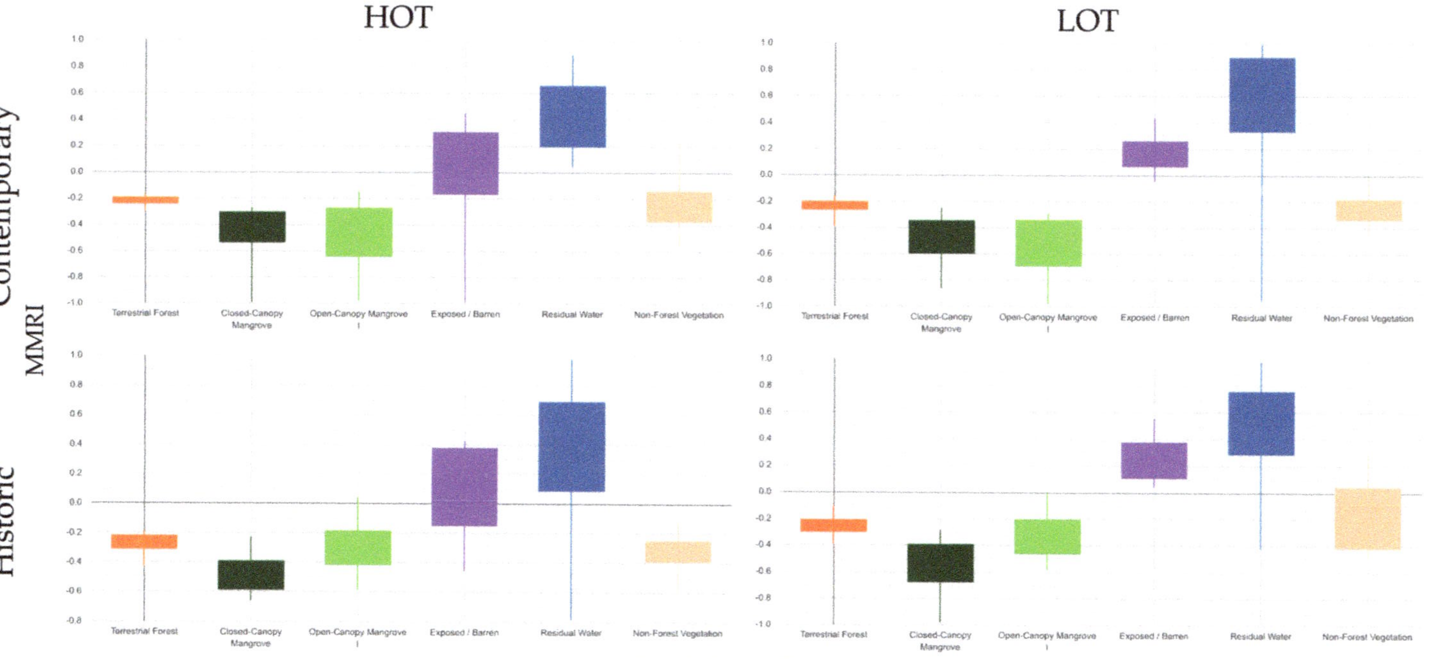

Figure 8. Example of index-specific overview of spectral values by land-cover class as represented by CRAs. MMRI is shown for each the historic and contemporary HOT and LOT datasets. The bar-whisker plots represent the min, max, and interquartile range (IQR) for each class.

Table 7. Final Historic and Contemporary Validation Error Matrices, using validation CRAs pixels.

Historic Classification Validation Error Matrix						
	Terrestrial Forest	Mangrove	Exposed/Barren	Residual Water	Non-Forest Vegetation	Total	User's Accuracy
Terrestrial Forest	24	0	0	0	0	24	100.0
Mangrove	0	54	0	0	0	54	100.0
Exposed/Barren	0	0	25	0	0	25	100.0
Residual Water	0	0	0	33	0	33	100.0
Non-Forest Vegetation	0	4	1	0	26	31	83.9
Total	24	58	26	33	26	167	
Producer's Accuracy	100.0	93.1	96.2	100.0	100.0		
Overall Accuracy						162/167	**97.0**

Contemporary Classification Validation Error Matrix						
	Terrestrial Forest	Mangrove	Exposed/Barren	Residual Water	Non-Forest Vegetation	Total	User's Accuracy
Terrestrial Forest	77	0	0	0	2	79	97.5
Mangrove	1	122	0	0	0	123	99.2
Exposed/Barren	0	0	33	0	0	33	100.0
Residual Water	0	0	0	24	0	24	100.0
Non-Forest Vegetation	2	0	0	0	71	73	97.3
Total	80	122	33	24	73	327	
Producer's Accuracy	96.3	100.0	100.0	100.0	97.3		
Overall Accuracy						327/332	**98.5**

3.2.3. Module 3—Dynamics and QAA

Classification results indicate that circa 2004–2008, Myanmar contained 995,412 ha of mangroves. In contrast, by 2014–2018, Myanmar contained 642,659 ha of mangroves. These results suggest that from 2004–2008 to 2014–2018 there was 551,570.99 ha of loss and 198,818.42 ha of gain (i.e., net loss 352,752.57 ha or 35.4%) (Figures 4 and 9, Table 5). As compared to Estoque et al. [56] and Giri et al. [70], estimated rates of loss are within reported trends and ranges; however, other studies reported lower rates of loss often coinciding with lower total estimates of mangrove cover (i.e., Bunting et al. [74]; Hamilton and Casey [78]; Richards and Friess [47]; Estoque et al. [56]). Figure 9 shows LPG from GEEMMM results within the loss hotspots identified through existing literature (i.e., Figure 1).

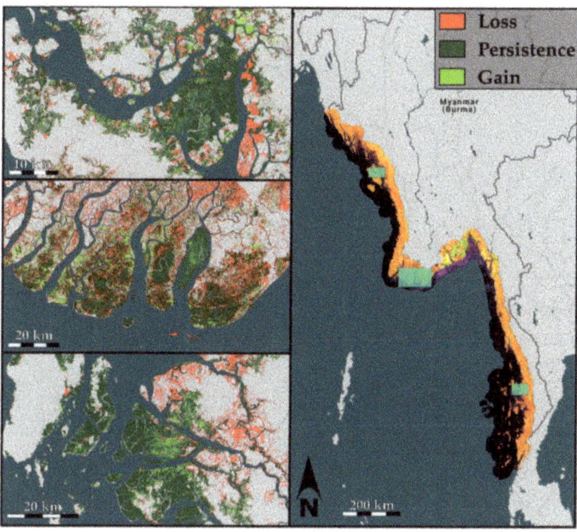

Figure 9. (**Left**) panels: Known mangrove loss hotspots (Figure 1). (**Top left**) shows loss, persistence, and gain (LPG) from 2004–2008 to 2014–2018 in Rakhine State; (**middle left**) panel shows the Ayeyarwady Region; (**bottom left**) shows Tanintharyi Region. (**Right**) Panel: contemporary high tide (HOT) image composite, false colour (R: NIR G: SWIR B: Red) with boundaries of left panels highlighted in cyan.

While there was a substantial net loss based on GEEMMM results, the reported gain seems relatively high. Portions of this likely reflect actual natural processes and increases in mangrove extent; however, the overall gain estimate is likely an overestimation. Exaggerated gain likely reflects the desk-based process of deriving CRAs. Clearly *any* classification is only as good as the examples used to calibrate the algorithm, and a limitation of this pilot was no direct access to field observations or ground truth, and constrained access to historical high spatial resolution satellite imagery. Disproportionate mangrove gain therefore likely reflects an underrepresentation of lower stature, less dense mangroves in the historic classification, which in turn exaggerates the amount of supposed gain (i.e., many of these areas were likely actually mangroves in both dates). Extensive field work and ground verification is required to confirm.

The GEEMMM QAA was conducted for the contemporary map, then repeated for the historic map. As part of the contemporary QAA, spot-checks were conducted over 108 sub-grid cells across Myanmar (Figure 1). The mangrove class was generally well-represented; however, at-times under-represented in favor of classes depicting portions of areas in the variable agricultural mosaic, i.e., non-forest vegetation, and exposed/barren. In both the contemporary, and less so the historic map, the agricultural mosaic was depicted as a patchwork of these two classes, on a pixel-by-pixel basis, given the inherent variability within the seasonal window. This resulted in some confusion between the two classes, and to some extent an under-representation of mangroves. In the contemporary map, sparser mangroves at the ecosystem periphery were at-times misclassified as non-forest vegetation, thereby under-representing mangrove and over-representing non-forest vegetation. Terrestrial forest was also at-times over-represented, occasionally at the expense of actual mangrove areas. Overall, the contemporary classification appeared to best represent Myanmar's south (i.e., the Tanintharyi coastline). The historic QAA, while not quite as comprehensive as the contemporary QAA (mainly due to the absence of historic imagery in GEE), found the mangrove class to be generally well-represented, though at-times over-represented at the expense of classes depicting the agricultural mosaic—an inverse to the contemporary map. Some portions of the agricultural mosaic were also found to be misclassified as terrestrial forest. As with the contemporary map, Myanmar's southern Tanintharyi coastline seemed best represented. Notably, most existing studies did not provide standard quantitative accuracy assessments, and no existing studies went beyond these and further qualitatively assessed resulting maps. While quantitative accuracy assessments should be a standard part of reporting, QAAs also help further assess resulting maps and identify areas for improvement. As such, the GEEMMM goes beyond standard accuracy—for which GEEMMM results were very high—and allows users to more closely examine actual distributions and subsequently dynamics.

3.2.4. Dissemination and Improvement

The code is available in a GitHub repository (see Supplementary Materials Section), with a GNU GPLv3 license permitting free use, modification, and sharing, provided that the source is disclosed and not used for commercial purposes. The code runs based on provided links, or is copied-and-pasted into GEE, which remains available for free non-profit and educational use. The tool itself continues to be adjusted and updated, as the GEE library evolves and as new mangrove remote sensing techniques become available.

While the tool performs well there are always potential improvements. Notably, CRAs are a key input for the workflow, and highly influence the outcome of the classifications. Future applications of the GEEMMM would benefit from direct access to field-based ground truth when deriving CRAs, particularly when it comes to confidently using mangrove sub-types as final map classes. While the need for and merit of isolating tidal conditions is proven, which tidal conditions are best requires further exploration—we used combined HOT and LOT in this pilot, whereas HOT or LOT on its own could also be employed. Furthermore, the choice of tidal condition depends on the intended application. For example, mangrove carbon projects may favor using HOT composites on their own for more conservative estimates of mangrove extent and change.

While going beyond standard accuracy metrics, the QAA is a somewhat complex component requiring significant user interaction; however, it too will evolve as the GEEMMM is further tested with other settings and applied to other AOIs. GEE itself also has notable limitations: the AOI can be as large or as small as the user requires but GEE has computational limits. Google shares its cloud processing among all GEE users, which means that if the task requested to process is too large (e.g., a long complex coastline, with collections containing hundreds of images) the user's allocated capacity may be exceeded and error(s) returned. Additionally, the functioning of this tool requires a relatively stable and reasonably strong internet connection, especially to view images and products within the GUI. If internet connectivity is limited, there may be latency issues loading data or even time-out errors. One of the benefits of working within the GEE environment however is that once a data product export has begun it will be completed on Google's server side. This means that internet access can be interrupted while using the tool, and it will continue to run. It was this feature of GEE, the server-side image/vector data exporting that drove the current configuration of three modules, where intermediate data products are exported to the user's assets, effectively saving their progress through the tool.

The GEEMMM is currently designed around the use of Landsat data—this was a conscious choice based around data availability. Sentinel imagery—which is also available through GEE—offers an increased revisit time (i.e., higher temporal resolution) and finer spatial resolution; however, it remains limited by a 2015-present temporal window. In contrast, the Landsat archive in GEE offers >35 years of imagery which facilitates more historically meaningful and robust dynamics assessments while also providing enough imagery to draw from multiple years to produce composites within preferred seasonal windows. Given the added benefits—especially once the archive spans 10+ years—future versions of the GEEMMM should also offer the choice of Sentinel imagery to users as an option.

4. Conclusions

We present a new tool—the GEEMMM—for mapping and monitoring mangrove ecosystems. By leveraging GEE, this new tool circumvents many traditional barriers to conventional methods. In addition, it presents an internal, image-based approach for tidal calibration. The GEEMMM—including the well commented source code—is available online and is ready to be used by practitioners anywhere mangrove ecosystems exist; please see information in Supplementary Material Section on how to access the GEEMMM.

While operational, the GEEMMM is not without its limitations: the larger the area the more complex the mapping task, particularly when it comes to creating optimal imagery composites within defined seasonal windows. In addition, the upper limits of GEE and internet connectivity present a challenge in terms of the time associated with and reliability of running the GEEMMM; however, when compared to the conventional processing times associated with standalone workstations it remains much faster, and once a part of the GEEMMM starts running it will continue to run even if the internet connection is lost. In any application, the resulting maps and dynamics assessments will only ever be as good as the examples of target map classes provided. Coastal managers will normally have such information available to them and GEEMMM provides them with a framework through which to capitalizes on this local knowledge, rather than relying on external datasets, which allow little to no customization, to map and monitor their mangroves.

The GEEMMM makes a significant and ready-to-go contribution toward accessible mangrove mapping and monitoring. It also remains a living tool wherein non-profit users are encouraged by the authors to make useful suggestions for modifications or additions, or modify the tool directly themselves to meet their own customized needs. While piloting the GEEMMM for Myanmar is an important first step, additional applications and tests are required, particularly for smaller areas of interest, wherein the GEEMMM can help fill a critical sub-national mapping gap. The authors welcome the opportunity to receive feedback from and work with users to more comprehensively assess the tool and gauge areas for improvement. A series of in-person and online instructional materials will go a long way toward ensuring the maximum and optimal utility of the GEEMMM. This first iteration of the GEEMMM further sets the stage for a comparatively more automated and even more accessible version to be deployable completely on mobile devices.

Supplementary Materials: The GEEMMM tool is freely available within the GitHub repository: https://github.com/Blue-Ventures-Conservation/GEEMMM.

Author Contributions: The GEEMMM was conceived of by, T.G.J., S.R.G., L.G., C.F., and J.M.M.Y. Contributions to the methodology were made by T.G.J., S.R.G., and C.F., with J.M.M.Y. developing the key tidal detection methods. J.M.M.Y. wrote all of the code in GEE for the GEEMMM tool, with the work reviewed by C.F. The results of this paper were validated by J.M.M.Y., T.G.J., S.R.G., C.F., and A.L. Formal analysis was performed by J.M.M.Y., T.G.J., and S.R.G., using the analysis tools developed by J.M.M.Y. Investigation for this work was conducted by J.M.M.Y., T.G.J., S.R.G., C.F., and A.L., J.M.M.Y. performed all of the data curation for this paper. The original manuscript writing was conducted by T.G.J. and J.M.M.Y.; with T.G.J. writing the introduction, discussion points, and conclusion and J.M.M.Y. writing the bulk of the methods and results. All authors, T.G.J., J.M.M.Y., S.R.G., C.F., A.L., and, L.G. were involved in writing—review and editing. Visualizations were generated by J.M.M.Y., A.L., and S.R.G. The project was administrated and supervised by T.G.J. and L.G. All authors have read and agreed to the published version of the manuscript.

Funding: This research was funded by Blue Ventures Conservation, with support from the UK Government's International Climate Fund, part of the UK commitment to developing countries to help them address the challenges presented by climate change and benefit from the opportunities.

Acknowledgments: We thank the following authors of studies referenced in this paper: Ake Rosenqvist, of solo Earth Observation (soloEO), for provision of and support regarding GMW data; J. Ronald Eastman and James Toledano, of Clark Labs, for guidance regarding the Clark Labs Aquaculture dataset; Chandra Giri, of United States Environmental Protection Agency, for provision of data and associated guidance; Edward L. Webb, of National University Singapore, for data provision and guidance; and Ate Poortinga, of Spatial Informatics Group, for provision of and guidance regarding SERVIR-Mekong data.

Conflicts of Interest: The authors declare no conflict of interest.

Appendix A

Table A1. Correlation matrices for the calculated spectral indices for both HOT and LOT historic and contemporary imagery extracted from the CRAs.

Contemporary HOT Index Band Correlation													
	SR	**NDVI**	**NDWI**	**MNDWI**	**CMRI**	**MMRI**	**SAVI**	**OSAVI**	**EVI**	**MRI**	**SMRI**	**LSWI**	**NDTI**	**EBBI**
SR	1	0.654	−0.61	−0.42	0.256	−0.46	0.654	0.654	0.13	0.018	−0.14	0.498	0.719	−0.71
NDVI	0.654	1	−0.98	−0.85	0.177	−0.7	0.999	0.999	0.203	−0.11	−0.38	0.163	0.755	−0.76
NDWI	−0.61	−0.98	1	0.907	0.005	0.679	−0.98	−0.98	−0.18	0.118	0.392	−0.06	−0.69	0.698
MNDWI	−0.42	−0.85	0.907	1	0.222	0.475	−0.85	−0.85	−0.14	0.222	0.342	0.328	−0.43	0.448
CMRI	0.256	0.177	0.005	0.222	1	−0.18	0.177	0.177	0.14	0	8.709 *	0.562	0.407	−0.4
MMRI	−0.46	−0.7	0.679	0.475	−0.18	1	−0.7	−0.7	−0.08	−0.04	0.154	−0.31	−0.72	0.617
SAVI	0.654	0.999	−0.98	−0.85	0.177	−0.7	1	0.999	0.203	−0.11	−0.38	0.163	0.755	−0.76
OSAVI	0.654	0.999	−0.98	−0.85	0.177	−0.7	0.999	1	0.203	−0.11	−0.38	0.163	0.755	−0.76
EVI	0.13	0.203	−0.18	−0.14	0.14	−0.08	0.203	0.203	1	−0.01	−0.09	0.032	0.145	−0.15
MRI	0.018	−0.11	0.118	0.222	0	−0.04	−0.11	−0.11	−0.01	1	0.033	0.227	0.067	0.085
SMRI	−0.14	−0.38	0.392	0.342	8.709 *	0.154	−0.38	−0.38	−0.09	0.033	1	0.028	−0.27	0.175
LSWI	0.498	0.163	−0.06	0.328	0.562	−0.31	0.163	0.163	0.032	0.227	0.028	1	0.566	−0.6
NDTI	0.719	0.755	−0.69	−0.43	0.407	−0.72	0.755	0.755	0.145	0.067	−0.27	0.566	1	−0.79
EBBI	−0.71	−0.76	0.698	0.448	−0.4	0.617	−0.76	−0.76	−0.15	0.085	0.175	−0.6	−0.79	1

*—Denotes an error output from the GEE servers for the index correlations.

Contemporary LOT Index Band Correlation													
	SR	**NDVI**	**NDWI**	**MNDWI**	**CMRI**	**MMRI**	**SAVI**	**OSAVI**	**EVI**	**MRI**	**SMRI**	**LSWI**	**NDTI**	**EBBI**
SR	1	0.718	−0.67	−0.44	0.286	−0.47	0.718	0.718	0.158	0.054	−0.14	0.509	0.717	−0.76
NDVI	0.718	1	−0.97	−0.79	0.234	−0.78	0.999	0.999	0.223	−0.1	−0.25	0.267	0.75	−0.79
NDWI	−0.67	−0.97	1	0.871	−0.02	0.774	−0.97	−0.97	−0.23	0.119	0.236	−0.14	−0.67	0.714
MNDWI	−0.44	−0.79	0.871	1	0.27	0.492	−0.79	−0.79	−0.22	0.251	0.221	0.337	−0.3	0.374
CMRI	0.286	0.234	−0.02	0.27	1	−0.13	0.234	0.234	−0.03	0.033	−0.1	0.612	0.459	−0.46
MMRI	−0.47	−0.78	0.774	0.492	−0.13	1	−0.78	−0.78	−0.21	−0.04	0.137	−0.37	−0.69	0.65
SAVI	0.718	0.999	−0.97	−0.79	0.234	−0.78	1	0.999	0.223	−0.1	−0.25	0.267	0.75	−0.79
OSAVI	0.718	0.999	−0.97	−0.79	0.234	−0.78	0.999	1	0.223	−0.1	−0.25	0.267	0.75	−0.79
EVI	0.158	0.223	−0.23	−0.22	−0.03	−0.21	0.223	0.223	1	−0.02	−0.04	0.006	0.126	−0.17
MRI	0.054	−0.1	0.119	0.251	0.033	−0.04	−0.1	−0.1	−0.02	1	0.033	0.253	0.081	−0.02
SMRI	−0.14	−0.25	0.236	0.221	−0.1	0.137	−0.25	−0.25	−0.04	0.033	1	−0.01	−0.14	0.134
LSWI	0.509	0.267	−0.14	0.337	0.612	−0.37	0.267	0.267	0.006	0.253	−0.01	1	0.716	−0.67
NDTI	0.717	0.75	−0.67	−0.3	0.459	−0.69	0.75	0.75	0.126	0.081	−0.14	0.716	1	−0.89
EBBI	0.76	−0.79	0.714	0.374	−0.46	0.65	−0.79	−0.79	−0.17	−0.02	0.134	−0.67	−0.89	1

Table A1. *Cont.*

| Historic HOT Index Band Correlation | | | | | | | | | | | | | |
	SR	NDVI	NDWI	MNDWI	CMRI	MMRI	SAVI	OSAVI	EVI	MRI	SMRI	LSWI	NDTI	EBBI
SR	1	0.844	−0.78	−0.56	0.603	−0.74	0.844	0.844	0.406	−0.17	−0.26	0.123	0.373	−0.83
NDVI	0.844	1	−0.98	−0.83	0.418	−0.82	0.999	0.999	0.38	−0.25	−0.45	−0.26	0.248	−0.69
NDWI	−0.78	−0.98	1	0.894	−0.26	0.773	−0.98	−0.98	−0.34	0.232	0.49	0.38	−0.2	0.595
MNDWI	−0.56	−0.83	0.894	1	0.027	0.618	−0.83	−0.83	−0.26	0.239	0.435	0.718	−0.06	0.244
CMRI	0.603	0.418	−0.26	0.027	1	−0.54	0.418	0.418	0.341	−0.21	0.03	0.56	0.341	−0.77
MMRI	−0.74	−0.82	0.773	0.618	−0.54	1	−0.82	−0.82	−0.45	0.184	0.279	0.031	−0.33	0.719
SAVI	0.844	0.999	−0.98	−0.83	0.418	−0.82	1	0.999	0.38	−0.25	−0.45	−0.26	0.248	−0.69
OSAVI	0.844	0.999	−0.98	−0.83	0.418	−0.82	0.999	1	0.38	−0.25	−0.45	−0.26	0.248	−0.69
EVI	0.406	0.38	−0.34	−0.26	0.341	−0.45	0.38	0.38	1	−0.09	−0.02	0.064	0.151	−0.38
MRI	−0.17	−0.25	0.232	0.239	−0.21	0.184	−0.25	−0.25	−0.09	1	−0.16	0.104	−0.02	0.231
SMRI	−0.26	−0.45	0.49	0.435	0.03	0.279	−0.45	−0.45	−0.02	−0.16	1	0.267	0.089	0.16
LSWI	0.123	−0.26	0.38	0.718	0.56	0.031	−0.26	−0.26	0.064	0.104	0.267	1	0.23	−0.45
NDTI	0.373	0.248	−0.2	−0.06	0.341	−0.33	0.248	0.248	0.151	−0.02	0.089	0.23	1	−0.38
EBBI	−0.83	−0.69	0.595	0.244	−0.77	0.719	−0.69	−0.69	−0.38	0.231	0.16	−0.45	−0.38	1

| Historic LOT Index Band Correlation | | | | | | | | | | | | | |
	SR	NDVI	NDWI	MNDWI	CMRI	MMRI	SAVI	OSAVI	EVI	MRI	SMRI	LSWI	NDTI	EBBI
SR	1	0.852	−0.77	−0.42	0.529	−0.77	0.852	0.852	0.135	0.158	−0.3	0.412	0.714	−0.82
NDVI	0.852	1	−0.97	−0.74	0.325	−0.86	0.999	0.999	0.183	0.037	−0.39	0.084	0.705	−0.69
NDWI	−0.77	−0.97	1	0.842	−0.11	0.824	−0.97	−0.97	−0.18	0.056	0.398	0.072	−0.63	0.574
MNDWI	−0.42	−0.74	0.842	1	0.28	0.539	−0.74	−0.74	−0.14	0.27	0.24	0.586	−0.28	0.086
CMRI	0.529	0.325	−0.11	0.28	1	−0.38	0.325	0.325	0.041	0.429	−0.08	0.724	0.485	−0.71
MMRI	−0.77	−0.86	0.824	0.539	−0.38	1	−0.86	−0.86	−0.14	−0.15	0.341	−0.25	−0.67	0.728
SAVI	0.852	0.999	−0.97	−0.74	0.325	−0.86	1	0.999	0.183	0.037	−0.39	0.084	0.705	−0.69
OSAVI	0.852	0.999	−0.97	−0.74	0.325	−0.86	0.999	1	0.183	0.037	−0.39	0.084	0.705	−0.69
EVI	0.135	0.183	−0.18	−0.14	0.041	−0.14	0.183	0.183	1	0.004	−0.01	0.008	0	−0.11
MRI	0.158	0.037	0.056	0.27	0.429	−0.15	0.037	0.037	0.004	1	−0.16	0.414	0.167	−0.32
SMRI	−0.3	−0.39	0.398	0.24	−0.08	0.341	−0.39	−0.39	−0.01	−0.16	1	−0.11	−0.25	0.322
LSWI	0.412	0.084	0.072	0.586	0.724	−0.25	0.084	0.084	0.008	0.414	−0.11	1	0.405	−0.71
NDTI	0.714	0.705	−0.63	−0.28	0.485	−0.67	0.705	0.705	0	0.167	−0.25	0.405	1	−0.72
EBBI	−0.82	−0.69	0.574	0.086	−0.71	0.728	−0.69	−0.69	−0.11	−0.32	0.322	−0.71	−0.72	1

References

1. Saenger, P. *Mangrove Ecology, Silviculture and Conservation*; Springer Science & Business Media: Berlin, Germany, 2002; Volume 3, ISBN 9789048160501.
2. Food and Agriculture Organization of the United Nations; Forestry Department (Rome). *Global Forest Resources Assessment 2000: Main Report*; Food and Agriculture Organization of the United Nations: Rome, Italy, 2001.
3. Xia, Q.; Qin, C.Z.; Li, H.; Huang, C.; Su, F.Z.; Jia, M.M. Evaluation of submerged mangrove recognition index using multi-tidal remote sensing data. *Ecol. Indic.* **2020**, 113. [CrossRef]
4. Lugo, A.E. Mangrove Ecosystems: Successional or Steady State? *Biotropica* **1980**, *12*, 65–72. [CrossRef]
5. Spalding, M.; Kainuma, M.; Collins, L. *World Atlas of Mangroves*; Earthscan: New York, NY, USA, 2010; ISBN 978-1-84407-657-4.
6. Duke, N.; Nagelkerken, I.; Agardy, T.; Wells, S.; van Lavieren, H. *The Importance of Mangroves to People: A Call to Action*; van Bochove, J.-W., Sullivan, E., Nakamura, T., Eds.; United Nations Environment Programme World Conservation Monitoring Centre: Cambridge, UK, 2014; ISBN 9789280733976.
7. Scales, I.R.; Friess, D.A.; Glass, L.; Ravaoarinorotsihoarana, L. Rural livelihoods and mangrove degradation in south-west Madagascar: Lime production as an emerging threat. *Oryx* **2018**, *52*, 641–645. [CrossRef]
8. Blue Ventures Conservation. Value Chain Analysis: The wild capture mud crab fishery of Madagascar's Menabe region. Available online: https://blueventures.org/publication/value-chain-analysis-the-wild-capture-mud-crab-fishery-of-madagascars-menabe-region/ (accessed on 1 October 2020).
9. Aye, W.N.; Wen, Y.; Marin, K.; Thapa, S.; Tun, A.W. Contribution of mangrove forest to the livelihood of local communities in Ayeyarwaddy Region, Myanmar. *Forests* **2019**, *10*, 414. [CrossRef]
10. Nagelkerken, I.; Blaber, S.J.M.; Bouillon, S.; Green, P.; Haywood, M.; Kirton, L.G.; Meynecke, J.O.; Pawlik, J.; Penrose, H.M.; Sasekumar, A.; et al. The habitat function of mangroves for terrestrial and marine fauna: A review. *Aquat. Bot.* **2008**, *89*, 155–185. [CrossRef]
11. Gopal, B.; Chauhan, M. Biodiversity and its conservation in the Sundarban mangrove ecosystem. *Aquat. Sci.* **2006**, *68*, 338–354. [CrossRef]
12. Gardner, C.J. Use of Mangroves by Lemurs. *Int. J. Primatol.* **2016**, *37*, 317–332. [CrossRef]
13. Alongi, D.M. Carbon cycling and storage in mangrove forests. *Annu. Rev. Mar. Sci.* **2014**, *6*, 195–219. [CrossRef]
14. Donato, D.C.; Kauffman, J.B.; Murdiyarso, D.; Kurnianto, S.; Stidham, M.; Kanninen, M. Mangroves among the most carbon-rich forests in the tropics. *Nat. Geosci.* **2011**, *4*, 293–297. [CrossRef]
15. Sanderman, J.; Hengl, T.; Fiske, G.; Solvik, K.; Adame, M.F.; Benson, L.; Bukoski, J.J.; Carnell, P.; Cifuentes-Jara, M.; Donato, D.; et al. A global map of mangrove forest soil carbon at 30 m spatial resolution. *Environ. Res. Lett.* **2018**, 13. [CrossRef]
16. Rakotomahazo, C.; Ravaoarinorotsihoarana, L.A.; Randrianandrasaziky, D.; Glass, L.; Gough, C.; Boleslas Todinanahary, G.G.; Gardner, C.J. Participatory planning of a community-based payments for ecosystem services initiative in Madagascar's mangroves. *Ocean Coast. Manag.* **2019**, *175*, 43–52. [CrossRef]
17. Ahmed, N.; Glaser, M. Coastal aquaculture, mangrove deforestation and blue carbon emissions: Is REDD+ a solution? *Mar. Policy* **2016**, *66*, 58–66. [CrossRef]
18. Valiela, I.; Bowen, J.L.; York, J.K. Mangrove forests: One of the world's threatened major tropical environments. *BioScience* **2001**, *51*, 807–815. [CrossRef]
19. Gandhi, S.; Jones, T.G. Identifying mangrove deforestation hotspots in South Asia, Southeast Asia and Asia-Pacific. *Remote Sens.* **2019**, *11*, 728. [CrossRef]
20. Richards, D.R.; Thompson, B.S.; Wijedasa, L. Quantifying net loss of global mangrove carbon stocks from 20 years of land cover change. *Nat. Commun.* **2020**, *11*, 4260. [CrossRef]
21. Global Mangrove Watch Mangrove Atlas. Available online: https://www.globalmangrovewatch.org/ (accessed on 8 September 2020).
22. Friess, D.A.; Rogers, K.; Lovelock, C.E.; Krauss, K.W.; Hamilton, S.E.; Lee, S.Y.; Lucas, R.; Primavera, J.; Rajkaran, A.; Shi, S. The State of the World's Mangrove Forests: Past, Present, and Future. *Ann. Rev. Environ. Resour.* **2019**, *44*, 89–115. [CrossRef]
23. Kuenzer, C.; Bluemel, A.; Gebhardt, S.; Quoc, T.V.; Dech, S. Remote sensing of mangrove ecosystems: A review. *Remote Sens.* **2011**, *3*, 878–928. [CrossRef]

24. UN Millennium Project. *Investing in Development: A Practical Plan to Achieve the Millennium Development Goals. Overview*; United Nations Development Programme: Washington, DC, USA, 2005; ISBN 1844072177.

25. Ramsar Convention Secretariat. The Fourth RAMSAR Strategic Plan 2016–2024. Available online: https://www.ramsar.org/the-ramsar-strategic-plan-2016--24 (accessed on 1 October 2020).

26. Wang, L.; Jia, M.; Yin, D.; Tian, J. A review of remote sensing for mangrove forests: 1956–2018. *Remote Sens. Environ.* **2019**, 231. [CrossRef]

27. Rogers, K.; Lymburner, L.; Salum, R.; Brooke, B.P.; Woodroffe, C.D. Mapping of mangrove extent and zonation using high and low tide composites of Landsat data. *Hydrobiologia* **2017**, *803*, 49–68. [CrossRef]

28. Zhang, X.; Treitz, P.M.; Chen, D.; Quan, C.; Shi, L.; Li, X. Mapping mangrove forests using multi-tidal remotely-sensed data and a decision-tree-based procedure. *Int. J. Appl. Earth Obs. Geoinf.* **2017**, *62*, 201–214. [CrossRef]

29. Xia, Q.; Qin, C.Z.; Li, H.; Huang, C.; Su, F.Z. Mapping mangrove forests based on multi-tidal high-resolution satellite imagery. *Remote Sens.* **2018**, *10*, 1343. [CrossRef]

30. Sagar, S.; Roberts, D.; Bala, B.; Lymburner, L. Extracting the intertidal extent and topography of the Australian coastline from a 28 year time series of Landsat observations. *Remote Sens. Environ.* **2017**, *195*, 153–169. [CrossRef]

31. Bishop-Taylor, R.; Sagar, S.; Lymburner, L.; Beaman, R.J. Between the tides: Modelling the elevation of Australia's exposed intertidal zone at continental scale. *Estuar. Coast. Shelf Sci.* **2019**, *223*, 115–128. [CrossRef]

32. Murray, N.J.; Phinn, S.R.; DeWitt, M.; Ferrari, R.; Johnston, R.; Lyons, M.B.; Clinton, N.; Thau, D.; Fuller, R.A. The global distribution and trajectory of tidal flats. *Nature* **2019**, *565*, 222–225. [CrossRef]

33. Martin, P.J.; Smith, S.R.; Posey, P.G.; Dawson, G.M.; Riedlinger, S.H. *Use of the Oregon State University Tidal Inversion Software (OTIS) to Generate Improved Tidal Prediction in the East-Asian Seas*; Stennis Space Center: Hancock, MS, USA, 2009.

34. Gorelick, N.; Hancher, M.; Dixon, M.; Ilyushchenko, S.; Thau, D.; Moore, R. Google Earth Engine: Planetary-scale geospatial analysis for everyone. *Remote Sens. Environ.* **2017**, *202*, 18–27. [CrossRef]

35. U.S. Geological Survey. *Landsat 8 Collection 1 (C1) Land Surface Reflectance Code (LaSRC) Product Guide*; LSDS-1368; U.S. Geological Survey, Dempartment of the Interior: Sioux Falls, SD, USA, 2020.

36. Wulder, M.A.; Coops, N.C.; Roy, D.P.; White, J.C.; Hermosilla, T. Land cover 2.0. *Int. J. Remote Sens.* **2018**, *39*, 4254–4284. [CrossRef]

37. Hansen, M.C.; Potapov, P.V.; Moore, R.; Hancher, M.; Turubanova, S.A.; Tyukavina, A.; Thau, D.; Stehman, S.V.; Goetz, S.J.; Loveland, T.R.; et al. High-Resolution Global Maps of 21st-Century Forest Cover Change. *Science* **2013**, *342*, 850–853. [CrossRef]

38. Vos, K.; Splinter, K.D.; Harley, M.D.; Simmons, J.A.; Turner, I.L. CoastSat: A Google Earth Engine-enabled Python toolkit to extract shorelines from publicly available satellite imagery. *Environ. Model. Softw.* **2019**, *122*, 104528. [CrossRef]

39. Chen, B.; Xiao, X.; Li, X.; Pan, L.; Doughty, R.; Ma, J.; Dong, J.; Qin, Y.; Zhao, B.; Wu, Z.; et al. A mangrove forest map of China in 2015: Analysis of time series Landsat 7/8 and Sentinel-1A imagery in Google Earth Engine cloud computing platform. *ISPRS J. Photogramm. Remote Sens.* **2017**, *131*, 104–120. [CrossRef]

40. Pimple, U.; Simonetti, D.; Sitthi, A.; Pungkul, S.; Leadprathom, K.; Skupek, H.; Som-ard, J.; Gond, V.; Towprayoon, S. Google Earth Engine Based Three Decadal Landsat Imagery Analysis for Mapping of Mangrove Forests and Its Surroundings in the Trat Province of Thailand. *J. Comput. Commun.* **2018**, *6*, 247–264. [CrossRef]

41. Tieng, T.; Sharma, S.; Mackenzie, R.A.; Venkattappa, M.; Sasaki, N.K.; Collin, A. Mapping mangrove forest cover using Landsat-8 imagery, Sentinel-2, Very High Resolution Images and Google Earth Engine algorithm for entire Cambodia. In Proceedings of the IOP Conference Series: Earth and Environmental Science, 4th International Forum on Sustainable Future in Asia/4th NIES International Forum, Pan Pacific Hanoi, Vietnam, 23–24 January 2019; IOP Publishing: Bristol, UK, 2019; Volume 266.

42. Mondal, P.; Trzaska, S.; de Sherbinin, A. Landsat-derived estimates of mangrove extents in the Sierra Leone coastal landscape complex during 1990–2016. *Sensors* **2018**, *18*, 12. [CrossRef] [PubMed]

43. Diniz, C.; Cortinhas, L.; Nerino, G.; Rodrigues, J.; Sadeck, L.; Adami, M.; Souza-Filho, P.W.M. Brazilian mangrove status: Three decades of satellite data analysis. *Remote Sens.* **2019**, *11*, 808. [CrossRef]

44. Giri, C.; Ochieng, E.; Tieszen, L.L.; Zhu, Z.; Singh, A.; Loveland, T.; Masek, J.; Duke, N. Status and distribution of mangrove forests of the world using earth observation satellite data. *Glob. Ecol. Biogeogr.* **2011**, *20*, 154–159. [CrossRef]

45. Ellison, A.M.; Farnsworthf, E.J.; Merkt, R.E. Origins of Mangrove Ecosystems and the Mangrove Biodiversity Anomaly. *Glob. Ecol. Biogeogr.* **1999**, *8*, 95–115.

46. Food and Agricultural Organization (FAO). Loss of Mangroves Alarming. Available online: http://www.fao.org/newsroom/en/news/2008/1000776/index.html (accessed on 30 August 2020).

47. Richards, D.R.; Friess, D.A. Rates and drivers of mangrove deforestation in Southeast Asia, 2000–2012. *Proc. Natl. Acad. Sci. USA* **2016**, *113*, 344–349. [CrossRef]

48. Farnsworth, E.J.; Ellison, A.M. The global conservation status of mangroves. *AMBIO* **1997**, *26*, 328–334. [CrossRef]

49. Primavera, J.H. Development and conservation of Philippine mangroves: Institutional issues. *Ecol. Econ.* **2000**, *35*, 91–106. [CrossRef]

50. Dahdouh-Guebas, F. The use of remote sensing and GIS in the sustainable management of tropical coastal ecosystems. *Environ. Dev. Sustain.* **2002**, *4*, 93–112. [CrossRef]

51. Primavera, J.H. Mangroves, Fishponds, and the Quest for Sustainability. *Science* **2005**, *310*, 57–59. [CrossRef]

52. Primavera, J.H. Overcoming the impacts of aquaculture on the coastal zone. *Ocean Coast. Manag.* **2006**, *49*, 531–545. [CrossRef]

53. Gilman, E.L.; Ellison, J.; Duke, N.C.; Field, C. Threats to mangroves from climate change and adaptation options: A review. *Aquat. Bot.* **2008**, *89*, 237–250. [CrossRef]

54. Walters, B.B.; Rönnbäck, P.; Kovacs, J.M.; Crona, B.; Hussain, S.A.; Badola, R.; Primavera, J.H.; Barbier, E.; Dahdouh-Guebas, F. Ethnobiology, socio-economics and management of mangrove forests: A review. *Aquat. Bot.* **2008**, *89*, 220–236. [CrossRef]

55. Webb, E.L.; Jachowski, N.R.A.; Phelps, J.; Friess, D.A.; Than, M.M.; Ziegler, A.D. Deforestation in the Ayeyarwady Delta and the conservation implications of an internationally-engaged Myanmar. *Glob. Environ. Chang.* **2014**, *24*, 321–333. [CrossRef]

56. Estoque, R.C.; Myint, S.W.; Wang, C.; Ishtiaque, A.; Aung, T.T.; Emerton, L.; Ooba, M.; Hijioka, Y.; Mon, M.S.; Wang, Z.; et al. Assessing environmental impacts and change in Myanmar's mangrove ecosystem service value due to deforestation (2000–2014). *Glob. Chang. Biol.* **2018**, *24*, 5391–5410. [CrossRef]

57. De Alban, J.D.T.; Jamaludin, J.; Wong De Wen, D.; Than, M.M.; Webb, E.L. Improved estimates of mangrove cover and change reveal catastrophic deforestation in Myanmar. *Environ. Res. Lett.* **2020**, *15*, 034034. [CrossRef]

58. Alongi, D.M. Present state and future of the world's mangrove forests. *Environ. Conserv.* **2002**, *29*, 331–349. [CrossRef]

59. Alongi, D.M. Mangrove forests: Resilience, protection from tsunamis, and responses to global climate change. *Estuar. Coast. Shelf Sci.* **2008**, *76*, 1–13. [CrossRef]

60. Alongi, D.M. The Impact of Climate Change on Mangrove Forests. *Curr. Clim. Chang. Rep.* **2015**, *1*, 30–39. [CrossRef]

61. Sitoe, A.A.; Mandlate, L.J.C.; Guedes, B.S. Biomass and carbon stocks of Sofala Bay mangrove forests. *Forests* **2014**, *5*, 1967–1981. [CrossRef]

62. Field, C.D. Impact of expected climate change on mangroves. *Hydrobiologia* **1995**, *295*, 75–81. [CrossRef]

63. Krauss, K.W.; Lovelock, C.E.; McKee, K.L.; López-Hoffman, L.; Ewe, S.M.L.; Sousa, W.P. Environmental drivers in mangrove establishment and early development: A review. *Aquat. Bot.* **2008**, *89*, 105–127. [CrossRef]

64. Chan, H.T.; Baba, S. *Manual on Guidlines for Rehabilitation of Coastal Forests Damaged by Natural Hazards in the Asia-Pacific Region*; International Society for Mangrove Ecosystems and International Tropical Timber Organization: Okinawa, Japan, 2009; ISBN 9784906584130.

65. Suzuki, T.; Zijlema, M.; Burger, B.; Meijer, M.C.; Narayan, S. Wave dissipation by vegetation with layer schematization in SWAN. *Coast. Eng.* **2012**, *59*, 64–71. [CrossRef]

66. Di Nitto, D.; Neukermans, G.; Koedam, N.; Defever, H.; Pattyn, F.; Kairo, J.G.; Dahdouh-Guebas, F. Mangroves facing climate change: Landward migration potential in response to projected scenarios of sea level rise. *Biogeosciences* **2014**, *11*, 857–871. [CrossRef]

67. Giesen, W.; Wulffraat, S.; Zieren, M.; Scholten, L. *Mangrove guidebook for Southeast Asia*; Food and Agriculture Organization of the United Nations, Regional Office for Asia and the Pacific: Bangkok, Thailand, 2006.

68. Thomas, N.; Bunting, P.; Lucas, R.; Hardy, A.; Rosenqvist, A.; Fatoyinbo, T. Mapping mangrove extent and change: A globally applicable approach. *Remote Sens.* **2018**, *10*, 1466. [CrossRef]

69. Romañach, S.S.; DeAngelis, D.L.; Koh, H.L.; Li, Y.; Teh, S.Y.; Raja Barizan, R.S.; Zhai, L. Conservation and restoration of mangroves: Global status, perspectives, and prognosis. *Ocean Coast. Manag.* **2018**, *154*, 72–82. [CrossRef]

70. Giri, C.; Zhu, Z.; Tieszen, L.L.; Singh, A.; Gillette, S.; Kelmelis, J.A. Mangrove Forest Distributions and Dynamics (1975-2005) of the Tsunami-Affected Region of Asia. *J. Biogeogr.* **2008**, *35*, 519–528. [CrossRef]

71. Giri, C.; Long, J.; Abbas, S.; Murali, R.M.; Qamer, F.M.; Pengra, B.; Thau, D. Distribution and dynamics of mangrove forests of South Asia. *J. Environ. Manag.* **2015**, *148*, 101–111. [CrossRef]

72. Bhattarai, B. Assessment of mangrove forests in the Pacific region using Landsat imagery. *J. Appl. Remote Sens.* **2011**, *5*, 053509. [CrossRef]

73. Saah, D.; Tenneson, K.; Poortinga, A.; Nguyen, Q.; Chishtie, F.; Aung, K.S.; Markert, K.N.; Clinton, N.; Anderson, E.R.; Cutter, P.; et al. Primitives as building blocks for constructing land cover maps. *Int. J. Appl. Earth Obs. Geoinf.* **2020**, *85*, 101979. [CrossRef]

74. Bunting, P.; Rosenqvist, A.; Lucas, R.M.; Rebelo, L.M.; Hilarides, L.; Thomas, N.; Hardy, A.; Itoh, T.; Shimada, M.; Finlayson, C.M. The global mangrove watch—A new 2010 global baseline of mangrove extent. *Remote Sens.* **2018**, *10*, 1669. [CrossRef]

75. Clark Labs. Coastal Habitat Mapping: Mangrove and Pond Aquaculture Conversion. Available online: https://clarklabs.org/aquaculture/ (accessed on 30 August 2020).

76. Stibig, H.-J.; Belward, A.S.; Roy, P.S.; Rosalina-Wasrin, U.; Agrawal, S.; Joshi, P.K.; Hildanus; Beuchle, R.; Fritz, S.; Mubareka, S.; et al. A Land-Cover Map for South and Southeast Asia Derived from SPOT-VEGETATION Data. *J. Biogeogr.* **2007**, *34*, 625–637. [CrossRef]

77. Blasco, F.; Aizpuru, M.; Gers, C. Depletion of the mangroves of Continental Asia. *Wetl. Ecol. Manag.* **2001**, *9*, 245–256. [CrossRef]

78. Hamilton, S.E.; Casey, D. Creation of a high spatio-temporal resolution global database of continuous mangrove forest cover for the 21st century (CGMFC-21). *Glob. Ecol. Biogeogr.* **2016**, *25*, 729–738. [CrossRef]

79. *Google Earth Pro*; Google LLC: Mountain View, CA, USA, 2020.

80. Global LSIB Polygons Detailed 2017. Available online: https://catalog.data.gov/dataset/global-lsib-polygons-detailed-2017dec29 (accessed on 22 September 2020).

81. Tadono, T.; Ishida, H.; Oda, F.; Naito, S.; Minakawa, K.; Iwamoto, H. Precise Global DEM Generation by ALOS PRISM. *ISPRS Ann. Photogramm. Remote Sens. Spat. Inf. Sci.* **2014**, *II-4*, 71–76. [CrossRef]

82. GADM Maps and Data. Available online: https://www.gadm.org/ (accessed on 22 September 2020).

83. Li, P.; Feng, Z.; Jiang, L.; Liao, C.; Zhang, J. A review of swidden agriculture in Southeast Asia. *Remote Sens.* **2014**, *6*, 1654–1683. [CrossRef]

84. Streets, D.G.; Yarber, K.F.; Woo, J.-H.; Carmichael, G.R. Biomass burning in Asia: Annual and seasonal estimates and atmospheric emissions. *Glob. Biogeochem. Cycles* **2003**, *17*. [CrossRef]

85. Davis, J.H. *The Forests of Burma*; New York Botanical Garden: New York, NY, USA, 1964.

86. U.S. Geological Survey. *Landsat 8 (L8) Data Users Handbook*; LSDS-1574; U.S. Geological Survey, Department of the Interior: Sioux Falls, SD, USA, 2019.

87. Xu, H. Modification of normalised difference water index (NDWI) to enhance open water features in remotely sensed imagery. *Int. J. Remote Sens.* **2006**, *27*, 3025–3033. [CrossRef]

88. Hogarth, P.J. Mangrove Ecosystems. In *Encyclopedia of Biodiversity*, 2nd ed.; Elsevier: Amsterdam, The Netherlands, 2013; pp. 10–22. ISBN 9780123847195.

89. Eastman, J.R.; Crema, S.C.; Sangermano, F.; Cunningham, S.; Xiao, X.; Zhou, Z.; Hu, P.; Johnson, C.; Arakwiye, B.; Crone, N. *A Baseline Mapping of Aquaculture and Coastal Habitats in Thailand, Cambodia and Vietnam*; Aquaculture and Coastal Habitats Report No. 1; Clark Labs: Worcester, MA, USA, 2015.

90. Htway, O.; Matsumoto, J. Climatological onset dates of summer monsoon over Myanmar. *Int. J. Climatol.* **2011**, *31*, 382–393. [CrossRef]

91. Jones, T.G.; Glass, L.; Gandhi, S.; Ravaoarinorotsihoarana, L.; Carro, A.; Benson, L.; Ratsimba, H.R.; Giri, C.; Randriamanatena, D.; Cripps, G. Madagascar's mangroves: Quantifying nation-wide and ecosystem specific dynamics, and detailed contemporary mapping of distinct ecosystems. *Remote Sens.* **2016**, *8*, 106. [CrossRef]

92. Younes Cárdenas, N.; Joyce, K.E.; Maier, S.W. Monitoring mangrove forests: Are we taking full advantage of technology? *Int. J. Appl. Earth Obs. Geoinf.* **2017**, *63*, 1–14. [CrossRef]

93. Zhang, X.; Tian, Q. A mangrove recognition index for remote sensing of mangrove forest from space. *Curr. Sci.* **2013**, *105*, 1149–1154.

94. U.S. Geological Survey. *Landsat 4-7 Collection 1 (C1) Surface Reflectance (LEDAPS) Product Guide*; LSDS-1370; U.S. Geological Survey, Department of the Interior: Sioux Falls, SD, USA, 2020.

95. Jordan, C.F. Derivation of Leaf-Area Index from Quality of Light on the Forest Floor. *Ecology* **1969**, *50*, 663–666. [CrossRef]

96. Tarpley, J.D.; Schneider, S.R.; Money, R.L. Global Vegetation Indices from the NOAA-7 Meteorological Satellite. *J. Appl. Meteorol. Climatol.* **1984**, *23*, 491–494. [CrossRef]

97. Gao, B.-C. NDWI A Normalized Difference Water Index for Remote Sensing of Vegetation Liquid Water from Space. *Remote Sens. Envrion.* **1996**, *58*, 257–266. [CrossRef]

98. Gupta, K.; Mukhopadhyay, A.; Giri, S.; Chanda, A.; Datta Majumdar, S.; Samanta, S.; Mitra, D.; Samal, R.N.; Pattnaik, A.K.; Hazra, S. An index for discrimination of mangroves from non-mangroves using LANDSAT 8 OLI imagery. *MethodsX* **2018**, *5*, 1129–1139. [CrossRef]

99. Huete, A.R. A Soil-Adjusted Vegetation Index (SAVI). *Remote Sens. Environ.* **1988**, *25*, 295–309. [CrossRef]

100. Rondeaux, G.; Steven, M.; Baret, F. Optimization of soil-adjusted vegetation indices. *Remote Sens. Environ.* **1996**, *55*, 95–107. [CrossRef]

101. Huete, A.R.; Didan, K.; van Leeuwen, W. *Modis Vegetation Index (MOD 13) Algorithm Theoretical Basis Document*; University of Arizona: Tucson, AZ, USA, 1999; Version 3.

102. Chandrasekar, K.; Sesha Sai, M.V.R.; Roy, P.S.; Dwevedi, R.S. Land Surface Water Index (LSWI) response to rainfall and NDVI using the MODIS vegetation index product. *Int. J. Remote Sens.* **2010**, *31*, 3987–4005. [CrossRef]

103. Van Deventer, A.P.; Ward, A.D.; Gowda, P.M.; Lyon, J.G. Using thematic mapper data to identify contrasting soil plains and tillage practices. *Photogramm. Eng. Remote Sens.* **1997**, *63*, 87–93.

104. As-syakur, A.R.; Adnyana, I.W.S.; Arthana, I.W.; Nuarsa, I.W. Enhanced built-UP and bareness index (EBBI) for mapping built-UP and bare land in an urban area. *Remote Sens.* **2012**, *4*, 2957–2970. [CrossRef]

105. Weber, S.J.; Keddell, L.; Kemal, M. Myanmar Ecological Forecasting: Utilizing NASA Earth Observations to Monitor, Map, and Analyze Mangrove Forests in Myanmar for Enhanced Conservation. Available online: https://ntrs.nasa.gov/citations/20140006912 (accessed on 1 October 2020).

106. Breiman, L. Random forests. *Mach. Learn.* **2001**, *45*, 5–32. [CrossRef]

107. Akoglu, H. User's guide to correlation coefficients. *Turk. J. Emerg. Med.* **2018**, *18*, 91–93. [CrossRef] [PubMed]

108. Goodwin, L.D.; Leech, N.L. Understanding correlation: Factors that affect the size of r. *J. Exp. Educ.* **2006**, *74*, 249–266. [CrossRef]

109. Xie, Y.; Sha, Z.; Yu, M. Remote sensing imagery in vegetation mapping: A review. *J. Plant Ecol.* **2008**, *1*, 9–23. [CrossRef]

110. Giri, C.P. *Remote Sensing of Land Use and Land Cover: Principles and Applications*; CRC Press: Boca Raton, FL, USA, 2016; ISBN 9781420070750.

111. Li, W.; El-Askary, H.; Qurban, M.A.; Li, J.; ManiKandan, K.P.; Piechota, T. Using multi-indices approach to quantify mangrove changes over the Western Arabian Gulf along Saudi Arabia coast. *Ecol. Indic.* **2019**, *102*, 734–745. [CrossRef]

112. Jones, T.G.; Ratsimba, H.R.; Ravaoarinorotsihoarana, L.; Glass, L.; Benson, L.; Teoh, M.; Carro, A.; Cripps, G.; Giri, C.; Gandhi, S.; et al. The dynamics, ecological variability and estimated carbon stocks of mangroves in Mahajamba Bay, Madagascar. *J. Mar. Sci. Eng.* **2015**, *3*, 793–820. [CrossRef]

113. Chen, N. Mapping mangrove in Dongzhaigang, China using Sentinel-2 imagery. *J. Appl. Remote Sens.* **2020**, *14*, 1. [CrossRef]

114. Stehman, S.V.; Foody, G.M. Accuracy Assessment. In *The SAGE Handbook of Remote Sensing*; Warner, T.A., Nellis, M.D., Foody, G.M., Eds.; SAGE Publications Ltd.: London, UK, 2009; pp. 297–314. ISBN 9780857021052.

115. Heumann, B.W. Satellite remote sensing of mangrove forests: Recent advances and future opportunities. *Prog. Phys. Geogr.* **2011**, *35*, 87–108. [CrossRef]

116. Rosenfield, G.H. Analysis of thematic map classification error matrices. *Photogramm. Eng. Remote Sens.* **1986**, *52*, 681–686.

117. Lonneville, B.; Schepers, L.; Fernández Bejarano, S.; Vanhoorne, B.; Tyberghein, L. Marine Regions. Available online: https://www.marineregions.org/ (accessed on 22 September 2020).
118. Jones, T.G.; Ratsimba, H.R.; Carro, A.; Ravaoarinorotsihoarana, L.; Glass, L.; Teoh, M.; Benson, L.; Cripps, G.; Giri, C.; Zafindrasilivonona, B.; et al. The Mangroves of Ambanja and Ambaro Bays, Northwest Madagascar: Historical Dynamics, Current Status and Deforestation Mitigation Strategy. In *Estuaries of the World*; Diop, S., Scheren, P., Ferdinand Machiwa, J., Eds.; Springer: Cham, Switzerland, 2016; pp. 67–85.
119. Jones, T.G.; Ratsimba, H.R.; Ravaoarinorotsihoarana, L.; Cripps, G.; Bey, A. Ecological Variability and Carbon Stock Estimates of Mangrove Ecosystems in Northwestern Madagascar. *Forests* **2014**, *5*, 177–205. [CrossRef]

Publisher's Note: MDPI stays neutral with regard to jurisdictional claims in published maps and institutional affiliations.

 *remote sensing*

Technical Note

Gap-Free Monitoring of Annual Mangrove Forest Dynamics in Ca Mau Province, Vietnamese Mekong Delta, Using the Landsat-7-8 Archives and Post-Classification Temporal Optimization

Leon T. Hauser [1,*], Nguyen An Binh [2], Pham Viet Hoa [2], Nguyen Hong Quan [3,4] and Joris Timmermans [1,5]

[1] Department of Environmental Biology, Institute of Environmental Sciences, Leiden University, P.O. Box 9518, 2300 RA Leiden, The Netherlands; j.timmermans@cml.leidenuniv.nl

[2] Ho Chi Minh City Institute of Resources Geography, Vietnam Academy of Science and Technology, Ho Chi Minh City 700000, Vietnam; nabinh@hcmig.vast.vn (N.A.B.); pvhoa@hcmig.vast.vn (P.V.H.)

[3] Institute for Circular Economy Development, Vietnam National University, Ho Chi Minh City 700000, Vietnam; nh.quan@iced.org.vn

[4] Center of Water Management and Climate Change, Institute for Environment and Resources, Vietnam National University, Ho Chi Minh City 700000, Vietnam

[5] Biogeography & Macroecology Lab, Department Theoretical and Computational Ecology, Institute for Biodiversity and Ecosystem Dynamics, University of Amsterdam (UvA), 1090 GE Amsterdam, The Netherlands

* Correspondence: l.t.hauser@cml.leidenuniv.nl

Received: 19 September 2020; Accepted: 6 November 2020; Published: 13 November 2020

Abstract: Ecosystem services offered by mangrove forests are facing severe risks, particularly through land use change driven by human development. Remote sensing has become a primary instrument to monitor the land use dynamics surrounding mangrove ecosystems. Where studies formerly relied on bi-temporal assessments of change, the practical limitations concerning data-availability and processing power are slowly disappearing with the onset of high-performance computing (HPC) and cloud-computing services, such as in the Google Earth Engine (GEE). This paper combines the capabilities of GEE, including its entire Landsat-7 and Landsat-8 archives and state-of-the-art classification approaches, with a post-classification temporal analysis to optimize land use classification results into gap-free and consistent information. The results demonstrate its application and value to uncover the spatio-temporal dynamics of mangrove forests and land use changes in Ngoc Hien District, Ca Mau province, Vietnamese Mekong delta. The combination of repeated GEE classification output and post-classification optimization provides valid spatial classification (94–96% accuracy) and temporal interpolation (87–92% accuracy). The findings reveal that the net change of mangroves forests over the 2001–2019 period equals −0.01% annually. The annual gap-free maps enable spatial identification of hotspots of mangrove forest changes, including deforestation and degradation. Post-classification temporal optimization allows for an exploitation of temporal patterns to synthesize and enhance independent classifications towards more robust gap-free spatial maps that are temporally consistent with logical land use transitions. The study contributes to a growing body of work advocating full exploitation of temporal information in optimizing land cover classification and demonstrates its use for mangrove forest monitoring.

Keywords: data fusion; forest monitoring; Google Earth Engine; Landsat; mangrove forests; multi-temporal analysis; remote sensing; satellite earth observation; time series analysis; Vietnam

---

1. Introduction

Ecosystem services offered by mangrove forests are facing severe risks. Within the transition zone of land and sea of (sub)tropical coastal regions, mangroves have carved out a distinct niche to flourish and thereby provide vital services to mankind. Specifically, mangrove ecosystems have shown to be one of the world's most productive in terms of carbon sequestration, shelter and breeding grounds for aquatic species, and as important physical barriers against tides and ocean surges [1–4]. Despite the multitude of crucial ecosystem services these coastal forests offer to communities in coastal regions of more than 124 countries, the status of mangrove forests in many regions is under pressure due to forest loss and land degradation, caused by overexploitation and land use change driven by human development [5–7]. Due to the inaccessible, ever-changing, and extensive nature of these mangroves, remote sensing has become a primary instrument to monitor the health and dynamics of these ecosystems [8–10].

The field of Satellite Remote Sensing has moved into an era in which a tremendous wealth of earth observation (EO) data are gathered at increasing spectral, spatial, and temporal resolutions—supporting the wide-spread application of satellite data for studying global changes [11]. Orbiting EO satellites allow us to repeatedly revisit areas of interest to study temporal changes and facilitate time series analysis. The iconic Landsat-7 and Landsat-8 missions both offer average revisit intervals of 16 days and observations that go back as early as the year 2000. The later Landsat-8 mission collected over 3.35 Petabyte of scenes over the course of a single year in 2017 [12]. These data collections hold great potential to improve our monitoring efforts of mangrove ecosystems and study changes over time.

A critical review by Younes Cardenas et al. (2017) on using satellite remote sensing to monitor mangrove ecosystems points out that the majority of studies conducted—reviewing 55 recent peer-reviewed articles using Landsat/Aster imagery—are not making full use of the wealth of EO data available [13]. The authors specify that most studies between 2001–2016 used fewer than 10 images and longitudinal studies often analyze temporal changes with 7–11 years between scenes which leaves much of the potential of current satellite archives unlocked [13]. Yet, mangrove forests are frequently part of fast-changing landscapes driven by land use change at the interplay of volatile aquaculture markets, policy-making, and the biophysical dynamics of erosion, sedimentation, and changing tides [14–16]. This raises the question of how we can better unlock the potential of available satellite imagery archives to facilitate high temporal resolution monitoring of the fast-paced land use processes surrounding mangrove forest ecosystems.

The advances in high-performance computing (HPC) in combination with cloud-computing services, such as provided by the Google Earth Engine platform (GEE), allow us to address the major challenges of processing and handling enormous EO datasets and turning these into comprehensible information [13,17–21]. The GEE platform provides straightforward HPC cloud access to many of the major satellite archives as well as numerous image classifiers for mapping applications, including Classification and Regression Trees (CART) and Random Forests (RF) approaches. Illustrative of its capabilities, Hansen et al. (2013) mapped global forest cover change products from over 650 thousand Landsat-7 scenes [22]. Following this, a large body of regional studies has demonstrated high mapping accuracies using GEE's land use classifiers (CART) with Landsat images [19,23,24]. More specifically, we observe an increasing use and successful implementation (accuracies between 92% and 97%) of GEE-based land use classification for mangrove mapping [25–27].

Through GEE, we can efficiently organize longitudinal time series from satellite observations and independent classification efforts can be repeated over time with ease. These time series can be valuable to study and monitor temporal changes in land use. Conversely, the temporal dependencies of each time point in the series can also be used to further optimize the time series in terms of missing information and consistency. In other words, the temporal domain can facilitate post-classification optimization of GEE output towards maps that are gap-free and temporally consistent with logical land use transitions as well as provide a means of cross-validation. This is particularly crucial in cases with hampering climatic conditions (clouds, snow, dust, and aerosols), instrumentation errors, losses of image data during data transmission, or high uncertainties in information processing [28].

Temporal gap-filling and smoothing approaches are common practice in remote sensing of phenology and cropping cycles through continuous parameters, such as vegetation indices (e.g., NDVI, EVI) and surface parameters (Land Surface Temperature) [28–32]. However, in discrete land cover classification exercises, this practice remains less common, including in combination with the GEE platform [33,34]. Current studies tend to focus on gap-filling based on spatially neighboring pixels [35,36], spectral similarity, and/or multi-sensor (source) data fusion [34,37,38], rather than temporal integration. As such, few land use studies have taken full advantage of temporal dependencies to reduce both information gaps and inconsistent land use transitions [13,39–41]. This is a particularly rare undertaking for the monitoring of mangrove forests land use changes, whereas consistent and gap-free time series are crucial to closely monitor mangrove deforestation, degradation, and disturbance [13,15]. Land use changes tend to follow logical temporal land use transitions which can guide the optimization of classification maps [13,40].

The main objective of our study is to deploy high-performance computing techniques to monitor mangrove forest cover changes in our case study area; the mangrove-rich Ngoc Hien District, Ca Mau province in the Vietnamese Mekong delta. Rather than a single land use classification approach, we demonstrate how independent land use classifications conducted in GEE can be combined to optimize classification results in terms of completeness and consistency. As such, the study exploits both; (1) the computational capacity of GEE to deal with the entire Landsat-7 and -8 archives and (2) the temporal element of a longitudinal time series to optimize land use classification results into "gap-free" and temporally consistent information. This can help us better understand the spatio-temporal dynamics of mangrove forests, in terms of extent, distribution, and land use change and disturbances that threaten their conservation.

2. Materials and Methods

2.1. Study Area

The study area focuses on Vietnam's southernmost district, Ngoc Hien, Ca Mau province, located in the Southern Mekong Delta between latitude 8°33′–8°45′N and longitude 104°42′45″–105°3′54″E, spanning an area of 743 km^2 (See Figure 1). The district has been well-studied for its importance as a major aquaculture hub and its significant reserves of Vietnam's largest and last remaining old-growth mangrove forests, including the internationally acknowledged RAMSAR site of Mui Ca Mau (2012) and UNESCO Biosphere Reserve (2009) [42–44]. The landscape supports both ecologically important mangrove ecosystems and the economic livelihoods based on aquaculture.

2.2. Remote Sensing Data Pre-Processing

This study makes use of the archives of Landsat's later missions embodied by the Landsat-7 and Landsat-8 multispectral imagery available through GEE's public data catalogue of atmospherically corrected surface reflectance data. We have made use of all available 30 m spatial resolution bands of both missions, this implies: two short-wave infrared (SWIR) bands and four/five visible and near-infrared (VNIR) bands for Landsat-7 and Landsat-8, respectively. The study area is centered within the path-rows of 125–054 and 126–064 with an average 16-day revisit time. The Landsat-7 and -8 Quality Assessment bands and calculated F-mask were used to filter out pixels containing clouds, cirrus, cloud shadow, and atmospheric contamination of the reflectance signal.

Table 1 gives an overview of the available images for each year. Based on the spectral reflectance of all available images annual median composites are compiled that provide consistent cloud-free median images of spectral reflectance [45]. The malfunction of the Scan Line Corrector (SLC) of the Landsat-7 imager has resulted in that on average around a quarter of the data in a Landsat-7 scene is missing from the 31st of May 2003 onwards [46]. These products hence have considerable data gaps, but still maintain the same radiometric and geometric corrections as data collected prior to the SLC failure. The combination of high cloudiness of the region with SLC failure results in limited usability of the years 2003–2012. Therefore, our analysis deploys a cautionary interpretation of these years. Furthermore, years characterized by SLC failure are combined bi-annually to increase data availability, coverage of the region, and to lower uncertainties.

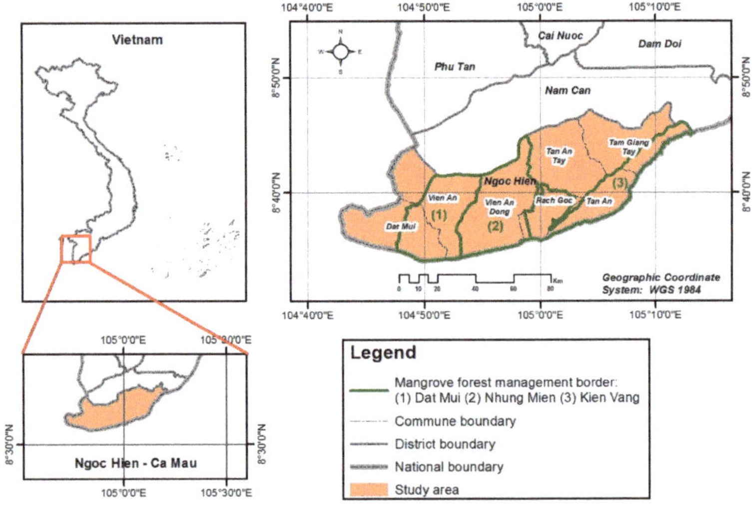

Figure 1. Location of study area.

Table 1. Overview of used images, per sensor (Landsat-7 and Landsat-8) per scene pathway, and missing pixels and anomalous land use transitions assessed with temporal data fusion. Years impeded by the SLC-malfunction are shaded in grey.

Year	Available Images				Missing Pixels for Gap Filling		LUC Anomalies Detected	
	LS-7		LS-8		No. of Pixels	%	No. of Pixels	%
	125–054	126–064	125–054	126–064				
2001–2002	19	23			1	0.0	0	0.0
2003–2004 †	22	19			0	0.0	449	0.0
2005–2006 †	24	17			257	0.0	1321	0.1
2007–2008 †	10	7			1352	0.1	3220	0.3
2009–2010 †	12	11			78	0.0	4352	0.5
2011–2012 †	11	9			6846	0.7	5112	0.5
2013	(10)	(8)	7	13	6468	0.7	8918	0.9
2014			21	19	143	0.0	2131	0.2
2015	16 *	12 *	17 *	19 *	5	0.0	976	0.1
2016			14	17	60	0.0	1372	0.1
2017			18	14	438	0.0	2964	0.3
2018			17	19	1026	0.1	2028	0.2
2019			17	20	3914	0.4	3720	0.4

* = Training year, † = Scan Line Corrector (SLC) malfunction.

2.3. Land Cover Classification

After pre-processing, the resulting cloud-free median multispectral annual composites are used to characterize land use, and the land use changes over time. The land use classification scheme of our study takes into consideration four dominant land uses within the Ngoc Hien district, namely (1) Dense Mangrove Forest, (2) Sparse Mangroves, (3) Aquaculture/Waterbodies, and (4) Built-up and Barren lands. Dense mangrove forest is defined by a minimum of 30% canopy cover. Vegetated mangrove areas that are 10–30% crown cover are classified as sparse mangroves.

We conducted a supervised classification to develop land use maps. There are several classification algorithms available within GEE, including; Classification and Regression Trees (CART), Random Forest (RF), Naïve Bayes, and Support-Vector Machine (SVM). Our study opted for the commonly used CART classifier which has produced relatively high accuracies when applied to Landsat data in numerous settings [19,23,24,26]. More specifically, several studies have reported the highest accuracy for CART land use classification of coastal wetlands and mangroves using GEE compared to other classifiers [25,27]. Most importantly, we ran trails in the study area for both CART and RF in which the first yielded the highest classification accuracy (94–96% for CART, against 89–94% for RF, respectively). GEE code implementations of both approaches and its validation against test data can be found in the Supplementary Materials Table S1.

Within CART, a decision tree (DT) classifier was instantiated and trained on field data using GEE's default parameters. The CART algorithm runs through a series of nodes that recursively split the input data in such a way that there is a decrease in entropy and an increase in information gain after the split. GEE's CART uses the Gini Impurity Index to decide the input features which will provide the best split at each node. A tabular overview of the exact decision rules for building the model can be found in Supplementary Materials Table S2. One disadvantage of the DT classifier is the considerable sensitivity to the training dataset. A small change to the training data can result in a very different set of subsets and can result in overfitting [19,47]. Nevertheless, our training and validation data relies on extensive fieldwork, including 514 georeferenced points gathered in-situ across the Ngoc Hien district in 2015, subdivided into the four classes; dense mangroves ($n = 247$), sparse mangroves ($n = 72$), waterbodies/aquaculture ponds ($n = 120$), and built-up and barren lands ($n = 75$). We used 70% of the field data for training and the remaining 30% for validation, thereby estimating the classification errors independently.

Following the initial training of the classifier, it is then deployed backward (LS-7) or forward (LS-8) through the time series based on spectral/change information of the surface reflectance data available in the composite datasets. Based on this method, land cover maps are generated from the surface reflectance of pre-processed yearly median composites between 2001 and 2019. The workflow of GEE pre-processing and land use classification is presented schematically in Figure 2. GEE code can be accessed through the URLs published in Supplementary Materials Table S1.

2.4. Post-Classification Optimization through Time Series Temporal Data Fusion

The longitudinal temporal data of the Landsat archives enabled the use of neighboring time points to cross-validate findings, fill in missing data through temporal data fusion, detect and revise illogical land use changes in the post-classification analysis [33,39–41,46,48].

Gap-filling through consideration of the temporal dimension in years preceding and following a missing value has allowed us to interpolate missing pixels to assure spatial and temporal continuity over the study area. The approach for temporal interpolation follows an adaptation of inverse distance weighting methods applied to discrete land use classification data, the basic assumption is that a temporally distributed variable at short distance is generally more similar than at larger distance [49,50]. The applied approach integrates land use classes of adjacent years weighted by a power (p = 1.5) of the distance (d) to the year of interpolation, formulated in an equation as:

$$\hat{z}_{i=0} = \max_z\left\{\sum_{i=1}^{n}\frac{1}{d_{z,i}^{p}}\right\}$$

in which i = 0 indicates the time point to be interpolated with predicted land use ($\hat{z}$), and which $i = n$ indicate the n years adjacent where land use (z) has been observed. As such, pixels missing valid observations are estimated by taking into account a seven-year pattern and scoring neighboring time points, in which observations nearest in time weight the heaviest. The seven-year time window corresponds with the consecutive years with full data availability (2013–2019). The land use class scored the highest will be used for gap-filling.

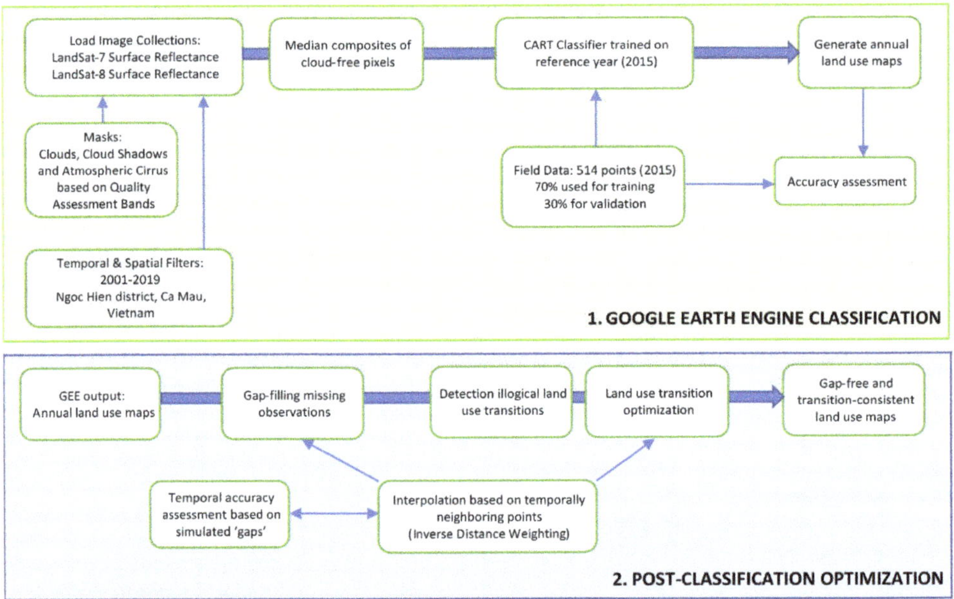

Figure 2. Data processing chain and workflow of the study separated in a repeated (1) GEE land use classification process and (2) post-classification optimization based on the temporal integration of land use classification output.

Similarly, further optimization of classification results can be achieved by taking into account that land use changes usually occur characterized by a logical transition [40,41]. Land use changes and transitions follow ecological rules [40,41]. For instance, the growth of dense mangrove forests takes at least multiple years. Understanding these land use transitions can help setting rules determined by ecology and feasibility to detect illogical land use transition from remotely sensed time series.

In our study, we have opted for a post-classification approach to detect and revise illogical land use (See Table 2). The assessment of uncertainties and reclassification by the CART classifier based on the temporal exclusion of certain illogical transition rules (e.g., maximum a posteriori (MAP) classification rule in a Bayesian framework [41]) has not been embedded within the GEE platform. Therefore, a post-classification approach that scores the likelihood of land use classes to occur based on temporal context is deployed. Table 2 provides an overview of the anomalies detected. Similar to the gap-filling of missing pixels, illogical land use transitions are revised by distance weighting of neighboring time points (analyzing a seven-year pattern) of the pixel under scrutiny to determine what land cover time is most probable to be in place. The entire workflow from pre-processing, land cover classification, post-classification optimization is presented schematically in Figure 2's diagram.

Table 2. Scheme of logical inter-annual land use changes. Color scheme indicates a classification of mangrove forest land use change trends. The color scheme also applies multi-annually.

Year 1 ↓ Year 2 →	Dense Mangroves	Sparse Mangroves	Waterbodies/Paddies	Built Env/Mudlands
Dense Mangroves	✓	✓	✓	✓
Sparse Mangroves	✓	✓	✓	✓
Waterbodies/Paddies	✗	✓	✓	✓
Built Env/Mudlands	✗	✓	✓	✓

Deforestation; Forest Degradation; Reforestation; Unchanged Forest.

We tested the gap-filling algorithm's predictive performance using a k-fold leave-one-out strategy. In these tests, we purposely removed a datapoint to run the gap-filling algorithm to predict the land use based on temporal neighbors. We assessed the percentage of correctly predicted land use classifications. We have conducted this for all available pixels and years in the analysis.

Taken together, temporal integration of GEE's individual land use classification maps into a gap-free sequence of maps that follow logical land use transitions enabled us to analyze the land use trends occurring throughout the available data. Land use changes were grouped into four categories illustrated through the color scheme in Table 2. Only mangrove forest changes were considered in the trend maps (other land use changes remain white). Moreover, we used 150×150 m box averages to highlight general trends and reduce uncertainties by spatial regularization.

3. Results

Temporal and Spatial Accuracy Assessment

Classification of surface reflectance, using a CART classifier ran within GEE and trained separately for both Landsat-7, and -8 on collected field data, produces 95.6% and 94.1% accuracy confiners for LS-7 and LS-8, respectively, for the reference year 2015 (Table 3). The gap-filling algorithm's performance to predict land use type based on the seven-year pattern surrounding missing pixels hovers between 87 to 92% depending on the year (Table 4).

Table 3. Validation results of land use classification for Landsat-7 and Landsat-8.

	Land Cover Type	Dense Mangrove		Sparse Mangrove		Water Bodies/Ponds (LS7/LS8)		Built Environment /Mudflats (LS7/LS8)		Total Classified Pixels (LS7/LS8)		User Accuracy (%) (LS7/LS8)	
		LS7	LS8	LS7	LS8	LS7	LS8	LS7	LS8	LS7	LS8	LS7	LS8
Classification Results	Dense mangrove	72	84	1	5	0	0	0	0	73	89	98.63	94.38
	Sparse mangrove	1	0	20	16	2	2	0	0	23	18	86.96	88.89
	Water bodies/Ponds	0	0	1	2	36	38	0	0	37	40	97.30	95.00
	Built environment/Mudflats	1	0	0	1	1	0	24	21	26	22	92.31	95.45
	Total ground truth pixel	74	84	22	24	39	40	24	21	159	169		
	Producer accuracy (%)	97.30	100.0	90.91	66.67	92.31	95.00	100.0	100.0				
	Overall accuracy (%) (LS7/LS8)											95.60	94.08

Ground Truth from Field Survey

Table 4. Validation results of land use classification for Landsat-7 and Landsat-8. Years with SLC malfunction are shaded in grey considering the higher uncertainties.

Year	Temporal Gap-Filling Accuracy (%)
2001–2002	-
2003–2004	85.67
2005–2006	85.19
2007–2008	86.64
2009–2010	85.43
2011–2012	87.16
2013	90.41
2014	91.68
2015	91.80
2016	86.92
2017	87.40
2018	86.89
2019	-

An overview of all land use classification maps after gap-filling and post-processing for logical land use transitions is presented in Figure 3. The maps depict all years with full data-availability and therefore the highest reliability ranging from 2001–2002, and 2013–2019. A visual inspection of the temporal dynamics indicates that the central regions in Ngoc Hien have been subject to a high frequency of land use changes whereas towards the coast, specifically the Western Cape, stable havens of dense mangrove forest have to large extent been minimally subjected to forest clearance and land use change. Over the years, three core mangrove forests areas can be identified (See Figure 1, for commune locations); (1) the cluster in the south-east of Ngoc Hien, including the eastern part of Tan An commune and adjacent southern Tam Giang Tay commune, (2) the mangrove forest strip along the southern coastline, and (3) the largest core mangrove area is located in the western tips of Ngoc Hien district; Dat Mui commune which encloses the Ca Mau Cape National Park and the north-western part of Vien An commune.

The total land use changes over time are highlighted in Figure 4. These results show that the total shares of land use cover between four classes have remained relatively stable with yearly fluctuations. In the span of two decades, we find sparse mangrove cover fluctuating between 33,000 ha and 40,000 ha and dense mangrove forests hovering between 18,000 ha and 24,000 ha. The average net change of dense mangroves over the 2001–2019 equals −0.01% annually. In recent years since 2013, we observe upward trends in dense mangrove forest cover as well as built environment and mudflats. Sparse mangrove cover has been decreasing in that same period. Waterbodies/ponds have remained more or less equal.

Despite these overall trends, the figure conceals the spatial dynamics behind the distribution of mangrove forests and land use changes that characterize the region. To gain insight into the spatial distribution of land use changes, an aggregated trend map was created (Figure 5) to identify hotspots of mangrove forest change while leaving changes of non-forest classes out of consideration. Generally, the dense mangrove patches in the Eastern regions have shrunk in size and extent, whereas the core mangrove forests on the Western Cape have seen an expansion through sedimentation and seaward colonization. Ngoc Hien's central communes have seen an increased integration of dense mangroves patches in the mosaic of waterbodies/ponds and sparse mangrove cover. Areas most prone to mangrove forest loss and degradation are the eastern central regions, possibly due to its vicinity to districts with higher urban development, infrastructure, and aquaculture production [51].

The trend map presented in Figure 5 gives insight into the spatial distribution of temporal dynamics of mangrove forest loss and gains. The map highlights how coastal erosion along the southward coastline and sedimentation along the westward frontier have resulted in losses and gains of mangrove forests. The strip of dense mangrove forests serving as a protective buffer along Ngoc Hien's southern coastline has seen shifting land use changes; on the one hand, inland regrowth and increased connectivity are observed, whereas on the other hand coastal erosion is becoming increasingly severe leading to seaward losses of dense mangrove forest. In Ngoc Hien's Western regions, Dat Mui and Vien An communes (see Figure 1, for commune locations), the remaining core mangrove forests have seen relatively few land use changes except for seaward mangrove expansion and colonization caused by coastal sedimentation (Figure 5).

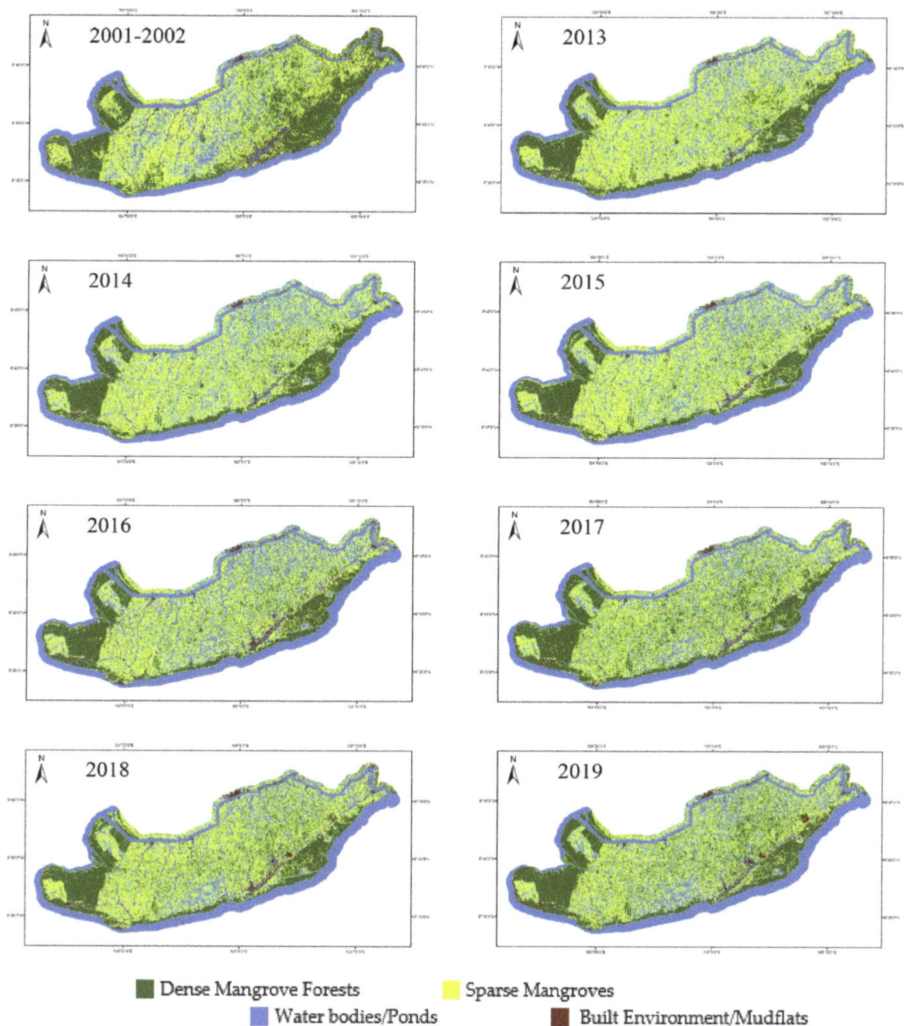

Figure 3. Time series of land use maps of Ngoc Hien District from 2001–2002 and 2013–2019, displayed are all years with full data availability (non-SLC malfunction).

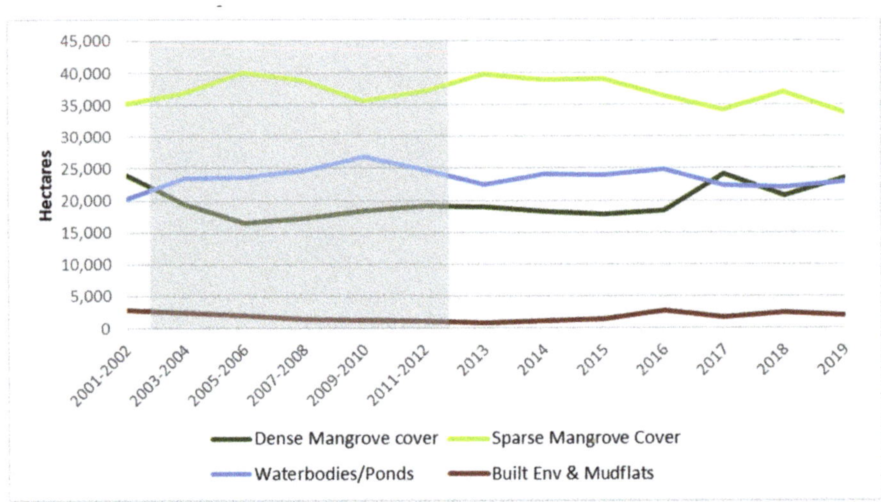

Figure 4. Temporal changes in total land cover surface across four land use classes in Ngoc Hien district. Years with SLC malfunction are shaded in grey considering the higher uncertainties.

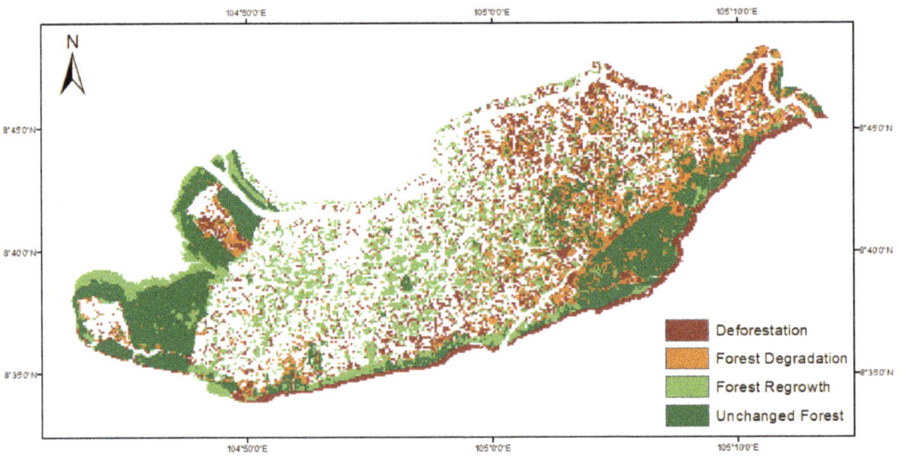

Figure 5. Mangrove forest trend map (2001–2019) of 150 × 150 m box averages.

4. Discussion

4.1. Overcoming Observation Gaps in Mangrove Monitoring

In the critical review by Younes Cárdenas et al. (2017) based on 55 recent peer-reviewed journal articles using Landsat or ASTER images to monitor mangrove forests, the authors conclude that the majority of multi-temporal studies focus on only a fraction of available satellite imagery with on average 7–11 years between scenes of multitemporal analysis [13]. Yet, high temporal change detection of mangrove forests is vital in mapping threats from aquaculture expansion and coastal development as well as to understand cyclic processes such as logging in production areas and seasonal biomass fluctuations [15]. In our study, 446 unique scenes were processed into annual median composites to study the temporal dynamics of mangrove forests in one of Vietnam's most prominent regions for mangrove conservation, Ngoc Hien district in the Ca Mau province. Despite a large number of scenes per composite (15.4 scenes on average), it still resulted in a high number of pixels with no observations and/or illogical land use transitions for different years between 2001–2019 (See Table 1). Following temporal gap-filling and post-classification optimization, the resulting optimized time series allowed us to better monitor the state of mangrove forests in Ngoc Hien with spatial and temporal continuity and logical consistency in transitions.

4.2. Post-Classification Temporal Optimization

Accurate land use classification remains a challenging undertaking in landscapes dominated by aquaculture and mangrove forest land use [13]. In our study, the accuracy assessment for the reference year 2015 yielded assuring accuracy confiners (Table 3). We used 514 single-year reference data points observed in-situ for training the classifier and validation of the results. Ideally, we would have such data available for multiple years; however, this is difficult to acquire and organize. With limited availability of ground-truth data, reducing classification uncertainties and increasing temporal consistency is key to provide high-quality land use maps.

The use of yearly median composites allowed for fast computational processing and a comprehensible annual time series output for policy and decision-makers. Nevertheless, the application of median composites also poses disadvantages. Adequate composites still require the presence of sufficient high-quality observations. Moreover, a year-round even temporal distribution of scenes is required to facilitate an adequate median composite that is representative for the entire year. Knowledge regarding yearly phenology, the impact of tides of reflectance signals, trends in biophysical parameters (functional traits) of mangrove ecosystems is for a large part still lacking to make appropriate judgments on possible biases in median composites [13]. In other words, gaining more understanding regarding these temporal processes, also within yearly cycles, will help gain insight into the robustness of median composites for mangrove forest ecosystems.

Further challenges in accurate classifications of mangrove forests are raised by; (1) the fine-grained landscape mosaic with mangrove plots and aquaculture ponds often sized at sub-pixel (30 m × 30 m) measures, (2) the unknown implications of tidal effects on spectral signals and the high level of water vapor observed in these coastal regions, and (3) recent trends towards integrated mangrove-shrimp farming production systems which have made discrimination between mangrove, aquaculture paddy land use classes more ambivalent [9,13,15,52,53]. These challenges highlight the importance of making the most of the temporal information available to lower uncertainties in the final classification product. This is particularly important when noise in remote sensing signals is high—which is commonly the case in cloud-covered mangrove regions—and when multi-annual validation/training data are scarce. Several situations can cause illogical changes in a land use time series, such as classification errors, reflectance signal noise, and imperfect image co-registration.

Here, we demonstrated how the exploitation of information in the temporal domain allowed for additional optimization and a means of cross-validation of the GEE classification outputs. Studying temporal patterns and cross-validating land use changes in relation to the previous and following years help increase the robustness to noise and credibility of classification efforts. Specifically, the temporal information of land cover maps in a time series has been used to detect illogical land use changes and improve classification results [48]. The approach for temporal mangrove monitoring outlined is relatively easy to implement using GEE output and post-classification optimization. At the same time, it provides valid spatial classification (Table 3; 94–96% accuracy) and temporal interpolation (Table 4; 87 to 92% accuracy). Temporal interpolation follows a simple discrete inverse distance weighting algorithm, however other and more advanced statistical learning approaches can potentially be interesting alternatives [41,49,50,54–56].

Figure 5 illustrates the further implementation of our gap-free time series in studying multi-temporal mangrove land use trends in Ngoc Hien. Bi-temporal approaches risk highlighting observations that result from isolated instances that can introduce inconsistencies and uncertainties in classification, especially considering the, at times, unfavorable signal-to-noise ratios found in satellite remote sensing [13]. Instead of comparing two single timestamps, the integration of yearly gap-free land use classification enables temporal cross-validation along logical land use transitions and gap-filling based on temporally neighboring information. This temporally dense analysis allows us to fully assess the direction and frequency of land use changes affecting mangrove forests [57,58]. This is also important to ensure that the unchanged forest between two time points has remained undisturbed in the years in between. Moreover, the multi-annual approach allows us to assess and quantify observed land use changes, e.g., forest disturbances, at a high temporal frequency, thereby opening venues to better monitor and study mangrove forest disturbance regimes and mangrove degradation processes as compared to bi-temporal land use change.

The land use trend map (Figure 5) facilitates the detection of hotspots for mangrove forest change; deforestation, degradation, and regrowth. While the map is informative for deforestation and the land use change drivers behind it, the assessment of forest degradation remains arbitrary inherent to the operationalization of the land use classification scheme. This ties in with the challenges and complexity of defining forest degradation [59]. A large variety of existing definitions of forest degradation require different methods for assessment based on the objectives of the intended study [60]. The gap-free land use change maps presented here may help flag changes in mangrove forest extent. However, to overcome the arbitrariness of classes, complimentary maps on quantitative canopy/leaf traits and biomass can further enhance our ability to assess forest degradation and forest regrowth along a spectrum of ecologically relevant indicators [61–63].

4.3. Future Implementation

The case study presented here builds on and contributes to a growing body of work advocating that approaches that account for temporal information in optimizing land cover classification are superior to temporally independent time series classifications and other single-date methods [39,64]. The application of the post-classification optimization to stabilize land cover trajectories, mitigate unrealistic land cover transitions and overcome the limitations and costs of obtaining ancillary and field data are still rarely applied in land use time series studies [39,64]. GEE implementation of these methods within its standard functionality offers an opportunity to improve temporal consistency and promote temporally dense analyses as a common practice. In addition to the post-classification method presented here, other opportunities to fully benefit from the temporal information available would be a GEE implementation of methods combining the probability functions of land use trajectories and the propagation of classification uncertainties temporally to decide on the final classification outcome [33,39–41,46,54,65]. Furthermore, data fusion of multiple sensors have shown large potential to further increase the temporal resolution of analysis [34,37,38]. In particular, the integration of optical and LiDAR or synthetic aperture radar imaging offers opportunities for high-frequency temporal analyses that are independent of the weather and cloud cover [66].

5. Conclusions

Our study aimed to advance temporal mangrove forest monitoring efforts to benefit from the potential that currently available satellite earth observation data and cloud-based high-performance computing can offer. The temporal domain of these information-dense datasets opens opportunities to apply data fusion principles to optimize classification outputs to be gap-free and temporally consistent with logical land use transitions as well as provide a means of cross-validation. The results of our case study on mangrove forests demonstrate how this information can be valuable in understanding the spatio-temporal dynamics, processes, and trends of land use changes and improve decision-making with detailed information. Thereby, our study builds on and contributes to a growing body of work advocating that accounting for temporal information in optimizing land cover classification is superior to temporally independent classifications in time series and other single-date methods.

Despite growing awareness, most mangrove forest cover classification studies are yet to take full advantage of Earth observation's potential and the rich temporal information available from time series data. Implementation of temporal optimization, either post-classification or during the classification process, within future implementations or that can be automated on top of GEE's output as presented here, can hopefully contribute to advance mangrove monitoring studies towards fully unlocking the potential of data available as the field of earth observation keeps evolving. The integration of synthetic aperture radar remote sensing in addition to optical observations will be key in advancing the approach presented here and increase the temporal resolution by overcoming data gaps due to weather or lighting conditions.

Supplementary Materials: The following are available online at http://www.mdpi.com/2072-4292/12/22/3729/s1, Table S1: Links to the GEE code implementation used for this study, Table S2: Decision rules of the regression tree (CART) build for land use classification (2015) by Landsat-7 and Landsat-8 respectively.

Author Contributions: Conceptualization, L.T.H. and N.A.B.; methodology, L.T.H. and N.A.B.; coding, L.T.H. and N.A.B.; validation, N.A.B. and P.V.H.; investigation, N.A.B. and P.V.H.; resources, N.A.B. and P.V.H.; data curation, N.A.B.; writing—original draft preparation, L.T.H.; writing—review and editing, N.H.Q. and J.T.; visualization, L.T.H.; supervision, L.T.H. and P.V.H.; project administration, N.A.B. and P.V.H.; funding acquisition, P.V.H. All authors have read and agreed to the published version of the manuscript.

Funding: This research was funded by Vietnam Academy of Science and Technology, grant number ĐLTE00.06/20-21 and the APC was funded by the Institute of Environmental Sciences (CML), Leiden University.

Acknowledgments: This research was partly supported by Vietnam Academy of Science and Technology through the project ĐLTE00.06/20-21.

Conflicts of Interest: The authors declare no conflict of interest.

References

1. Clough, B. *Continuing the Journey amongst Mangroves*; ISME Mangrove Educational Book Series No.1; ISME Publications: Okinawa, Japan, 2013; ISBN 9784906584161.

2. Kuenzer, C.; Tuan, V.Q. Assessing the ecosystem services value of Can Gio Mangrove Biosphere Reserve: Combining earth-observation- and household-survey-based analyses. *Appl. Geogr.* **2013**, *45*, 167–184. [CrossRef]

3. Brander, L.M.; Wagtendonk, A.J.; Hussain, S.S.; McVittie, A.; Verburg, P.H.; De Groot, R.S.; Van Der Ploeg, S. Ecosystem service values for mangroves in Southeast Asia: A meta-analysis and value transfer application. *Ecosyst. Serv.* **2012**, *1*, 62–69. [CrossRef]

4. Siikamäki, J.; Sanchirico, J.N.; Jardine, S.L. Global economic potential for reducing carbon dioxide emissions from mangrove loss. *Proc. Natl. Acad. Sci. USA* **2012**, *109*, 14369–14374. [CrossRef]

5. Costanza, R.; De Groot, R.; Sutton, P.C.; Van Der Ploeg, S.; Anderson, S.J.; Kubiszewski, I.; Farber, S.; Turner, R.K. Changes in the global value of ecosystem services. *Glob. Environ. Chang.* **2014**, *26*, 152–158. [CrossRef]

6. Duke, N.C.; Meynecke, J.-O.; Dittmann, S.; Ellison, A.M.; Anger, K.; Berger, U.; Cannicci, S.; Diele, K.; Ewel, K.C.; Field, C.D.; et al. A World Without Mangroves? *Science* **2007**, *317*, 41b–42b. [CrossRef]

7. Giri, C.; Ochieng, E.; Tieszen, L.L.; Zhu, Z.; Singh, A.; Loveland, T.; Masek, J.; Duke, N. Status and distribution of mangrove forests of the world using earth observation satellite data. *Glob. Ecol. Biogeogr.* **2010**, *20*, 154–159. [CrossRef]

8. Heumann, B.W. Satellite remote sensing of mangrove forests: Recent advances and future opportunities. *Prog. Phys. Geogr.* **2011**, *35*, 87–108. [CrossRef]

9. Kuenzer, C.; Bluemel, A.; Gebhardt, S.; Quoc, T.V.; Dech, S. Remote Sensing of Mangrove Ecosystems: A Review. *Remote Sens.* **2011**, *3*, 878–928. [CrossRef]

10. Giri, C. Observation and Monitoring of Mangrove Forests Using Remote Sensing: Opportunities and Challenges. *Remote Sens.* **2016**, *8*, 783. [CrossRef]

11. Belward, A.S.; Skøien, J.O. Who launched what, when and why; trends in global land-cover observation capacity from civilian earth observation satellites. *ISPRS J. Photogramm. Remote Sens.* **2015**, *103*, 115–128. [CrossRef]

12. Soille, P.; Burger, A.; De Marchi, D.; Kempeneers, P.; Rodriguez, D.; Syrris, V.; Vasilev, V. A versatile data-intensive computing platform for information retrieval from big geospatial data. *Futur. Gener. Comput. Syst.* **2018**, *81*, 30–40. [CrossRef]

13. Cárdenas, N.Y.; Joyce, K.E.; Maier, S.W. Monitoring mangrove forests: Are we taking full advantage of technology? *Int. J. Appl. Earth Obs. Geoinf.* **2017**, *63*, 1–14. [CrossRef]

14. Hamilton, S.E.; Casey, D. Creation of a high spatio-temporal resolution global database of continuous mangrove forest cover for the 21st century (CGMFC-21). *Glob. Ecol. Biogeogr.* **2016**, *25*, 729–738. [CrossRef]

15. Hauser, L.T.; Vu, G.N.; Nguyen, B.A.; Dade, E.; Nguyen, H.M.; Nguyen, T.T.Q.; Le, T.Q.; Vu, L.H.; Tong, A.T.H.; Pham, H.V. Uncovering the spatio-temporal dynamics of land cover change and fragmentation of mangroves in the Ca Mau peninsula, Vietnam using multi-temporal SPOT satellite imagery (2004–2013). *Appl. Geogr.* **2017**, *86*, 197–207. [CrossRef]

16. Li, F.; Liu, K.; Tang, H.; Liu, L.; Liu, H. Analyzing Trends of Dike-Ponds between 1978 and 2016 Using Multi-Source Remote Sensing Images in Shunde District of South China. *Sustainability* **2018**, *10*, 3504. [CrossRef]

17. Azzari, G.; Lobell, D. Landsat-based classification in the cloud: An opportunity for a paradigm shift in land cover monitoring. *Remote Sens. Environ.* **2017**, *202*, 64–74. [CrossRef]

18. Gorelick, N.; Hancher, M.; Dixon, M.; Ilyushchenko, S.; Thau, D.; Moore, R. Google Earth Engine: Planetary-scale geospatial analysis for everyone. *Remote Sens. Environ.* **2017**, *202*, 18–27. [CrossRef]

19. Shelestov, A.; Lavreniuk, M.; Kussul, N.; Novikov, A.; Skakun, S. Exploring Google Earth Engine Platform for Big Data Processing: Classification of Multi-Temporal Satellite Imagery for Crop Mapping. *Front. Earth Sci.* **2017**, *5*, 17. [CrossRef]

20. Yang, C.; Yu, M.; Hu, F.; Jiang, Y.; Li, Y. Utilizing Cloud Computing to address big geospatial data challenges. *Comput. Environ. Urban Syst.* **2017**, *61*, 120–128. [CrossRef]

21. Yang, C.; Xu, Y.; Nebert, D. Redefining the possibility of digital Earth and geosciences with spatial cloud computing. *Int. J. Digit. Earth* **2013**, *6*, 297–312. [CrossRef]

22. Hansen, M.C.; Potapov, P.V.; Moore, R.; Hancher, M.; Turubanova, S.A.; Tyukavina, A.; Thau, D.; Stehman, S.V.; Goetz, S.J.; Loveland, T.R.; et al. High-resolution global maps of 21st-century forest cover change. *Science* **2013**, *342*, 850–853. [CrossRef] [PubMed]

23. Johansen, K.; Phinn, S.R.; Taylor, M. Mapping woody vegetation clearing in Queensland, Australia from Landsat imagery using the Google Earth Engine. *Remote Sens. Appl. Soc. Environ.* **2015**, *1*, 36–49. [CrossRef]

24. Thwal, N.S.; Ishikwawa, T.; Watanabe, H. Comparison of Random Forest, k-Nearest Neighbor, and Support Vector Machine Classifiers for Land Cover Classification Using Sentinel-2 Imagery. *Sensors* **2017**, *18*, 18. [CrossRef]

25. Farda, N.M. Multi-temporal Land Use Mapping of Coastal Wetlands Area using Machine Learning in Google Earth Engine. *IOP Conf. Ser. Earth Environ. Sci.* **2017**, *98*, 012042. [CrossRef]

26. Mondal, P.; Liu, X.; Fatoyinbo, T.E.; Lagomasino, D. Evaluating Combinations of Sentinel-2 Data and Machine-Learning Algorithms for Mangrove Mapping in West Africa. *Remote Sens.* **2019**, *11*, 2928. [CrossRef]

27. Li, W.; El-Askary, H.; Qurban, M.A.; Li, J.; Manikandan, K.; Piechota, T. Using multi-indices approach to quantify mangrove changes over the Western Arabian Gulf along Saudi Arabia coast. *Ecol. Indic.* **2019**, *102*, 734–745. [CrossRef]

28. Belda, S.; Pipia, L.; Morcillo-Pallarés, P.; Rivera-Caicedo, J.P.; Amin, E.; De Grave, C.; Verrelst, J. DATimeS: A machine learning time series GUI toolbox for gap-filling and vegetation phenology trends detection. *Environ. Model. Softw.* **2020**, *127*, 104666. [CrossRef]

29. Zeng, L.; Wardlow, B.; Xiang, D.; Hu, S.; Li, D. A review of vegetation phenological metrics extraction using time-series, multispectral satellite data. *Remote Sens. Environ.* **2020**, *237*, 111511. [CrossRef]

30. Weiss, D.J.; Atkinson, P.M.; Bhatt, S.; Mappin, B.; Hay, S.I.; Gething, P.W. An effective approach for gap-filling continental scale remotely sensed time-series. *ISPRS J. Photogramm. Remote Sens.* **2014**, *98*, 106–118. [CrossRef]

31. Noormets, A. Phenology of Ecosystem Processes. In *Climate Change 2013—The Physical Science Basis 53*; Springer Science & Business Media: New York, NY, USA, 2013; ISBN 978-1-107-05799-1.

32. Gutman, G.; Masek, J.G. Long-term time series of the Earth's land-surface observations from space. *Int. J. Remote Sens.* **2012**, *33*, 4700–4719. [CrossRef]

33. Wulder, M.A.; Coops, N.C.; Roy, D.P.; White, J.C.; Wulder, M.A.; Coops, N.C.; Roy, D.P.; White, J.C. Land cover 2.0. *Int. J. Remote Sens.* **2018**, *39*, 4254–4284. [CrossRef]

34. Ienco, D.; Interdonato, R.; Gaetano, R.; Minh, D.H.T. Combining Sentinel-1 and Sentinel-2 Satellite Image Time Series for land cover mapping via a multi-source deep learning architecture. *ISPRS J. Photogramm. Remote Sens.* **2019**, *158*, 11–22. [CrossRef]

35. Asare, Y.M.; Forkuo, E.K.; Forkuor, G.; Thiel, M. Evaluation of gap-filling methods for Landsat 7 ETM+ SLC-off image for LULC classification in a heterogeneous landscape of West Africa. *Int. J. Remote Sens.* **2019**, *41*, 2544–2564. [CrossRef]

36. Chen, J.; Zhu, X.; Vogelmann, J.E.; Gao, F.; Jin, S. A simple and effective method for filling gaps in Landsat ETM+ SLC-off images. *Remote Sens. Environ.* **2011**, *115*, 1053–1064. [CrossRef]

37. Dusseux, P.; Corpetti, T.; Hubert-Moy, L.; Corgne, S. Combined Use of Multi-Temporal Optical and Radar Satellite Images for Grassland Monitoring. *Remote Sens.* **2014**, *6*, 6163–6182. [CrossRef]

38. Adepoju, K.A.; Adelabu, S.A. Improving accuracy of Landsat-8 OLI classification using image composite and multisource data with Google Earth Engine. *Remote Sens. Lett.* **2020**, *11*, 107–116. [CrossRef]

39. Gómez, C.; White, J.C.; Wulder, M.A. Optical remotely sensed time series data for land cover classification: A review. *ISPRS J. Photogramm. Remote Sens.* **2016**, *116*, 55–72. [CrossRef]

40. Yang, G.; Fang, S.; Dian, Y.; Bi, C. Improving Seasonal Land Cover Maps of Poyang Lake Area in China by Taking into Account Logical Transitions. *ISPRS Int. J. Geo-Inf.* **2016**, *5*, 165. [CrossRef]

41. Cai, S.; Liu, D.; Sulla-menashe, D.; Friedl, M.A. Remote Sensing of Environment Enhancing MODIS land cover product with a spatial—Temporal modeling algorithm. *Remote Sens. Environ.* **2014**, *147*, 243–255. [CrossRef]

42. Tue, N.T.; Dung, L.V.; Nhuan, M.T.; Omori, K. Carbon storage of a tropical mangrove forest in Mui Ca Mau National Park, Vietnam. *Catena* **2014**, *121*, 119–126. [CrossRef]

43. Van, T.T.; Wilson, N.N.; Thanh-Tung, H.H.; Quisthoudt, K.K.; Quang-Minh, V.V.; Tuan, L.X.L.; Dahdouh-Guebas, F.; Koedam, N. Changes in mangrove vegetation area and character in a war and land use change affected region of Vietnam (Mui Ca Mau) over six decades. *Acta Oecol.* **2015**, *63*, 71–81. [CrossRef]

44. Ha, T.T.P.; Van Dijk, H.; Visser, L. Impacts of changes in mangrove forest management practices on forest accessibility and livelihood: A case study in mangrove-shrimp farming system in Ca Mau Province, Mekong Delta, Vietnam. *Land Use Policy* **2014**, *36*, 89–101. [CrossRef]

45. Moreno-Martínez, Á.; Camps-Valls, G.; Kattge, J.; Robinson, N.; Reichstein, M.; Van Bodegom, P.; Kramer, K.; Cornelissen, J.H.C.; Reich, P.; Bahn, M.; et al. A methodology to derive global maps of leaf traits using remote sensing and climate data. *Remote Sens. Environ.* **2018**, *218*, 69–88. [CrossRef]

46. Roy, D.P.; Ju, J.; Lewis, P.; Schaaf, C.; Gao, F.; Hansen, M.; Lindquist, E. Multi-temporal MODIS—Landsat data fusion for relative radiometric normalization, gap filling, and prediction of Landsat data. *Remote Sens. Environ.* **2008**, *112*, 3112–3130. [CrossRef]

47. Bishop, C.M. *Pattern Recognition and Machine Learning*; Information Science and Statistics; Springer: New York, NY, USA, 2016; ISBN 9781493938438.

48. Shen, H.; Li, X.; Cheng, Q.; Zeng, C.; Yang, G.; Li, H.; Zhang, L. Missing Information Reconstruction of Remote Sensing Data: A Technical Review. *IEEE Geosci. Remote Sens. Mag.* **2015**, *3*, 61–85. [CrossRef]

49. Knotters, M.; Hoogland, T. *A Disposition of Interpolation Techniques*; Statutory Research Tasks Unit for Nature and the Environment; WOt-Werkdocument 190: Wageningen, The Netherlands, 2010.

50. Lepot, M.; Aubin, J.-B.; Clemens, F.H. Interpolation in Time Series: An Introductive Overview of Existing Methods, Their Performance Criteria and Uncertainty Assessment. *Water* **2017**, *9*, 796. [CrossRef]

51. Son, N.-T.; Chen, C.-F.; Chang, N.-B.; Chen, C.-R.; Chang, L.-Y.; Thanh, B.-X. Mangrove Mapping and Change Detection in Ca Mau Peninsula, Vietnam, Using Landsat Data and Object-Based Image Analysis. *IEEE J. Sel. Top. Appl. Earth Obs. Remote Sens.* **2014**, *8*, 503–510. [CrossRef]

52. Tran, T.T.H.; Ha, T.; Bush, S.R.; Mol, A.P.J.; Dijk, H. Van Organic coasts? Regulatory challenges of certifying integrated shrimp e mangrove production systems in Vietnam. *J. Rural Stud.* **2012**, *28*, 631–639. [CrossRef]

53. Baumgartner, U.; Kell, S.; Nguyen, T.H. Arbitrary mangrove-to-water ratios imposed on shrimp farmers in Vietnam contradict with the aims of sustainable forest management. *Springerplus* **2016**, *5*, 1–10. [CrossRef]

54. Lahoz, W.A.; Schneider, P. Data assimilation: Making sense of Earth Observation. *Front. Environ. Sci.* **2014**, *2*, 1–28. [CrossRef]

55. Abercrombie, S.P.; Friedl, M.A. Improving the Consistency of Multitemporal Land Cover Maps Using a Hidden Markov Model. *IEEE Trans. Geosci. Remote Sens.* **2016**, *54*, 703–713. [CrossRef]

56. Gong, W.; Fang, S.; Yang, G.; Ge, M. Using a Hidden Markov Model for Improving the Spatial-Temporal Consistency of Time Series Land Cover Classification. *ISPRS Int. J. Geo-Inf.* **2017**, *6*, 292. [CrossRef]

57. Jia, K.; Liang, S.; Zhang, N.; Wei, X.; Gu, X.; Zhao, X.; Yao, Y.; Xie, X. Land cover classification of finer resolution remote sensing data integrating temporal features from time series coarser resolution data. *ISPRS J. Photogramm. Remote Sens.* **2014**, *93*, 49–55. [CrossRef]
58. Hilker, T.; Wulder, M.A.; Coops, N.C.; Linke, J.; McDermid, G.; Masek, J.G.; Gao, F.; White, J.C. A new data fusion model for high spatial- and temporal-resolution mapping of forest disturbance based on Landsat and MODIS. *Remote Sens. Environ.* **2009**, *113*, 1613–1627. [CrossRef]
59. Modica, G.; Merlino, A.; Solano, F.; Mercurio, R. An index for the assessment of degraded mediterranean forest ecosystems. *For. Syst.* **2015**, *24*, 1–13. [CrossRef]
60. Wu, B.; Meng, X.; Ye, Q.; Sharma, R.P.; Duan, G.; Lei, Y.; Fu, L. Method of estimating degraded forest area: Cases from dominant tree species from Guangdong and Tibet in China. *Forests* **2020**, *11*, 930. [CrossRef]
61. Lausch, A.; Erasmi, S.; King, D.J.; Magdon, P.; Heurich, M. Understanding forest health with remote sensing-Part I-A review of spectral traits, processes and remote-sensing characteristics. *Remote Sens.* **2016**, *8*, 29. [CrossRef]
62. Shrestha, S.; Miranda, I.; Kumar, A.; Pardo, M.L.E.; Dahal, S.; Rashid, T.; Remillard, C.; Mishra, D.R. Identifying and forecasting potential biophysical risk areas within a tropical mangrove ecosystem using multi-sensor data. *Int. J. Appl. Earth Obs. Geoinf.* **2019**, *74*, 281–294. [CrossRef]
63. Pham, T.D.; Le, N.N.; Ha, N.T.; Nguyen, L.V.; Xia, J.; Yokoya, N.; To, T.T.; Trinh, H.X.; Kieu, L.Q.; Takeuchi, W. Estimating mangrove above-ground biomass using extreme gradient boosting decision trees algorithm with fused sentinel-2 and ALOS-2 PALSAR-2 data in can Gio biosphere reserve, Vietnam. *Remote Sens.* **2020**, *12*, 777. [CrossRef]
64. Hermosilla, T.; Wulder, M.A.; White, J.C.; Coops, N.C.; Hobart, G.W.; Hermosilla, T.; Wulder, M.A.; White, J.C.; Coops, N.C. Disturbance-Informed Annual Land Cover Classification Maps of Canada's Forested Ecosystems for a 29-Year Landsat Time Series Disturbance-Informed Annual Land Cover Classification Maps of Canada's. *Can. J. Remote Sens.* **2018**, *44*, 67–87. [CrossRef]
65. Lewis, P.; Gómez-Dans, J.; Kaminski, T.; Settle, J.; Quaife, T.; Gobron, N.; Styles, J.; Berger, M. An Earth Observation Land Data Assimilation System (EO-LDAS). *Remote Sens. Environ.* **2012**, *120*, 219–235. [CrossRef]
66. Pham, T.D.; Yokoya, N.; Bui, D.T.; Yoshino, K.; Friess, D.A. Remote sensing approaches for monitoring mangrove species, structure, and biomass: Opportunities and challenges. *Remote Sens.* **2019**, *11*, 230. [CrossRef]

Publisher's Note: MDPI stays neutral with regard to jurisdictional claims in published maps and institutional affiliations.

remote sensing

Article

Identification before-after Forest Fire and Prediction of Mangrove Forest Based on Markov-Cellular Automata in Part of Sembilang National Park, Banyuasin, South Sumatra, Indonesia

Soni Darmawan [1,2,*], Dewi Kania Sari [1], Ketut Wikantika [2], Anggun Tridawati [3], Rika Hernawati [1] and Maria Kurniawati Sedu [1]

[1] Department of Geodesy Engineering, Faculty of Civil Engineering and Planning, Institut Teknologi Nasional Bandung, PHH Mustofa No. 23, Bandung 40124, West Java, Indonesia; dewiks@itenas.ac.id (D.K.S.); rikah@itenas.ac.id (R.H.); athysedu97@mhsw.itenas.ac.id (M.K.S.)

[2] Center for Remote Sensing, Institut Teknologi Bandung, Ganesha No. 10, Bandung 40132, West Java, Indonesia; ketut@gd.itb.ac.id

[3] Department of Geodesy Engineering, Faculty of Engineering, Universitas Lampung, Prof. Dr. Ir. Sumantri Brojonegoro No. 1, Bandar Lampung 35141, Indonesia; angguntridawati.30@gmail.com

* Correspondence: soni_darmawan@itenas.ac.id; Tel.: +62-822-1660-2544

Received: 30 September 2020; Accepted: 2 November 2020; Published: 11 November 2020

Abstract: In 1997, the worst forest fire in Indonesia occurred and hit mangrove forest areas including in Sembilang National Park Banyuasin Regency, South Sumatra. Therefore, the Indonesian government keeps in trying to rehabilitate the mangrove forest in Sembilang National Park. This study aimed to identify the mangrove forest changing and to predict on the future year. The situations before and after forest fire were analyzed. This study applied an integrated Markov Chain and Cellular Automata model to identify mangrove forest change in the interval years of 1989–2015 and predict it in 2028. Remote sensing technology is used based on Landsat satellite imagery (1989, 1998, 2002, and 2015). The results showed mangrove forest has decreased around 9.6% from 1989 to 1998 due to forest fire, and has increased by 8.4% between 1998 and 2002, and 2.3% in 2002–2015. Other results show that mangroves area has continued to increase from 2015 to 2028 by 27.4% to 31% (7974.8 ha). It shows that the mangrove ecosystem is periodically changing due to good management by the Indonesian government.

Keywords: mangrove; Markov chain; cellular automata

1. Introduction

Mangrove forests are located along sloping shores, river estuaries, deltas, bays influenced by tides, and generally found in tropical and subtropical areas [1,2]. As a defense for shore and marine ecosystems, mangroves are an essential link to maintaining the waters' biological cycle [3,4]. Mangrove forests have several benefits, among others, as a carbon storage [3,5], prevent abrasion [5], reduce the impact of tsunamis [6] and as habitat breeding fish [7].

Indonesia is a country that has the largest mangrove forest in the world, reaching 59.8% of the total area of mangrove forests in Southeast Asia [8]. The area of mangrove forests in Indonesia is around 4.5 million ha with the proportion (18–23%) exceeding Brazil (1.3 million ha), Nigeria (1.1 million ha), and Australia (0.97 million ha) [9]. South Sumatra Province is one of the provinces in Indonesia which has widespread mangrove forests. Based on the results of the inventory and description of mangrove forests implemented by the Musi Watershed Management Center in 2006, the area of mangrove forests in South Sumatra province is around 1,693,110.1 ha [10]. It is also consistent with the Decree of the Minister of Forestry Number 95/Kpts-II/2003 dated March 19, 2003, which declared that South Sumatra has a mangrove area of 202,896.3 ha, specifically in Sembilang National Park [11]. Sembilang National Park is dominated by mangrove forests due to its position on the coast of the Banyuasin peninsula. However, to the West and Northwest of Sembilang National Park, there is a large stretch of peat swamp forest which is an extension of the peat forest in the Berbak region of Jambi Province [12]. The condition of mangroves in Indonesia, especially in the National Park, is experiencing tremendous pressure, both from human activities and environmental factors [13]. Generally, the destruction of mangrove forests is caused by building materials, animals feed and forest fire [14].

The forest fire occurred in Indonesia during the dry weather in 1997. Firstly, forest fires in Indonesia were caused by human activities, such as: cultivation of deliberate slash and burn by farmers on peatland areas, land conversion, fishing, and logging, nevertheless, the extent of the respective causes are unknown [15,16]. Then, the fire quickly burns dry organic matter so that spreads over a large area and caused mega fire. Mega forest fires started in Southern Sumatra and Southern Kalimantan in early May and continued until the second week of November 1997 [17]. Mega forest fires in Indonesia have also been triggered by the El Niño climate phenomenon [18]. It is thought that during the 1997 El Niño fires in Indonesia, between 0.8 and 2.6 Gt of carbon was released into the atmosphere as a result of burning peat and vegetation. This amount is equivalent to 13–40% of global and carbon emissions [19,20]. On the other hand, this disaster also affected the health of the population in Sumatra, Kalimantan, and neighboring countries, and disrupted political stability [21]. Approximately 35 million people in Southeast Asia were affected [22]. The cost of smog pollution costs around USD 674–799 million and is associated with carbon emission losses of around USD 2.8 billion [23]. This occurrence was declared to be one of the worst environmental disasters of all time [24].

According to Figure 1 many fires started since early May 1997. A wave of land clearance fires moved from the north to the south of the island. Fire numbers peaked in Aceh, North Sumatra and Riau provinces in May and including July was the large number of fires, in Jambi from July to September, in South Sumatra in September [25]. September was the primary month of forests fire in South Sumatra [15,25,26]. The most significant concentrations were in four regions, there were Pampangan district, between Palembang city and Jambi province border, Pendopo district in the central-western part of South Sumatra, and Jambi east near to Berbak National Park and in Lampung province from August to October [27]. The number of hotspots decreased slightly at the end of October, and then, over two to three days starting from 6 November, all major fires discontinued, apparently after heavy rain [27]. All significant fires correlated with the lowest month of rainfall in each province, because the habit of seasonal rainfall controls the tendency for fires to occurred in Sumatra and is strongly influenced by the striking differences in land use types in each of the eight provinces. Despite significant regional differences in average annual rainfall in Sumatra, the climate is almost humid, and 85 per cent of the island has a dry season (mean monthly rainfall less than 100 mm) of less than two months [28].

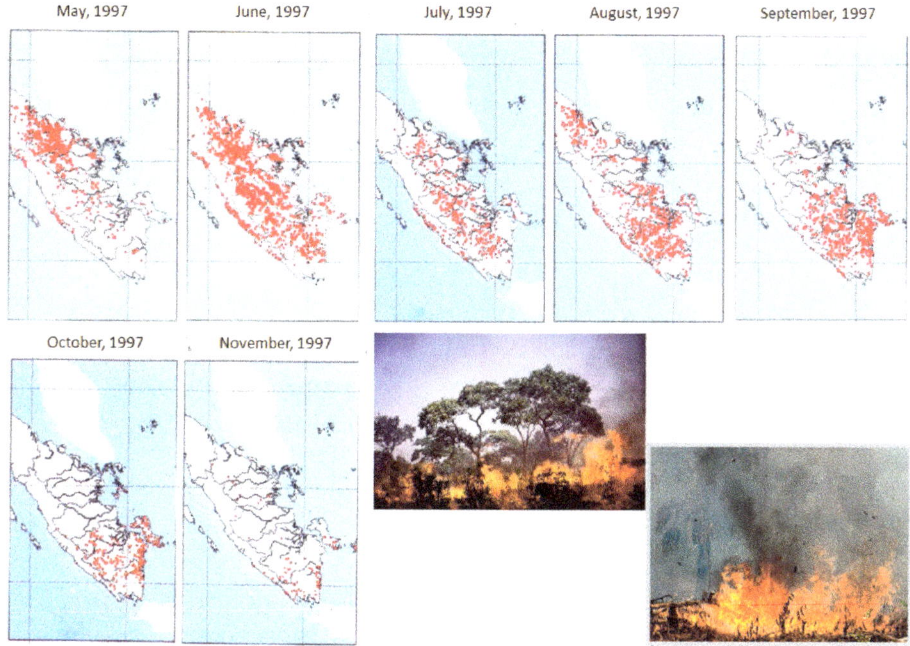

Figure 1. Distribution of hot spot in Sumatra 1997 [25].

The total area affected by the fire in Indonesia appears to have been excess of two mile ha, including Sumatra [29]. The main areas affected by forest fires in Sumatra is Sembilang National Park. Therefore, the Indonesian government keeps on trying to rehabilitate the mangrove forest in Sembilang National Park as the largest mangrove forest in Sumatra through publish policies related to forest fires by issuing rules and regulations relating to the prevention and control of forest fires which are regulated in Law no. 6 of 1990, Law No. 5 of 1994, Law No. 23 of 1997, Law No. 41 of 1999 and Government Regulation No. 4 of 2001 [30]. These regulations consist of prevention and control through coordinated extension activities, the prohibition of burning activities, improvement the skills of human resources both from the government and companies, and compliance, and procurement of fire-fighting equipment. Hence, the condition of the mangrove forests in the Sembilang National Park is very dynamic and changed every year. Mangrove forests need to be observed to control and rehabilitate. Moreover, the future of land use is also needed to support planning and policies [31]. This aspect can be approached through land cover change modeling as an instrument to support the analysis of the causes and consequences of land cover change.

Modeling land cover change depends upon the accurate extraction of both past and present land cover information [32], which the past and future scenarios are evaluated by model. Remote sensing has been widely proven to be essential in providing information regarding the land cover change [33,34], in which most studies use pixel-based image analysis methods [35–37]. Generally, the algorithm used is the maximum likelihood [38,39] and then the land cover change is evaluated and assessed by map algebra [40,41]. Meanwhile, another method used for predicting land cover change is based on multivariate analysis through image regression [42]. However, the limitations of this model are cannot quantify the change and aim to observe the temporal analysis [43]. Therefore, this condition takes Markov-Cellular Automata to overcome these limitations. Markov-Cellular Automata is an efficient, simple model and has an excellent ability to simulate and predict land cover change based on spatial data [44–46]. The Markov-Cellular Automata model is an integration of the Markov Chains and Cellular Automata models. The Markov Chain is a statistical model used to determine the probability (probabilistic) of change for each land class from two land data sets at different periods [47,48], while the Cellular Automata model is expressed as an automaton (raster data cell), which is having cell contents that can change or transfer at any time, according to the transition rules that are recognized in each cell [49–52]. The Markov-Cellular Automata model is a good application for identifying and predicting land cover change because it estimates spatial and temporal components [53].

Moreover, the application of a suitable classification algorithm is essential. Various classification algorithms have been developed for mangrove mapping such as ISODATA [54], Maximum likelihood [55–58], object based classification [59,60], and support vector machine [61–63]. Support vector machine was a reliable machine learning algorithm that provides acceptable accuracy for mangrove mapping [64,65]. Madanguit et al. (2017) [64] compared the support vector machine and QUEST classification algorithms for mangrove mapping. The results showed that the support vector machine algorithm provides higher accuracy than QUEST with 94.9% and 93.6%. Firmansyah et al. (2019) [65] also said that support vector machines could minimize mangrove mapping errors compared to decision trees. Feng et al., 2016 [63] calculated and simulated urban development in Shanghai-China using a machine learning-based Markov-Cellular Automata integration model. The support vector machine algorithm was used and compared with the conventional algorithms. His research results had indicated that the conventional algorithm was not good enough to simulate the complex boundaries between urban and non-urban areas, and the support vector machine algorithm provides accurate results. Referring to several previous studies, the objectives of this study are:

1. To identify the mangrove forest changes in Sembilang National Park, Banyuasin Regency in 1989, 1998, 2002, and 2015.
2. To predict the area of mangrove forest in Sembilang National Park, Banyuasin Regency in 2028.

The model uses Markov-Cellular Automata based on a support vector machine algorithm. It is assumed that it can be used by a study to establish policies, especially in anticipating negative impacts on environmental changes and mangrove planning and management purposes.

2. Materials and Methods

2.1. Study Area

Sembilang National Park (104°14′–104°54′E, 1°53′–2°27′S) with Berbak National Park (a Ramsar Site) to the north is part of the Greater Berbak-Sembilang Ecosystem on the Indonesian island of Sumatra [66]. The Sembilang National Park (Figure 2) location is in the west and is bounded by the Benue River and the provincial boundary with Jambi province in the north and the Lalan and Banyuasin rivers in the south [11]. The typical climate in Sembilang National Park are humid air to heavy rainfall from November to March and dry season during June-September [67]. The hydrology of the Sembilang National Park is characterized by the smooth transition of freshwater and brackish water habitats [66].

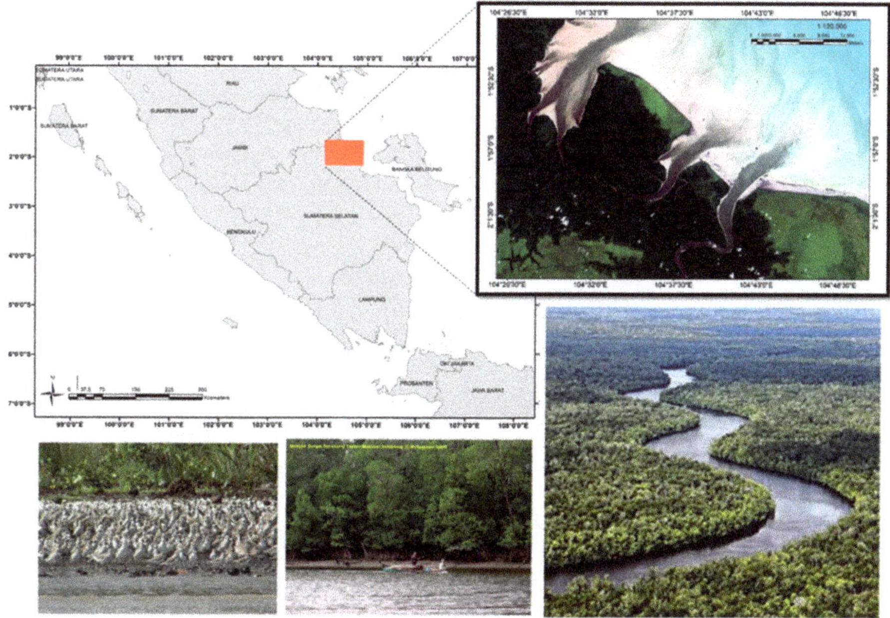

Figure 2. Location of the study area is in Sembilang National Park, Musi Banyuasin, South Sumatra Province, Indonesia. Selected remotely sensed image from Landsat 7 ETM+ (source from USGS) and administrative boundaries of the study area (source from Indonesian Geospatial Agency/BIG), and taken photograph from [68–70].

Sembilang National Park is the largest mangrove area in the Indo-Malaya region and the only mangrove area that still has an intact natural transition into the nearby freshwater forest and peat swamp [66]. Mangrove species that live in this area are *Rhizophora* (*Rhizophora apiculata* and *Rhizophora mucronata*), *Nepenthes ampullaria* which is an indicator species on deep peat, *Brugierra* (*Bruguierra gymnorrhiza, Bruguierra parviflora, Bruguierra sexangula,* and *Bruguierra corniculatum* and *Aegiceras*), *Kandelia candel, Sonneratia* (*Sonneratia caseolaris, Sonneratia alba,* and *Sonneratia Ovata*), *Avicennia* (*Avicennia marina, Avicennia alba,* and *Avicennia ofificinalis*), *Ceriops* (*Ceriops decandra* and *Ceriops taga*), *Xylocarpus* (*Xylocarpus granatum*) and *Xylocarpus Excoecaria agallocha* [71].

The park is biologically rich sites with more than two hundred species of birds, one hundred and forty species of fish and more than fifty species of mammals [68]. The area is estimated to support 70% of the significant coastal fisheries of South Sumatra in terms of breeding, spawning, and nursery areas [67]. Many of these species are endangered, such as the endangered Sumatran Tiger (*Panthera tigris sumatrae*), and the endangered Indian Elephant (*Elephas maximus*), the Storm Stork (*Ciconia stormi*), and the Malayan Giant Turtle (*Orlitia borneensis*) [67]. More than 43% of mangrove species in Indonesia are also found here [67]. Around 0.5–1 million shorebirds use the area and during the winter and almost 80,000–100,000 migratory birds feed and rest here [68]. It supports more than 1% of the population of Milky Stork (*Mycteria cinerea*), Asian Dowitcher (*Limnodromus semipalmatus*), Spotted Greenshank (*Tringa guttifer*), Far Eastern Curlew (*Numenius madagascariensis*) and Lesser Adjutant (*Leptoptilos javanicus*) [67,68].

As an area affected by forest fire in 1997, various efforts have been made to rehabilitate mangrove forests in Sembilang Nasional Park (Figure 3) by involving Government and various stakeholders. Several related activities have been carried out [17]; (1) Integrated Swamp Development Project (ISDP) between Japan International Cooperation Agency (JICA) and Forest Fire Prevention Management Project/FFPMP in Northern Sembilang Nasional Park in 1997 to 2001; this project produce a document of descriptions, programs and suggestions to help planning, conservation, and control of resource use in buffer zones, (2) Forest Fire Prevention Management Project (FFPMP) by JICA in Sei Rambut Village in Northern Sembilang Nasional Park in 1997 to 2000, (3) Global Environment Facility (GEF) Project Berbak-Sembilang in 2000 to 2004; the purpose of this project are spatial planning, assessment, monitoring, and capacity building and environmental awareness; and (4) Climate Change, Forest and Peatlands in Indonesia project (CCFPI) in 2002 to 2005; this project consists of community based activities and policy development activities related to the protection and rehabilitation of swamp forests and peatlands in Indonesia.

Figure 3. Implementation of restoration program through natural regeneration assistance, enrichment planting, and seeding around Sembilang National Park [12,72,73].

2.2. Methodology

The methodology in this study is divided into five steps (Figure 4) including (1) a suit of data Landsat 5, Landsat 7 ETM+, and Landsat 8 OLI multi-temporal include preprocessing, (2) image classification using support vector machine (3) measurements accuracy, (4) Mangrove change using the Markov Chain and Cellular Automata model, and (5) validation. The description of each step of the research is as follows:

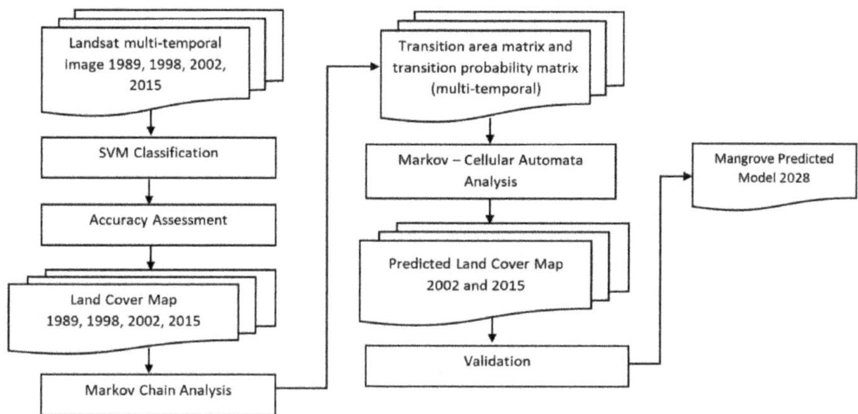

Figure 4. Methodology.

2.3. Data and Preprocessing

Landsat 5, Landsat 7 ETM+, and Landsat 8 OLI images acquired for 1989, 1998, 2002, and 2015 of 30×30 m^2 spatial resolution derived from USGS. Landsat 2015 was used both to predict land cover in 2028. We also used the administrative boundary data of 1:50,000. The images were registered to the geographic coordinate projection using World Geodetic System 1984 (WGS-84). And auxiliary data of administrative boundaries data was obtained from Geospatial Information Agency (BIG). The data collected can be seen in Figure 5. The information of collected data can be seen in Table 1.

Table 1. The Data Collected.

Dataset	Date and Scale	Source
Landsat 5	17 May 1989 and 24 April 1998	USGS
Landsat 7 ETM+	30 June 2002	USGS
Landsat 8 OLI	26 June 2015	USGS
The administrative boundaries	1:50,000	BIG

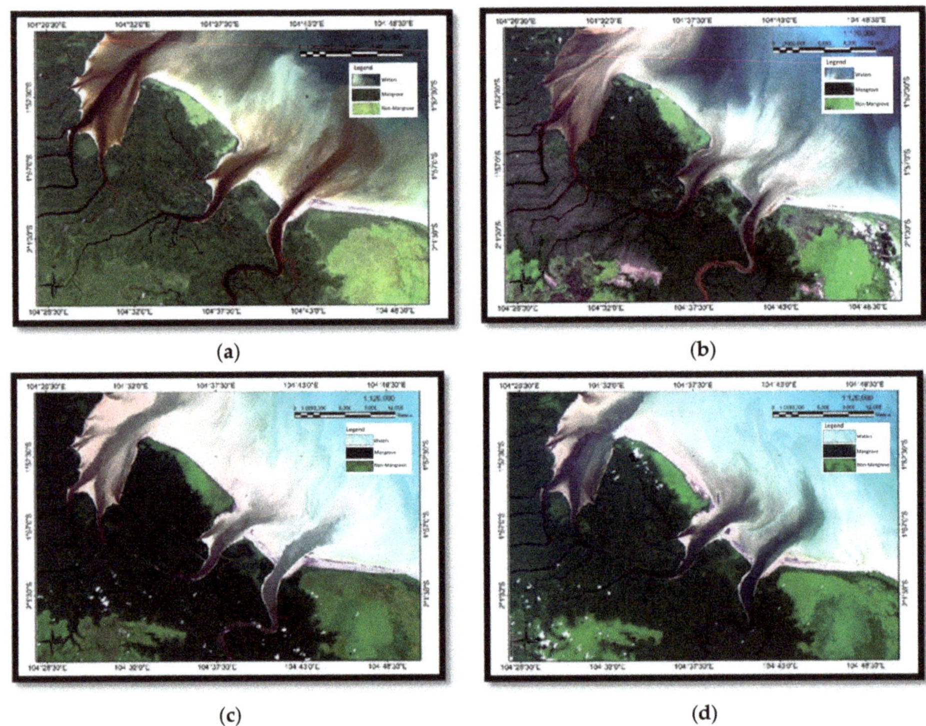

Figure 5. Landsat imageries acquired in: (**a**) Landsat 5 (1989), (**b**) Landsat 5 (1998), (**c**) Landsat 7 ETM+ (2002), and (**d**) Landsat 8 OLI (2015).

2.4. Classification

The support vector machine was used for classification of mangrove forests in study area. This classification was applied to the Landsat 5 images in 1989, Landsat 5 in 1998, Landsat 7 ETM+ in 2002, and Landsat 8 OLI images in 2015. This study was used support vector machine to fit an optimal separating hyperplane or set in a high or infinite-dimensional space to locate the optimal boundaries between classes. In this case the 3 classes defined previously there are mangroves, non-mangroves and water (Table 2).

Table 2. Description of LULC classes.

Class	Definition
Non-mangrove	Land, built-up land, bare ground, roads, shrubs, vegetation, other habitats of mangrove.
Mangrove	A shrub or small tree that grows in coastal saline or brackish water.
Waters	Water in certain areas, both static and dynamic, such as seas, rivers, lakes.

Based on statistical theory, support vector machines operate by classifying two or more classes by studying for the best hyperplane that utilizes data at the separation point (super vector) even for a limited number of samples [74,75]. The support vector machine equation can be seen in the following Equation (1):

$$SVM_{(F,\lambda)}(R) = sign \left(\sum_{i}^{N} y_i \alpha_i (f_R . f_i) + b \right) \tag{1}$$

where the SVM trainer of R is the class of region based on specific feature type F and specific scale λ. y_i is the support vector class and α_i ($i = 1, \ldots, N$) is decision coefficient with N is total of region. The support vectors are the f_R is the feature vector the region and f_i such that $\alpha_i > 0$, and b is a parameter found during the training.

2.5. Training Data Collection Scheme

The training data were based on two different acquisition methods and incorporate field survey image interpretation using satellite data. Considering the time and access to the area is not always possible, GPS measurements were also impossible. The majority classes in Sembilang National Park where is not built-up area, the land cover only mangrove, non-mangrove (such as brush area), and water area. The distribution of training data can be seen in Figure 6. This study, mapping mangrove forest using Landsat 5, Landsat 7 ETM+, and Landsat 8 OLI images which spatial resolution is 30 m therefore training data of mangrove, non-mangrove and water patched on the field is more than 0.009 ha.

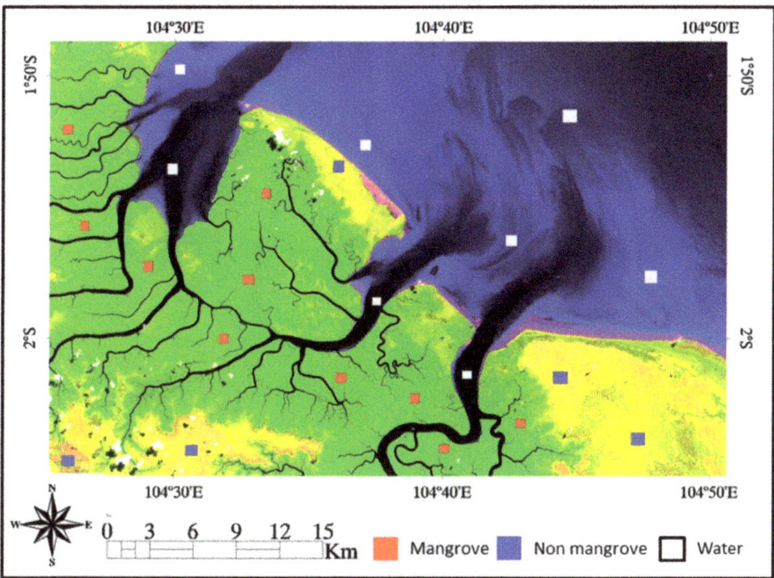

Figure 6. The distribution of training data.

2.6. Accuracy Assessment

Classification accuracy was assessed based on the classification accuracy statistics, the error matrix (user/producer's accuracy and omission/commission error), overall accuracy and kappa statistic [76]. Validation of the classification maps produced from the support vector machine implementation was performed against the set of validation pixels for each class collected following the procedure [77]. In addition to the classification statistics, the land cover classification was generated from the support vector machine algorithm and shows correspondence between the classification result and a reference data. The value of reference data was collected in 1458 pixels from the field in Equations (2)–(5).

$$\text{Kappa} = \frac{N\sum_{i=1}^{r} Xii - \sum_{i=1}^{r} Xi + (X+i)}{N^2 - \sum_{i=1}^{r} Xi + (X+i)} \times 100\% \tag{2}$$

$$\text{Overall Accuracy} = \frac{\sum_{i=1}^{r} Xii}{N} \times 100\% \tag{3}$$

$$\text{User's Accuracy} = \frac{Xii}{X+i} \times 100\% \tag{4}$$

$$\text{Producer's Accuracy} = \frac{Xii}{X+i} \times 100\% \tag{5}$$

where N is the number of all pixels used for observation, r is the number of rows in the error matrix (number of classes), Xii is diagonal values of the contingency matrix of row i and column i, $X + i$ is column pixel number i, and $Xi +$ row pixel number i.

2.7. Markov Chain

Modeling using Markov-Cellular Automata has been widely applied in several fields by researchers, including for the study of regional-scale land-use change, watershed management [78,79], regional monitoring cities [80–83], monitoring of plantation and agricultural areas [84], monitoring of erosion [85], simulating forest cover change [86], evaluating the integration of land use and climate change [87], and monitoring sand areas [88]. In 2015, Halmy et al. (2015) [88] used the Markov-Cellular Automata model to predict sand areas using Landsat TM 5 data, which yields 90% accuracy. The results show that the Markov-Cellular Automata model is a useful model for applying and predicting land cover.

Markov Chain determines how much land cover would be estimated to change from the latest date to the predicted date [89]. In this study, the Cellular Automata (CA) -Markov model was applied to predict the 2028 LULC in the Sembilang National Park area to identify variations in future mangrove land use. First, classified images from the period of 1989 and 1998; between 1998 to 2002, and 2002 until 2015 were selected as input into the model, to calculate matrix of conversion areas and conversion probabilities. The transition probability maps were used to produce maps of land use for the year of 2028. In an iterative process CA-Markov uses the transition probability maps of each land cover to establish the inherent suitability of each pixel to change from one land use type to another. The transition area matrix shows the total area (in cells) expected to change in the next period of 1989–1998, 1998–2002, and 2002–2015.

The prediction of land use changes is calculated by the following Equation (6) [89,90]:

$$S(t, t+1) = Pij \times S(t) \tag{6}$$

where $S(t)$ is the system status at time of t, $S(t + 1)$ is the system status at time of $t + 1$; Pij is the transition probability matrix in a state [90]. If P is transition probability, namely the probability of converting current state to another state in next period [91], the expression is as follows:

$$P_{ij} = \begin{bmatrix} P_{11} & P_{12} \cdots & P_{1n} \\ P_{21} & P_{22} \cdots & P_{2n} \\ \cdots & \cdots & \cdots \\ P_{n1} & P_{n2} & P_{nn} \end{bmatrix} \tag{7}$$

$$\left(0 \le P_{ij} \le 1\right) \tag{8}$$

where P is the transition probability; P_{ij} stands for the probability of converting from current state i to another state j in next time; P_n is the state probability of any time. Low transition will have a probability near 0 and high transition have probabilities near 1 [44].

2.8. Cellular Automata (CA)

The Cellular Automata (CA) is produces to determine iteration times, combining transition area matrix and potential transition maps as the CA local transition rule, land use map in the future could be simulated. In this study, Markov Chain results from data in the form of a transition probability matrix, transition area matrix, and a set of conditional probability images (1989–1998, 1998–2002) and actual land use maps in 2002 and 2015 were applied with the Cellular Automata model to obtain predictions of land cover in 2002 and 2015.

The CA model can be expressed as follows in Equation (9) [90]:

$$S(t, t+1) = f\left(S(t),\ N\right) \tag{9}$$

where S is the set of limited and discrete cellular states, N is the Cellular field, t and $t + 1$ indicate the different times, and f is the transformation rule of cellular states in local space.

2.9. Validation

In terms of validating the CA-Markov predictions and evaluating the applied models results, validating process map predictions based on actual maps can be achieved. In this study for validating the model, the condition of land cover in 2002 and 2015 was estimated and compared with actual land use maps [56,92]. The precision of simulation or classification image results, pixel-by-pixel, is accessed via the kappa accuracy index.

Kappa index of agreement provided *Kno, Klocation, KlocationStrata* and *Kstandard* index. This statistic ranges from −1 (significantly worse than random) to 1 (perfect), but it typically lies between 0 and 1. *Kno* (kappa for no ability) showed the proportion between the actual map and the prediction map accurately determine the location, *Klocation* (kappa for location) showed the proportion between the actual map and the prediction map based on input at a location, *Klocationstrata* showed the proportion between the actual map and prediction map based on the number, and *Kstandard* (kappa index) aims to compare the proportion that is observed to be correct with the proportion that is expected due to probability [93]. The formulas for the summary statistics following Equations (10)–(13) [93,94]:

$$K_{no} = \frac{(M\,(m)N\,(n))}{(P\,(p) - N\,(n))} \tag{10}$$

$$K_{location} = \frac{(M\,(m)\,N\,(m))}{(P\,(m) - N\,(m))} \tag{11}$$

$$K_{locationstrata} = \frac{(M\,(m)\,H(m))}{(K\,(m) - H\,(m))} \tag{12}$$

$$K_{standard} = \frac{(M\,(m)N\,(n))}{(P\,(p) - N\,(n))} \tag{13}$$

where no information is defined by $N(n)$, medium stratum level information by $H(m)$, medium grid cell level information by $M(m)$, perfect grid cell-level information given imperfect stratum-level information by $K(m)$ mean, and perfect grid cell-level information across the landscape by $P(p)$.

3. Results

3.1. Land Cover Classification in 1989, 1998, 2002, and 2015

The land cover classification was obtained from Landsat 5, Landsat 7 ETM+, and Landsat 8 OLI Images in 1989, 1998, 2002, and 2015 using the support vector machine algorithm in Sembilang National Park. Land cover classes were categorized into three: Mangroves, Non-mangroves, and Water.

We obtained the land cover map from the classification algorithm by use of the support vector machine (Figure 7). We identified three dominant classes in the study area including Mangrove, Non-mangrove, and Water. Based on Figure 7b, for the years 1989 to 1998 we can see visually, there was the decrease of Mangrove forest because of forest fire, while, from the years 1998 to 2015 there was the significant increase of Mangrove forest. The accuracy assessment of classification obtained by each year can be seen on Tables 3–6.

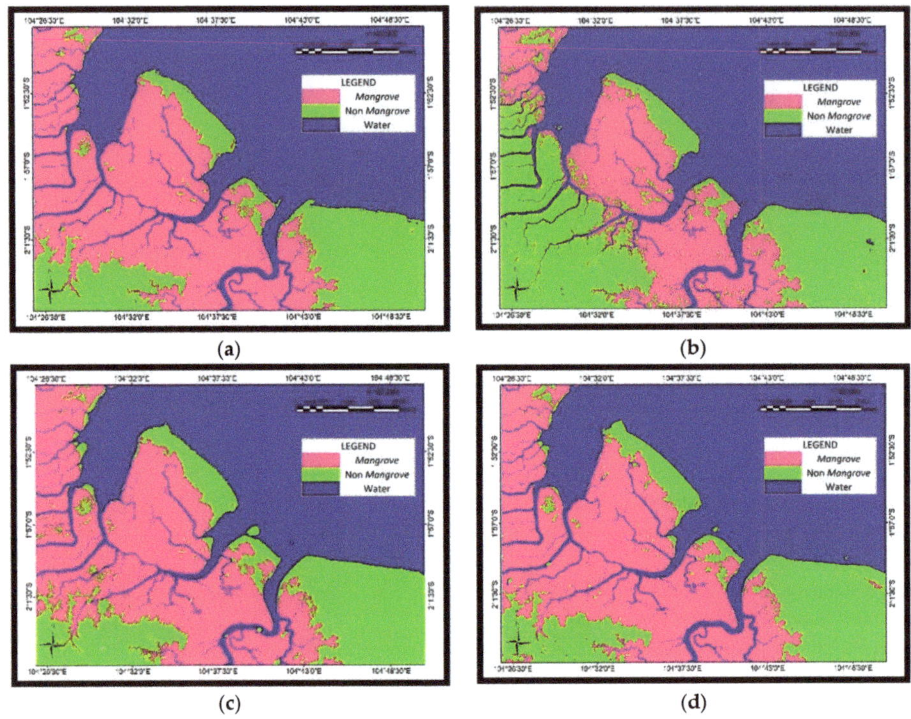

Figure 7. Land cover map as classification result in (**a**) 1989, (**b**) 1998, (**c**) 2002, (**d**) 2015.

Table 3. The confusion matrix of support vector machine classification in 1989.

	Classes	Training Data for Ground Check					
		Mangrove	Non-Mangrove	Water	Total	User Accuracy	Error Commission
Classification SVM	Mangrove	495	2	0	497	99.6	0.4
	Non-mangrove	0	538	0	538	100	0
	Water	0	0	405	405	100	0
	Total	495	540	405	1440		
	Producer Accuracy	100	99.6	100		OA	99.8
	Error omission	0	0.4	0		Kappa	0.9

Table 4. The confusion matrix of support vector machine classification in 1998.

	Classes	Training Data for Ground Check					
		Mangrove	Non-Mangrove	Water	Total	User Accuracy	Error Commission
Classification SVM	Mangrove	446	13	0	459	97.1	2.8
	Non-mangrove	4	649	0	653	97.4	0.6
	Water	0	4	342	346	100	1.1
	Total	450	666	342	1458		
	Producer Accuracy	99.1	97.4	100		OA	98.5
	Error omission	0.9	2.5	0		Kappa	0.9

Table 5. The confusion matric of support vector machine classification in 2002.

<table>
<tr><td rowspan="8">Classification SVM</td><td rowspan="2">Classes</td><td colspan="6">Training Data for Ground Check</td></tr>
<tr><td>Mangrove</td><td>Non-Mangrove</td><td>Water</td><td>Total</td><td>User Accuracy</td><td>Error Commission</td></tr>
<tr><td>Mangrove</td><td>533</td><td>17</td><td>1</td><td>551</td><td>96.7</td><td>3.2</td></tr>
<tr><td>Non-mangrove</td><td>7</td><td>607</td><td>0</td><td>614</td><td>98.8</td><td>1.1</td></tr>
<tr><td>Water</td><td>0</td><td>6</td><td>359</td><td>365</td><td>98.3</td><td>1.6</td></tr>
<tr><td>Total</td><td>540</td><td>630</td><td>360</td><td>1530</td><td></td><td></td></tr>
<tr><td>Producer Accuracy</td><td>98.7</td><td>96.3</td><td>99.7</td><td></td><td>OA</td><td>97.9</td></tr>
<tr><td>Error omission</td><td>1.3</td><td>3.6</td><td>0.3</td><td></td><td>Kappa</td><td>0.9</td></tr>
</table>

Table 6. The confusion matric of support vector machine classification in 2015.

<table>
<tr><td rowspan="8">Classification SVM</td><td rowspan="2">Classes</td><td colspan="6">Training Data for Ground Check</td></tr>
<tr><td>Mangrove</td><td>Non-Mangrove</td><td>Water</td><td>Total</td><td>User Accuracy</td><td>Error Commission</td></tr>
<tr><td>Mangrove</td><td>540</td><td>7</td><td>0</td><td>547</td><td>98.7</td><td>1.3</td></tr>
<tr><td>Non-mangrove</td><td>0</td><td>623</td><td>0</td><td>623</td><td>100</td><td>0</td></tr>
<tr><td>Water</td><td>0</td><td>0</td><td>360</td><td>360</td><td>100</td><td>0</td></tr>
<tr><td>Total</td><td>540</td><td>630</td><td>360</td><td>1530</td><td></td><td></td></tr>
<tr><td>Producer Accuracy</td><td>100</td><td>98.1</td><td>100</td><td></td><td>OA</td><td>99.5</td></tr>
<tr><td>Error omission</td><td>0</td><td>1.1</td><td>0</td><td></td><td>Kappa</td><td>0.9</td></tr>
</table>

The confusion matrix shows accuracy assessment of classifications for the years 1989, 1998, 2002, and 2015 (Tables 3–6). The classification provided from the use of training data indicated that land cover reliability mapped, showed that all categories had above 90% rate of overall accuracy and kappa statistic. The discrimination of mangrove class also shows good accuracy with a commission error in years 1989, 1998, 2002, and 2015 of 0.4%, 2.8%, 3.2%, and 1.3%, respectively. In addition, the commission error in 1989, 1998, 2002, and 2015 of 0%, 0.9%, 1.3%, and 0%, respectively.

3.2. Land Cover Change Classification

The support vector machine classification produces an area of land-use classes (Mangrove, Non-mangrove, and Water). Overall, each land-use class area for years 1989, 1998, 2002, and 2015 can be seen in Table 7.

Table 7. Area Land-use (years) 1989, 1998, 2002, and 2015.

Classes	Area (ha) and Percentages (%)	1989	1998	2002	2015
Mangrove	area (ha)	58,145.5	36,847.4	55,548.3	60,697.5
	(%)	26.2	16.6	25.1	27.4
Non-mangrove	area (ha)	53,265.4	73,327.4	58,419.1	51,965.8
	(%)	24.1	33.1	26.3	23.4
Water	area (ha)	109,886	111,122	107,329	108,633
	(%)	49.6	50.2	48.5	49.1

The area occupied by each class in 1989 was: Mangrove 58,145.5 ha (26.2%), Non-mangrove 53,265.4 ha (24.1%), and 109,886 ha (49.6%). In 1998, Mangrove area was decreased to 36,847.4 ha (16.6%), Non-mangrove area was increased to 73,327.4 ha (33.1%) and the Water area was increased to 111,112 ha (50.2%). In contrast to 1989, mangrove areas show increased trends both in 2002 and 2015. In 2002, that the Mangrove area was increased to 55,548.3 ha (25.1%), Non-mangrove has decreased to 58,419.1 ha (26.3%), and the water area has decreased to 107,329 ha (48.5%). In 2015, the Mangrove areas has increased to 60,697.5 ha (27.4%), Non-mangrove areas has decrease to 51,965.8 ha (23.4%), and the Waters area has increased to 108,633 ha (49.1%).

3.3. Transition Matrix and Transition Probability Matrix for the Land Cover

The transition matrix area and transition probability matrix for years 1989–1998, 1998–2002, and 2002–2015 can be seen in Tables 8 and 9. The transition probability matrix from 1989 to 2015 then used to predict mangrove forest change in the year 2028. Validation of Markov-Cellular Automata identify map was carried out by comparing prediction Mangrove forest map of the year 2028 with the support vector machine 2015 classified Mangrove forest map.

Table 8. Transition Probability matrix in 1989–1998, 1998–2002 and 2002–2015.

Period	Land Cover	Mangrove	Non-Mangrove	Waters
	Mangrove	0.6	0.3	0
1989–1998	Non-mangrove	0.1	0.8	0.1
	Water	0.1	0.1	0.8
	Mangrove	0.7	0.2	0.1
1998–2002	Non-mangrove	0.7	0.3	0.1
	Water	0.1	0.2	0.7
	Mangrove	0.8	0.1	0.1
2002–2015	Non-mangrove	0.2	0.7	0.1
	Water	0.1	0.1	0.8

Table 9. Transition matrix area in 1989–1998, 1998–2002 and 2002–2015.

Period	Land Cover	Mangrove	Non-mangrove	Waters
	Mangrove	266,660	142,756	0
1989–1998	Non-mangrove	84,159	668,662	61,928
	Water	141,087	46,333	1,047,267
	Mangrove	446,200	123,355	47,648
1998–2002	Non-mangrove	431,900	202,641	14,561
	Water	100,020	208,094	884,435
	Mangrove	547,050	99,300	28,066
2002–2015	Non-mangrove	128,752	414,686	33,960
	Water	87,208	104,804	1,015,025

The transition probability matrix for years 1989–1998, 1998–2002, and 2002–2015 show that changing mangroves to non-mangroves from 1989 to 1998 is 0.3 and probability decreased by 0.2 in 1998–2002 (Table 8). When transition of non-mangrove into mangrove was observed, the transition probability was very low in 1989–1998 of 0.1. The probability of changing non-mangroves to mangrove areas in 2002–2015 decreased by 0.2. Mangrove, non-mangrove, and waters classes for the period 1989–1998, 1998–2002, and 2002–2015 have a probability value above 0 can change to the other classes, indicating the possibilities appropriate for analyzing changes in existing land cover.

Land-cover change was shown as transition matrix area in 1989–1998, 1998–2002 and 2002–2015 (Table 9). The probability transition shows that from 1989 to 1998, mangroves had the highest chance of becoming non-mangroves with a prediction of pixel allocations of 142,756 pixels or equivalent with 12,848.4 ha (area pixels = 900 m^2). In addition, in the same period, waters had the most significant chance of becoming mangrove with the prediction of pixel allocations of 141,087 equivalent with 12,697.8 ha. The 1998–2002 period shows that mangroves, non-mangroves, and water have the highest chance of becoming changed with several pixel allocations.

3.4. Prediction of Land Cover Change in Year 2002 and 2015

The simulation results of the Markov-CA model to predict land cover in 2002 and 2015 were shown in Figure 8, while the area of land cover prediction map in 2002 and 2015 can be seen in Table 10.

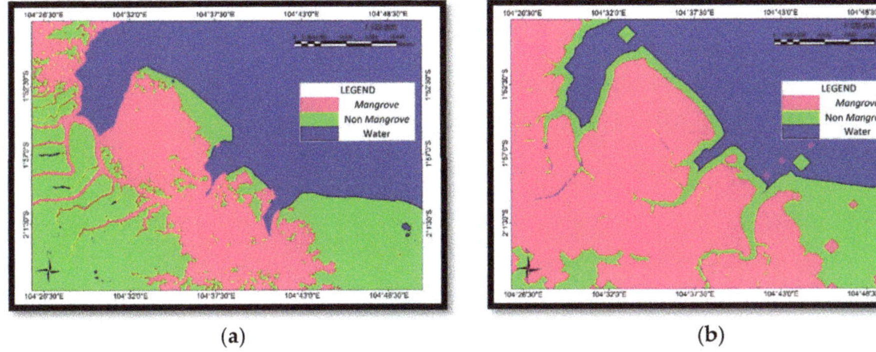

(a)   (b)

Figure 8. Land cover prediction map in (**a**) 2002 and (**b**) 2015.

Table 10. Land cover area of prediction map in 2002 and 2015.

Classes	Area (ha) and Percentages (%)	Prediction (2002)	Prediction (2015)
Mangrove	area (ha)	42,224.1	86,245.2
	(%)	19.1	38.9
Non-mangrove	area (ha)	77,012.6	47,087.1
	(%)	34	21.2
Water	area (ha)	10,059.9	87,964.2
	(%)	46.1	39.7

The area occupied by each class in 2002 (prediction) was: Mangrove of 42,224.1 ha (19.1%), Non-mangrove of 77,012.6 ha (34%), and Water of 10,059.9 ha (46.1%) (Table 10). In contrast, 2015 (prediction), Mangrove area was increased to 86,245.2 ha (38.9%), while both Non-mangrove area and Water were decreased to 47,087.1 ha (21.2%) and 87,964.2 ha (39.7%), respectively.

3.5. Kappa Index Agreement

Kappa evaluate how well classification or modeling performs excluding chance agreement [93]. In this study, kappa was used to assess the agreement between the 2002 and 2015 actual land cover maps and simulations. Kappa index agreement of prediction years 2002 and 2015 can be seen in Table 11. We found that the lowest kappa index agreement of 2002 (prediction) is 0.7 as *Kstandard*, while the highest kappa index agreement of 2002 is 0.8 as *Klocation* and *KlocationStrata*. In addition, the lowest kappa index agreement of 2015 (prediction) is 0.7 as *Kstandard*, while the highest kappa index agreement of 2015 (prediction) is 0.8 as *Klocation* and *KlocationStrata*. According to Gwet (2014) [95], *Kno* and *Kstandard* obtained from both prediction map in 2002 and 2015 indicated substantial agreement, while *Klocation* and *KlocationStrata* obtained from both prediction map in 2002 and 2015 indicated almost perfect agreement.

Table 11. Kappa Index Agreement in prediction years 2002 and 2015.

Kappa Index of Agreement	2002 (Prediction)	2015 (Prediction)
Kno	0.7	0.7
Klocation	0.8	0.8
KlocationStrata	0.8	0.8
Kstandard	0.7	0.7

3.6. Prediction of Land Cover Change in 2028

Land cover change years 2028 prediction using land over obtained from land cover year 2015. The 2015 land cover data were used for the base map, the potential transition map, and a transition area matrix for 2002–2015. The future land cover models are predicted, as shown in Figure 9 while the prediction area for 2028 is seen in Table 12.

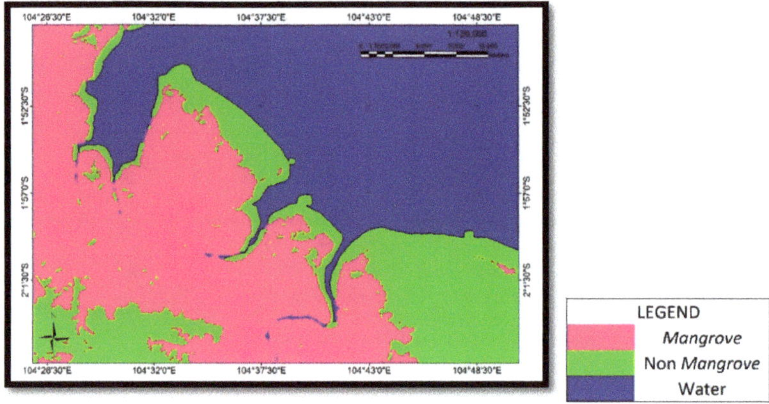

Figure 9. Prediction map for 2028.

Table 12. Land cover change prediction.

Classes	Area (in ha) and Percentages (%)	1989	1998	2002	2015	2028
Mangrove	area (ha)	58,145.5	36,847.4	55,548.3	60,697.5	68,672.3
	(%)	26.2	16.6	25.1	27.4	31
Non-mangrove	area (ha)	53,265.4	73,327.4	58,419.1	51,965.8	55,691.1
	(%)	24.1	33.1	26.3	23.4	25.1
Water	area (ha)	109,886	111,122	107,329	108,633	96,933.3
	(%)	49.6	50.2	48.5	49.1	43.8

Table 12 shows the period between 2015 and 2028, area of Mangroves increased from 27.4% to 31% or 7974.8 ha, while the area of Non-mangrove also increased from 23.4% to 25.1% or 3725.3 ha, and Water area decreased from 49.1% to 43.8% or 11,696.7 ha. In general, the area of classification results in 1989, 1998, 2002, and 2015 and the predicted results for 2028 can be seen in Table 12.

4. Discussion

4.1. Imagery Data on This Study

Landsat data was used in this study because it maps general land cover classes at a spatial resolution of 30 m for large areas. The identification of mangrove cover based on satellite images is not free from ambiguity. The problems usually occur related to the class representation, mixed pixel effects, and tidal effects [96]. For selecting the images, different seasons and atmospheric conditions must be considered [97]. According to Darmawan et al. (2015) mangrove identification and estimation of above-ground mangrove forest biomass are influenced by tidal height [98]. Therefore, the acquisition of Landsat 5 was carried out on 17 May 1989, at 02:39:43 UTC, Landsat 5 on 24 April 1998, at 02:48:21, Landsat 7 ETM+ on 30 June 2002, at 02:59:41, and Landsat 8 OLI on 26 June 2015, at 03:10:34 occurs at the same tides which are affected by the tidal height (Table 13), and also in April–June in South Sumatra has entered the dry season. In Indonesia, there are only two seasons, namely the rainy season and the dry season. It means that for each tidal effect, the Landsat image and atmospheric conditions are at the same conditions.

Table 13. The suitability time of Landsat imagery and tidal height.

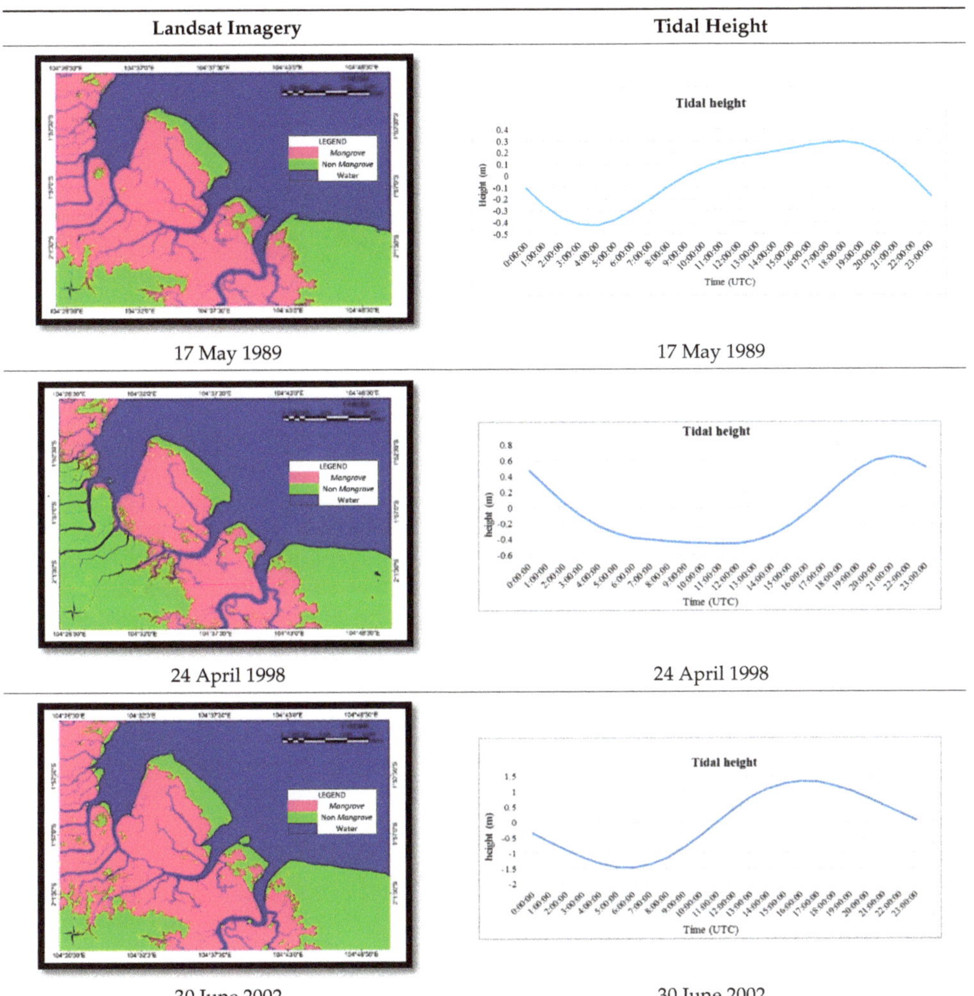

Landsat Imagery	Tidal Height
17 May 1989	17 May 1989
24 April 1998	24 April 1998
30 June 2002	30 June 2002

Table 13. *Cont.*

Landsat Imagery	Tidal Height

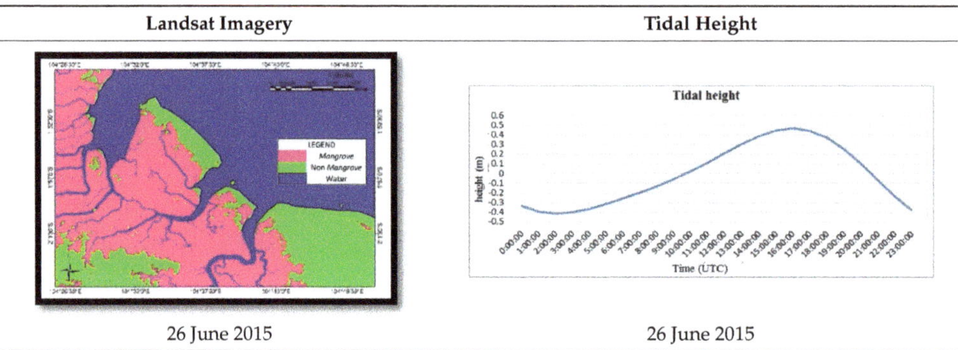

26 June 2015	26 June 2015

4.2. Accuracy Assessment

The sufficient number of training data and the selection of classification approaches are essential factors in successful classification [99,100]. In this study, training data were collected from more acceptable spatial resolution imagery. We collected a total of 1042 training pixels covering the Sembilang National Park of the area mapped. In this study, the amount both of the training and testing used are unbalance data. There are 432 training pixels with the primary class of mangrove, 333 training pixels labeled as land, and 277 training pixels labeled as water. The number of training data was following [101]. The training data size should not be smaller than 10 to 30 times the number of bands for each object. All pixels were selected randomly following to uniform in ground truth data. According to some researchers, balancing samples in classification is a controversial topic [102–104].

In some cases, unbalanced data was inevitable due to the complexity and heterogeneous landscapes in the study area when choosing training data [105]. Therefore, this problem can be handled using the appropriate classifications approach, such as a vector machine. The support vector machine for handling unbalanced data in the classification process is the best choice, proven in high accuracy (Table 14).

Table 14. The classification accuracy.

Years	Overall Accuracy	Kappa Statistics	Mangrove User Accuracy	Mangrove Producer Accuracy
1989	99.8	0.9	99.6	100
1998	98.5	0.9	97.2	99.1
2002	97.9	0.9	96.7	98.7
2015	99.5	0.9	98.7	100

The classification accuracy obtained from Landsat 1989, 1998, 2002, and 2015 indicated that the support vector machine was the right option for mangrove mapping based on an unbalance training sample. In this study, the overall accuracy of the land cover maps for 1989 (99.8%), 1998 (98.5%), 2002 (97.9%), and 2015 (99.5%) were achieved. All accuracy indicators of overall accuracy, kappa statistics, Mangrove user accuracy, and Mangrove producer accuracy were above 90%. Table 14 shows that the support vector machine classifier's performance can dramatically decrease with a relatively small number of mislabeled examples [76]. According to Mountrakis (2011) [76], support vector machines are not relatively sensitive to training sample size, and some literature has improved support vector machines to work successfully with limited quantity and quality of training data [76].

4.3. Matrix Probability Transition

The probability matrix is a factor that sets the trend of change in surrounding cells as a function of cell conditions themselves. The CA-Markov was applied for simulation of mangrove cover changed. Three intervals period were used in this study, including an interval of 9 years (1989–1998), an interval of 4 years (1998–2002), and 13 years (2002–2015). Each interval represents each land cover category projections, while the third interval is determined by the results of project accuracy in 2028.

The probability transition matrix had been done for the interval in 9 years (1989–1998), an interval of 4 years (1998–2002), and 13 years (2002–2015) (Figure 10). Figure 10, the interval of 9 years (1989–1998), indicated that more than half of the Mangrove was changed into Non-mangrove areas (34.8%). This result was related to the Forestry Research and Development Agency (2013) that South Sumatra experienced relatively high deforestation during the 1990, resulting in a decrease in mangrove forest cover due to forest fires in 1997–1998. The worst of them occurred in 1997 during the dry weather fostered by El Niño [29]. At the same time, there was no change in Mangrove becoming Water. In the same interval years, 82% of Non-mangrove remained unconverted. So, the water 84.8% remained unconverted. The four-year (1998–2002) interval showed a positive change concerning the Mangrove forest recovery, which is showed 66.4% Non-mangrove and 8.3% Water shifted to mangrove areas. The increasing Mangrove areas resulted in better management by the government in Sembilang National Park. It is also proven from 13 years (2002–2015), indicating that 81.1% of Mangroves remain unconverted. Totally 22.3% Non-mangrove and 7.2% Water changed to Mangrove areas. The increase of Mangrove areas indicated better management in Sembilang National Park.

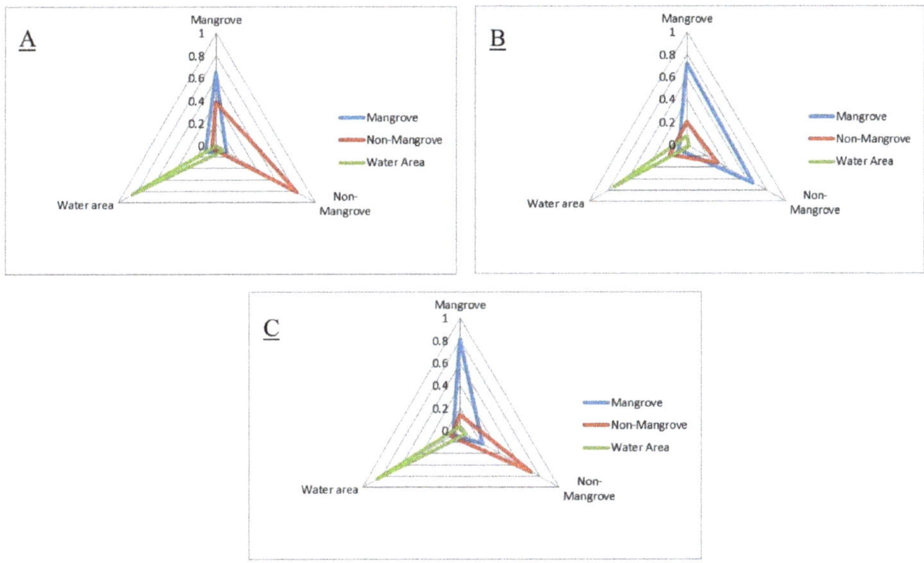

Figure 10. The probability matrix: (**A**) an interval of 9 years (1989–1998), (**B**) an interval of 4 years (1998–2002), and (**C**) an interval of 13 years (2002–2015).

4.4. Land Cover Area in Years 1989, 1998, 2002, 2015 and Predicted 2028

The statistics area (ha) trend was derived from the support vector machine classification for 1989, 1998, 2002, 2015, and prediction 2028 (Figure 11). In 1989–2018, the trend of mangrove land cover fluctuated indicated the lowest mangrove area is in 1998 of 36,847.4 ha, and the highest mangrove area predicted in 2028 of 68,672.3 ha. In the interval years of 1989–1998, the loss mangroves in Sembilang National Park are 21,298.1 ha. The decrease of mangrove was caused by the worst forest fires in

Indonesia in 1997 [29]. However, in 1998–2028, the increase of mangrove forest became 55,548.3 ha in 2002, 60,697.5 ha in 2015, and 68,672.3 ha in 2028. The highest rate of increasing mangrove forests was in 1998–2002 of 8.4%. The increase of the mangrove forest area indicated that the government had succeeded in mangrove rehabilitation management.

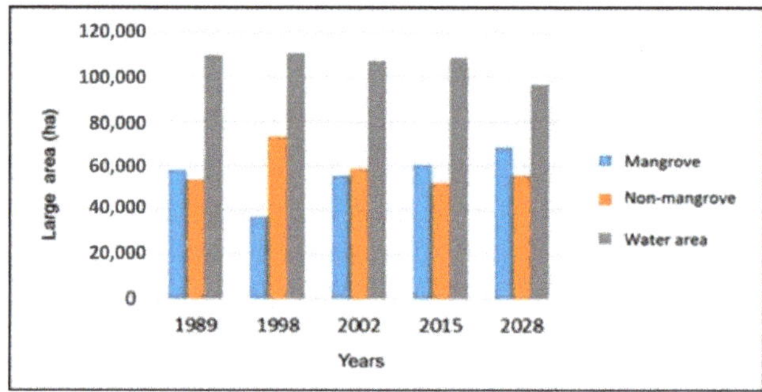

Figure 11. Area statistics (ha) of land cover class for the years 1989, 1998, 2002, 2015, and predicted 2028.

5. Conclusions

In this study, we calculated mangrove forest changes areas in Sembilang National Park, Banyuasin, Indonesia from 1989, 1998, 2002, 2015, and predicted mangrove forest areas in 2028 using the support vector machine algorithm and the Markov-Cellular Automata model. Based on the historical land cover from 1989 to 2015, the study has attempted to detect, simulate and predict the future expansion trends up to 2028, with a specific empirical focus on changes in the mangrove forest area. The results showed that the mangrove forest area from 1989 to 1998 has decreased around 9.6%. The worst forest fires caused the decrease of mangrove forests in Indonesia in 1997. However, the Indonesian government keeps trying to rehabilitate mangrove forests in Sembilang National Park as the largest mangrove forest. It was proven that the mangrove area has increased by 8.4% between 1998 and 2002, and 2.3% in 2002–2015. Other results showed that the mangroves area has continued to increase from 2015 to 2028 by 27.4% to 31% (7974.8 ha). The increase of the mangrove forest area indicated that the government had succeeded in mangrove rehabilitation management.

Author Contributions: Conceptualization, S.D., D.K.S. and K.W.; Data curation, M.K.S.; Formal analysis, S.D. and R.H.; Methodology, S.D., D.K.S., K.W. and R.H.; Validation, M.K.S.; Visualization, A.T. and M.K.S.; Writing—original draft, S.D. and A.T.; Writing—review & editing, R.H. All authors have read and agreed to the published version of the manuscript.

Funding: This research was funded by Ministry of Research and Technology/National Agency for Research and Innovation (Kemenristek/BRIN), Number: 378/B.05/LPPM-Itenas/IV/2020.

Acknowledgments: In addition, the researcher would like to acknowledge the funding support from Ministry of Research and Technology/National Agency for Research and Innovation (Kemenristek/BRIN) and LPPM Itenas Bandung.

Conflicts of Interest: The authors declare no conflict of interest.

References

1. Nybakken, J.W.; Eidman, H.M. *Biologi Laut: Suatu Pendekatan Ekologis*; PT Gramedia Pustaka Utama: Jakarta, Indonesia, 1992; ISBN 979403018X 9789794030189.
2. Romimohtarto, K. *Biologi Laut: Ilmu Pengetahuan Tentang Biota Laut/Kasijan Romimohtarto, Sri Juwana*; Djambatan: Jakarta, Indonesia, 2001; ISBN 9794284009.
3. Donato, D.C.; Kauffman, J.B.; Murdiyarso, D.; Kurnianto, S.; Stidham, M.; Kanninen, M. Mangroves among the most carbon-rich forests in the tropics. *Nat. Geosci.* **2011**, *4*, 293–297. [CrossRef]

4. Paillon, C.; Wantiez, L.; Kulbicki, M.; Labonne, M.; Vigliola, L. Extent of Mangrove Nursery Habitats Determines the Geographic Distribution of a Coral Reef Fish in a South-Pacific Archipelago. *PLoS ONE* **2014**, *9*, e105158. [CrossRef] [PubMed]

5. Barbier, E.B.; Hacker, S.D.; Kennedy, C.; Koch, E.W.; Stier, A.C.; Silliman, B.R. The value of estuarine and coastal ecosystem services. *Ecol. Monogr.* **2011**, *81*, 169–193. [CrossRef]

6. Brown, D. *Mangrove: Nature's Defences Against Tsunamis*; Environmental Justice Foundation: London, UK, 2004.

7. Walters, B.B.; Rönnbäck, P.; Kovacs, J.M.; Crona, B.; Hussain, S.A.; Badola, R.; Primavera, J.H.; Barbier, E.; Dahdouh-Guebas, F. Ethnobiology, socio-economics and management of mangrove forests: A review. *Aquat. Bot.* **2008**, *89*, 220–236. [CrossRef]

8. Giesen, W. *Indonesia's Mangrove: An Update on Remaining Area and Main Management Issues*; Asian Wetland Bureau (AWB): Wageningen, The Netherlands, 1993.

9. Spalding, M.; Blasco, F.; Field, C. *World Mangrove Atlas*; FAO: Rome, Italy, 1997; p. 178.

10. Suwignyo, R.A.; Ulqodry, T.Z.; Halimi, E.S.; Dwipa, H.S. Hutan Mangrove Pada Masyarakat. In Proceedings of the Lokakarya Pembentukan Kelompok Kerja Mangrove Daerah (KKMD) Provinsi Sumatera Selatan Balai Pengelolaan Hutan Mangrove Wilayah II Direktorat Jenderal Bina Pengelolaan Daerah Alir, Palembang, Indonesia, 26 May 2011.

11. *Ekosistem Sembilang—Taman Nasional Berbak*; TFCA Sumatera: Jakarta, Indonesia. Available online: http://tfcasumatera.org/bentang_alam/ekosistem-sembilang-taman-nasional-berbak/ (accessed on 4 October 2020).

12. Lubis, I.R.; Suryadiputra, I.N.N. Upaya pengelolaan terpadu hutan rawa gambut bekas terbakar diwilayah Berbak-Sembilang. In *Kebakaran di lahan rawa/gambut di Sumatera: Masalah dan Solusi. Prosiding Semiloka*; CIFOR: Bogor, Indonesia, 2004; pp. 105–119.

13. Tirtakusumah, R. Pengelolaan Hutan Mangrove Jawa Barat dan Beberapa Pemikiran untuk Tindak Lanjut. In Proceedings of the Dalam Prosiding Seminar V Ekosistem Mangrove, Jember, Indonesia, 3–6 August 1994; pp. 3–6.

14. Dahuri, R.; Rais, J.; Ginting, S.P.; Sitepu, M.J. *Integrated Coastal and Marine Resource Management*, 2nd ed.; PT. Pradnya Paramita: Jakarta, Indonesia, 2001; p. 328.

15. Anderson, I.P.; Bowen, R. *Fire Zones and the Threat to the Wetlands of Sumatra, Indonesia*; Report; European Union Ministry of Forestry: Palembang, Indonesia, 2000; pp. 1–46.

16. Barber, C.V.; Schweithelm, J. *Trial by Fire: Forest Fires and Forestry Policy in Indonesia's Era of Crisis and Reform*; Report; In collaboration with WWF-Indonesia and Telapak Indonesia Foundation; World Resources Institute (WRI), Forest Frontiers Initiative: South Jakarta, Indonesia, 2000; ISBN 1569734089.

17. Lubis, I.R.; Suryadiputra, I.N.N. *Upaya Pengelolaan Terpadu Hutan Rawa Gambut Bekas Terbakar di WIlayah Berbak-Sembilang*; CIFOR: Bogor, Indonesia, 2003; ISBN 9793361492.

18. Parameswaran, K.; Nair, S.K.; Rajeev, K. Impact of Indonesian forest fires during the 1997 El Nino on the aerosol distribution over the Indian Ocean. *Adv. Space Res.* **2004**, *33*, 1098–1103. [CrossRef]

19. Page, S.E.; Siegert, F.; Rieley, J.O.; Boehm, H.-D.V.; Jaya, A.; Limin, S. The amount of carbon released from peat and forest fires in Indonesia during 1997. *Nature* **2002**, *420*, 61–65. [CrossRef]

20. Harrison, M.E.; Page, S.E.; Limin, S.H. The global impact of Indonesian forest fires. *Biologist* **2009**, *56*, 156–163.

21. Boer, C. Forest and fire suppression in East Kalimantan, Indonesia. In *Proceedings of the International Conference on Community Involvement in FIRE Management*; Food and Agriculture Organization of the United Nations (FAO), Regional Office for Asia and the Pacific: Rome, Italy, 2002; pp. 69–71.

22. BAPPENAS (National Development Planning Agency); Asian Development Bank (ADB). *Causes, extent, impact and costs of 1997/98 fires and drought. Final Report, Annex 1 and 2*; ADB: Jakarta, Indonesia, 1999.

23. Tacconi, L. *Kebakaran Hutan di Indonesia: Penyebab, Biaya Dan Implikasi Kebijakan*; CIFOR: Bogor, Indonesia, 2003.

24. Glover, D.; Jessup, T. The Indonesian fires and haze of 1997: The economic toll. In Proceedings of the Economy and Environment Program for SE Asia (EEPSEA) Singapore and the World Wildlife Fund (WWF) Indonesia, Jakarta, Indonesia, May 1998.

25. Anderson, I.P.; Bowen, M.R.; Imanda, I.D. *Muhnandar Forest Fire Prevention and Control Project Forest Fire Prevention and Control Project Vegetation Fires in Indonesia: The Fire History of the Sumatra Provinces 1996–1998 As a Predictor of Future Areas At Risk*; Report; Balai Inventaris dan Perpetaan Hutan Wilayah II and Kanwil Kehutanan dan Perkebunan: Palembang, Indonesia, 1999.

26. CIFOR. *A Review of Fire Projects in Indonesia (1982–1998)*; CIFOR: Bogor, Indonesia, 1999; ISBN 9798764307.

27. Legg, C.A.; Laumonier, Y. Fires in Indonesia, 1997: A remote sensing perspective. *Ambio* **1999**, *28*, 479–485. [CrossRef]

28. Bowen, M.R.; Bompard, J.M.; Anderson, I.P.; Guizol, P.; Gouyon, A. *Anthropogenic Fires in Indonesia: A View from Sumatra*; CIFOR: Bogor, Indonesia, 2001; pp. 41–66.

29. Podgorny, I.A.; Li, F.; Ramanathan, V. Large aerosol radiative forcing due to the 1997 Indonesian forest fire. *Geophys. Res. Lett.* **2003**, *30*. [CrossRef]

30. Pemerintah Republik Indonesia. *Undang-Undang Nomor 5 Tahun 1990 Tentang Konservasi Sumber Daya Alam Hayati dan Ekosistemnya*; Pemerintah Republik Indonesia: Jakarta, Indonesia, 1990; Volume 1988, pp. 1–26.

31. Verburg, P.H.; Schot, P.P.; Dijst, M.J.; Veldkamp, A. Land use change modelling: Current practice and research priorities. *GeoJournal* **2004**, *61*, 309–324. [CrossRef]

32. Sohl, T.; Sleeter, B. Role of remote sensing for land-use and land-cover change modeling. In *Remote Sensing of Land Use and Land Cover: Principles and Applications*; Giri, C.P., Ed.; CRC Press: Boca Raton, FL, USA, 2012; pp. 225–239.

33. Rogan, J.; Chen, D. Remote sensing technology for mapping and monitoring land-cover and land-use change. *Prog. Plan.* **2004**, *61*, 301–325. [CrossRef]

34. Gómez, C.; White, J.C.; Wulder, M.A. Optical remotely sensed time series data for land cover classification: A review. *ISPRS J. Photogramm. Remote Sens.* **2016**, *116*, 55–72. [CrossRef]

35. Singh, A. Review article digital change detection techniques using remotely-sensed data. *Int. J. Remote Sens.* **1989**, *10*, 989–1003. [CrossRef]

36. Lu, D.; Mausel, P.; Brondizio, E.; Moran, E. Change detection techniques. *Int. J. Remote Sens.* **2004**, *25*, 2365–2401. [CrossRef]

37. Alqurashi, A.; Kumar, L. Investigating the use of remote sensing and GIS techniques to detect land use and land cover change: A review. *Adv. Remote Sens.* **2013**. [CrossRef]

38. Bolstad, P.; Lillesand, T.M. Rapid maximum likelihood classification. *Photogramm. Eng. Remote Sens.* **1991**, *57*, 67–74.

39. Franklin, J.; Woodcock, C.E.; Warbington, R. Multi-attribute vegetation maps of forest service lands in California supporting resource management decisions. *Photogramm. Eng. Remote Sens.* **2000**, *66*, 1209–1218.

40. Dewan, A.M.; Yamaguchi, Y.; Rahman, M.Z. Dynamics of land use/cover changes and the analysis of landscape fragmentation in Dhaka Metropolitan, Bangladesh. *GeoJournal* **2012**, *77*, 315–330. [CrossRef]

41. Guild, L.S.; Cohen, W.B.; Kauffman, J.B. Detection of deforestation and land conversion in Rondonia, Brazil using change detection techniques. *Int. J. Remote Sens.* **2004**, *25*, 731–750. [CrossRef]

42. Yuan, D.; Elvidge, C. NALC land cover change detection pilot study: Washington DC area experiments. *Remote Sens. Environ.* **1998**, *66*, 166–178. [CrossRef]

43. Lambin, E.F.; Baulies, X.; Bockstael, N.; Fischer, G.; Krug, T.; Leemans, R.; Moran, E.F.; Rindfuss, R.R.; Sato, Y.; Skole, D.; et al. *Land-Use and Land-Cover Change (LUCC): Implementation Strategy*; IGBP Report No. 48, IHDP Report No. 10; International Geosphere-Biosphere Programme (IGBP), International Human Dimension Programme on Global Environmental Change (IHDP): Stockholm, Sweden, 1999; p. 125.

44. Behera, M.D.; Borate, S.N.; Panda, S.N.; Behera, P.R.; Roy, P.S. Modelling and analyzing the watershed dynamics using Cellular Automata (CA)-Markov model—A geo-information based approach. *J. Earth Syst. Sci.* **2012**, *121*, 1011–1024. [CrossRef]

45. Zhang, Y.; Gong, H.; Zhao, W.; Li, X. Analyzing the mechanism of land use change in Beijing City from 1990 to 2000. *Resour. Sci.* **2007**, *29*, 206–213.

46. Memarian, H.; Balasundram, S.K.; Talib, J.B.; Teh, C.; Sung, B.; Sood, A.M.; Abbaspour, K. Validation of CA-Markov for Simulation of Land Use and Cover Change in the Langat Basin, Malaysia. *J. Geogr. Inf. Syst.* **2012**, *4*, 542–554. [CrossRef]

47. Huang, W.; Liu, H.; Luan, Q.; Jiang, Q.; Liu, J.; Liu, H. Detection and prediction of land use change in Beijing based on remote sensing and GIS. *Int. Arch. Photogramm. Remote Sens. Spat. Inf. Sci.* **2008**, *37*, 75–82.

48. Weng, Q. Land use change analysis in the Zhujiang Delta of China using satellite remote sensing, GIS and stochastic modelling. *J. Environ. Manag.* **2002**, *64*, 273–284. [CrossRef] [PubMed]

49. Benenson, I.; Torrens, P. *Geosimulation: Automata-Based Modeling of Urban Phenomena*; John Wiley & Sons Inc.: Hoboken, NJ, USA, 2004; ISBN 0470843497.

50. Candau, J.; Rasmussen, S.; Clarke, K.C. A coupled cellular automaton model for land use/land cover dynamics. In Proceedings of the Fourth International Conference on Integrating GIS and Environmental Modeling (GIS/EM4): Problems, Prospects and Research Needs, Banff, AB, Canada, 2–8 September 2000.

51. Koomen, E.; Rietveld, P.; Nijs, T. Modelling land-use change for spatial planning support. *Ann. Reg. Sci.* **2008**, *42*, 1–10. [CrossRef]

52. Torrens, P.M.; Benenson, I. Geographic automata systems. *Int. J. Geogr. Inf. Sci.* **2005**, *19*, 385–412. [CrossRef]

53. Houet, T.; Hubert-moy, L. Modelling and Projecting Land-Use and Land-Cover Changes with a Cellular Automaton in Considering Landscape Trajectories: An Improvement for Simulation of Plausible Future States. *EASeL eProceedings* **2006**, *5*, 63–76.

54. Yirsaw, E.; Wu, W.; Shi, X.; Temesgen, H.; Bekele, B. Land Use/Land Cover change modeling and the prediction of subsequent changes in ecosystem service values in a coastal area of China, the Su-Xi-Chang region. *Sustainability* **2017**, *9*, 1204. [CrossRef]

55. Fei, S.X.; Shan, C.H.; Hua, G.Z. Remote sensing of mangrove wetlands identification. *Procedia Environ. Sci.* **2011**, *10*, 2287–2293. [CrossRef]

56. Chen, C.F.; Son, N.T.; Chang, N.B.; Chen, C.R.; Chang, L.Y.; Valdez, M.; Centeno, G.; Thompson, C.A.; Aceituno, J.L. Multi-decadal mangrove forest change detection and prediction in honduras, central america, with landsat imagery and a markov chain model. *Remote Sens.* **2013**, *5*, 6408–6426. [CrossRef]

57. Aghighi, H.; Trinder, J.; Lim, S.; Tarabalka, Y. Improved adaptive Markov random field based super-resolution mapping for mangrove tree identification. *ISPRS Ann. Photogramm. Remote Sens. Spat. Inf. Sci.* **2014**, *2*, 61. [CrossRef]

58. Dan, T.T.; Chen, C.F.; Chiang, S.H.; Ogawa, S. Mapping and change analysis in mangrove forest by using Landsat imagery. *ISPRS Ann. Photogramm. Remote Sens. Spat. Inf. Sci.* **2016**, *3*, 109. [CrossRef]

59. Kux, H.J.; Souza, U.D. Object-based image analysis of WORLDVIEW-2 satellite data for the classification of mangrove areas in the city of São Luís, Maranhão State, Brazil. *ISPRS Ann. Photogramm Remote Sens. Spat. Inf. Sci* **2012**, 95–100. [CrossRef]

60. Kamal, M.; Phinn, S.; Johansen, K. Object-based approach for multi-scale mangrove composition mapping using multi-resolution image datasets. *Remote Sens.* **2015**, *7*, 4753–4783. [CrossRef]

61. Heenkenda, M.K.; Joyce, K.E.; Maier, S.W.; Bartolo, R. Mangrove species identification: Comparing WorldView-2 with aerial photographs. *Remote Sens.* **2014**, *6*, 6064–6088. [CrossRef]

62. Kanniah, K.D.; Sheikhi, A.; Cracknell, A.P.; Goh, H.C.; Tan, K.P.; Ho, C.S.; Rasli, F.N. Satellite images for monitoring mangrove cover changes in a fast growing economic region in southern Peninsular Malaysia. *Remote Sens.* **2015**, *7*, 14360–14385. [CrossRef]

63. Feng, Y.; Liu, Y.; Batty, M. Modeling urban growth with GIS based cellular automata and least squares SVM rules: A case study in Qingpu–Songjiang area of Shanghai, China. *Stoch. Environ. Res. Risk Assess.* **2016**, *30*, 1387–1400. [CrossRef]

64. Madanguit, C.J.G.; Oñez, P.J.L.; Tan, H.G.; Villanueva, M.D.; Ordaneza, J.E.; Aurelio, R.M.; Novero, A.U. Application of Support Vector Machine (SVM) and Quick Unbiased Efficient Statistical Tree (QUEST) Algorithms on Mangrove and Agricultural Resource Mapping using LiDAR Data Sets. *Int. J. Appl. Environ. Sci.* **2017**, *12*, 973–6077.

65. Firmansyah, S.; Gaol, J.L.; Susilo, S.B. Perbandingan Klasifikasi SVM dan Decision Tree untuk Pemetaan Mangrove Berbasis Objek Menggunakan Citra Satelit Sentinel-2B di Gili Sulat, Lombok Timur. *J. Pengelolaan Sumberd. Alam dan Lingkung. J. Nat. Resour. Environ. Manag.* **2019**, *9*, 746–757.

66. Finlayson, C.M.; Milton, G.R.; Prentince, R.C.; Davidson, N.C. *The Wetland Book II: Distribution, Description, and Conservation*; Springer: Dordrecht, The Netherlands, 2018; Volume 3, ISBN 9789400740013.

67. Silvius, M.J.; Noor, Y.R.; Lubis, I.R.; Giesen, W.; Rais, D. *Sembilang National Park: Mangrove Reserves of Indonesia BT—The Wetland Book: II: Distribution, Description, and Conservation*; Finlayson, C.M., Milton, G.R., Prentice, R.C., Davidson, N.C., Eds.; Springer: Dordrecht, The Netherlands, 2018; pp. 1819–1829. ISBN 978-94-007-4001-3.

68. Sembilang National Park | Service D'information sur les Sites Ramsar. Available online: https://rsis.ramsar.org/fr/ris/1945?language=fr (accessed on 1 October 2020).

69. Silvius, M.; Giesen, W.; Lubis, R.; Salathé, T. Ramsar Advisory Mission N° 85 Berbak National Park Ramsar Site N° 554 (with references to Sembilang National Park Ramsar Site N° 1945) Peat fire prevention through

green land development and conservation, peatland rewetting and public awareness. *Ramsar Conv. Rep. 85* **2018**, *554*, 1–60.

70. Sembilangan National Park. Available online: https://www.indonesia-tourism.com/south-sumatra/sembilangan.html (accessed on 17 October 2020).

71. Verheugt, W.J.M.; Purwoko, A.; Danielsen, F.; Skov, H.; Kadarisman, R. Integrating mangrove and swamp forests conservation with coastal lowland development; the Banyuasin Sembilang swamps case study, South Sumatra Province, Indonesia. *Landsc. Urban Plan.* **1991**, *20*, 85–94. [CrossRef]

72. Taman Nasional Sembilang | Technical Cooperation Projects | JICA. Available online: https://www.jica.go.jp/project/indonesian/indonesia/008/outline/05.html (accessed on 18 October 2020).

73. Giesen, W. Causes of peat swamp forest degradation in Berbak NP, Indonesia, and recommendations for restoration causes of peat swamp forest degradation in Berbak NP, Indonesia, and Recommendations Part of the project on "Promoting the river basin and ecosystem". *Tech. Rep.* **2004**, 125. [CrossRef]

74. Zhu, G.; Blumberg, D.G. Classification using ASTER data and SVM algorithms: The case study of Beer Sheva, Israel. *Remote Sens. Environ.* **2002**, *80*, 233–240. [CrossRef]

75. Mountrakis, G.; Im, J.; Ogole, C. Support vector machines in remote sensing: A review. *ISPRS J. Photogramm. Remote Sens.* **2011**, *66*, 247–259. [CrossRef]

76. Congalton, R.G. A review of assessing the accuracy of classifications of remotely sensed data. *Remote Sens. Environ.* **1991**, *37*, 35–46. [CrossRef]

77. Petropoulos, G.P.; Kontoes, C.; Keramitsoglou, I. Burnt area delineation from a uni-temporal perspective based on landsat TM imagery classification using Support Vector Machines. *Int. J. Appl. Earth Obs. Geoinf.* **2011**, *13*, 70–80. [CrossRef]

78. Kusratmoko, E.; Albertus, S.D.Y. Modelling land use/cover changes with markov-cellular automata in Komering Watershed, South Sumatera. In *Proceedings of the IOP Conference Series: Earth and Environmental Science*; IOP Publishing: Bristol, UK, 2017; Volume 54, p. 12103.

79. Ruben, G.B.; Zhang, K.; Dong, Z.; Xia, J. Analysis and projection of land-use/land-cover dynamics through scenario-based simulations using the CA-Markov model: A case study in guanting reservoir basin, China. *Sustainability* **2020**, *12*, 3747. [CrossRef]

80. Cetin, M.; Demirel, H. Modelling and simulation of urban dynamics. *Fresenius Environ. Bull.* **2010**, *19*, 2348–2353.

81. Li, C.L.; Liu, M.; Hu, Y.M.; Xu, Y.Y.; Sun, F.Y. Driving forces analysis of urban expansion based on boosted regression trees and Logistic regression. *Acta Ecol. Sin.* **2014**, *34*, 727–737.

82. Arsanjani, J.J.; Helbich, M.; Kainz, W.; Boloorani, A.D. Integration of logistic regression, Markov chain and cellular automata models to simulate urban expansion. *Int. J. Appl. Earth Obs. Geoinf.* **2013**, *21*, 265–275. [CrossRef]

83. Al-sharif, A.A.A.; Pradhan, B. Monitoring and predicting land use change in Tripoli Metropolitan City using an integrated Markov chain and cellular automata models in GIS. *Arab. J. Geosci.* **2014**, *7*, 4291–4301. [CrossRef]

84. Yang, X.; Zheng, X.Q.; Chen, R. A land use change model: Integrating landscape pattern indexes and Markov-CA. *Ecol. Model.* **2014**, *283*, 1–7. [CrossRef]

85. D'ambrosio, D.; Di Gregorio, S.; Gabriele, S.; Gaudio, R. A cellular automata model for soil erosion by water. *Phys. Chem. Earth Part B Hydrol. Ocean. Atmos.* **2001**, *26*, 33–39. [CrossRef]

86. Adhikari, S.; Southworth, J. Simulating forest cover changes of Bannerghatta National Park based on a CA-Markov model: A remote sensing approach. *Remote Sens.* **2012**, *4*, 3215–3243. [CrossRef]

87. Louca, M.; Vogiatzakis, I.N.; Moustakas, A. Modelling the combined effects of land use and climatic changes: Coupling bioclimatic modelling with Markov-chain Cellular Automata in a case study in Cyprus. *Ecol. Inform.* **2015**, *30*, 241–249. [CrossRef]

88. Halmy, M.W.A.; Gessler, P.E.; Hicke, J.A.; Salem, B.B. Land use/land cover change detection and prediction in the north-western coastal desert of Egypt using Markov-CA. *Appl. Geogr.* **2015**, *63*, 101–112. [CrossRef]

89. Hamad, R.; Balzter, H.; Kolo, K. Predicting land use/land cover changes using a CA-Markov model under two different scenarios. *Sustainability* **2018**, *10*, 3421. [CrossRef]

90. Sang, L.; Zhang, C.; Yang, J.; Zhu, D.; Yun, W. Simulation of land use spatial pattern of towns and villages based on CA-Markov model. *Math. Comput. Model.* **2011**, *54*, 938–943. [CrossRef]

91. Kumar, S.; Radhakrishnan, N.; Mathew, S. Land use change modelling using a Markov model and remote sensing. *Geomat. Nat. Hazards Risk* **2014**, *5*, 145–156. [CrossRef]
92. Wang, S.Q.; Zheng, X.Q.; Zang, X.B. Accuracy assessments of land use change simulation based on Markov-cellular automata model. *Procedia Environ. Sci.* **2012**, *13*, 1238–1245. [CrossRef]
93. Pontius, R.G. Quantification error versus location error in comparison of categorical maps. *Photogramm. Eng. Remote Sens.* **2000**, *66*, 1011–1016.
94. Pontius, R.G.; Huffaker, D.; Denman, K. Useful techniques of validation for spatially explicit land-change models. *Ecol. Model.* **2004**, *179*, 445–461. [CrossRef]
95. Gwet, K.L. *Handbook of Inter-Rater Reliability: The Definitive Guide to Measuring the Extent of Agreement among Raters*; Advanced Analytics, LLC.: San Francisco, CA, USA, 2014; ISBN 0970806280.
96. Rocchini, D.; Delucchi, L.; Bacaro, G.; Cavallini, P.; Feilhauer, H.; Foody, G.M.; He, K.S.; Nagendra, H.; Porta, C.; Ricotta, C.; et al. Calculating landscape diversity with information-theory based indices: A GRASS GIS solution. *Ecol. Inform.* **2013**, *17*, 82–93. [CrossRef]
97. Toosi, A.S.; Calbimonte, G.H.; Nouri, H.; Alaghmand, S. River basin-scale flood hazard assessment using a modified multi-criteria decision analysis approach: A case study. *J. Hydrol.* **2019**, *574*, 660–671. [CrossRef]
98. Darmawan, S.; Takeuchi, W.; Vetrita, Y.; Wikantika, K.; Sari, D.K. Impact of Topography and Tidal Height on ALOS PALSAR Polarimetric Measurements to Estimate Aboveground Biomass of Mangrove Forest in Indonesia. *J. Sens.* **2015**. [CrossRef]
99. Atkinson, P.M.; Aplin, P. Spatial variation in land cover and choice of spatial resolution for remote sensing. *Int. J. Remote Sens.* **2004**, *25*, 3687–3702. [CrossRef]
100. Atkinson, P.M.; Curran, P.J. Choosing an appropriate spatial resolution for remote sensing investigations. *Photogramm. Eng. Remote Sens.* **1997**, *63*, 1345–1351.
101. Van Niel, T.G.; McVicar, T.R.; Datt, B. On the relationship between training sample size and data dimensionality: Monte Carlo analysis of broadband multi-temporal classification. *Remote Sens. Environ.* **2005**, *98*, 468–480. [CrossRef]
102. Chen, D.; Stow, D. The effect of training strategies on supervised classification at different spatial resolutions. *Photogramm. Eng. Remote Sens.* **2002**, *68*, 1155–1162.
103. Landgrebe, D.A. *Signal Theory Methods in Multispectral Remote Sensing*; John Wiley & Sons Inc.: Hoboken, NJ, USA, 2003.
104. Mather, P.M. *Computer Processing of Remotely Sensed Images*; John Wiley & Sons Inc.: Hoboken, NJ, USA, 2004.
105. Gaertner, J.; Genovese, V.B.; Potter, C.; Sewake, K.; Manoukis, N.C. Vegetation classification of Coffea on Hawaii Island using WorldView-2 satellite imagery. *J. Appl. Remote Sens.* **2017**, *11*, 46005. [CrossRef]

Publisher's Note: MDPI stays neutral with regard to jurisdictional claims in published maps and institutional affiliations.

 *remote sensing*

Article

Land Cover Dynamics and Mangrove Degradation in the Niger Delta Region

Iliya Ishaku Nababa [1], Elias Symeonakis [1,*], Sotirios Koukoulas [2], Thomas P. Higginbottom [3], Gina Cavan [1] and Stuart Marsden [1]

[1] Department of Natural Sciences, Manchester Metropolitan University, Manchester M15 6BH, UK; ILIYA.I.NABABA@stu.mmu.ac.uk (I.I.N.); g.cavan@mmu.ac.uk (G.C.); s.marsden@mmu.ac.uk (S.M.)
[2] Department of Geography, University of the Aegean, 81100 Mytilene, Greece; skouk@geo.aegean.gr
[3] School of Mechanical, Aerospace, and Civil Engineering, University of Manchester, Manchester M13 9PL, UK; thomas.higginbottom@manchester.ac.uk
*   Correspondence: e.symeonakis@mmu.ac.uk; Tel.: +44-161-247-1587

Received: 30 September 2020; Accepted: 31 October 2020; Published: 4 November 2020

Abstract: The Niger Delta Region is the largest river delta in Africa and features the fifth largest mangrove forest on Earth. It provides numerous ecosystem services to the local populations and holds a wealth of biodiversity. However, due to the oil and gas reserves and the explosion of human population it is under threat from overexploitation and degradation. There is a pressing need for an accurate assessment of the land cover dynamics in the region. The limited previous efforts have produced controversial results, as the area of western Africa is notorious for the gaps in the Landsat archive and the lack of cloud-free data. Even fewer studies have attempted to map the extent of the degraded mangrove forest system, reporting low accuracies. Here, we map the eight main land cover classes over the NDR using spectral-temporal metrics from all available Landsat data centred around three epochs. We also test the performance of the classification when L-band radar data are added to the Landsat-based metrics. To further our understanding of the land cover change dynamics, we carry out two additional assessments: a change intensity analysis for the entire NDR and, focusing specifically on the mangrove forest, we analyse the fragmentation of both the healthy and the degraded mangrove land cover classes. We achieve high overall classification accuracies in all epochs (~79% for 1988, and 82% for 2000 and 2013) and are able to map the degraded mangroves accurately, for the first time, with user's accuracies between 77% and 87% and producer's accuracies consistently above 82%. Our results show that mangrove forests, lowland rainforests, and freshwater forests are reporting net and highly intense losses (mangrove net loss: ~500 km^2; woodland net loss: ~1400 km^2), while built-up areas have almost doubled in size (from 1990 km^2 in 1988 to 3730 km^2 in 2013). The mangrove forests are also consistently more fragmented, with the opposite effect being observed for the degraded mangroves in more recent years. Our study provides a valuable assessment of land cover dynamics in the NDR and the first ever accurate estimates of the extent of the degraded mangrove forest and its fragmentation.

Keywords: Niger Delta Region; mangroves; land cover dynamics; intensity analysis; fragmentation; spectral-temporal metrics; land degradation; Landsat; ALOS PALSAR-2; JERS-1; GLCM

---

1. Introduction

Deltas are economic and environmental hot spots [1]. They take up less than 1% of the Earth's surface but are home to more than ca. 7% of the global population—a density more than 10 times the average [2]. Deltas are able to support such high human populations thanks to the high productivity, biodiversity, and the ability to use the waterways for transport. They are key contributors to the production of agricultural goods and are, therefore, highly important in the fight against global food insecurity [3]. However, these important systems are highly delicate and vulnerable. Tropical delta regions, in specific, are under risk of numerous threats, including sea level rise, extreme floods, storm surges, erosion, subsidence, and salinity intrusion, amongst others, which are expected to increase both in frequency and magnitude with the climate crisis [4]. These problems have been proven to increase out-migration rates and human security risks in developing regions, often inhabited by some of the poorest populations in the world [3]. Given the importance and the vulnerability of tropical deltas, monitoring and understanding the land cover dynamics in these regions is vital for achieving efficient policy planning and progress toward achieving the Sustainable Development Goals [5].

The Niger River Delta (NRD) is the largest river delta in Africa [6] and home to a rapidly increasing human population. It features the largest mangrove forest in Africa, estimated to be ~5% of the global mangrove coverage and the fifth largest mangrove forest in the world [7]. It is recognised as a highly important resource for the local communities, as it is utilised for fisheries, fuelwood, construction material, flood protection, medicinal purposes, recreation, and tourism, and holds an important spiritual value [8–13]. Substantial oil and gas deposits are found under the mangrove ecosystem of the NRD. Over the last decades, this highly significant ecosystem is under threat of loss or degradation, mainly due to oil and gas exploration activities, the overexploitation of the mangroves for fuelwood, urbanisation, and the invasion of the Nipa palm species (*Nypa fruticans*) [11,14–19]. Climate change [13,20], sea level rise [21], and coastal erosion [22] are also threats to the mangrove system. Despite the importance of the NDR resources, and the perceived degradation from anthropogenic and environmental pressures, reliable information on land cover dynamics and, particularly, on the extent and condition of the mangrove forest, is still lacking.

Assessing land cover dynamics over large areas is only possible via Earth Observation technologies, which is commonly done with multi-temporal Landsat data. The Landsat archive is truly invaluable as it constitutes the only global medium-scale data available for ~50 years. More 'traditional' approaches have used image mosaics or single images from single-sensor data to map two (before and after) dates and assess change from these [9,23–27]. However, over certain parts of the world, e.g., western and eastern Africa, the data archive has significant gaps [27,28]. Moreover, the use of optical data for accurately mapping and monitoring land cover dynamics over the tropics can be problematic due to the extensive cloud contamination, which renders the creation of image mosaics over large areas an unachievable task [29–31].

Recent advances in data availability, computing power, cloud computing, and algorithm development (e.g., machine and deep learning) have given rise to new approaches to multi-temporal assessments of land cover, e.g., image compositing [32], and spectral-temporal metrics [33,34]. The combination of optical and radar data has also been hailed as an important advancement in regional-scale land cover mapping as certain land cover types, such as mangroves and savannah woody vegetation, are mapped successfully using radar backscatter data, taking advantage of their ability to 'see' through cloud [19,35–41]. Over the last decade, object-based image analysis (OBIA) approaches have also been tested to successfully separate mangrove species from other coastal vegetation [42], to map the Amazonian mangrove belt [38], and to assess long-term variations of forest loss, fragmentation, and degradation using a combination of OBIA and spatial autocorrelation indicators [43].

There has been a limited number of studies that mapped land cover dynamics in the NDR [9,23,27,44] as the area is one of the most affected worldwide from the gaps in the Landsat archive and a consistent cloud contamination. With the exception of Nwobi et al. [19], these have employed 'traditional' remote sensing approaches and results have been contradictory. Even fewer studies have attempted to estimate the spatial extent of the degraded mangrove cover. Kuenzer et al. [27] used mosaics of Landsat images to map land cover change in the NDR over three dates but reported low per class classification accuracies for both the "tall mangrove" and the "degraded mangrove" classes, making area calculations unreliable. Salami et al. [45] compared the accuracies achieved by using Landsat ETM+, ASTER and NigeriaSat-1 data to map the six main land cover classes. For the mapping of degraded mangrove, they reported high accuracies for all three platforms. However, their study covered a small fraction of the NRD.

Based on the initial assessment of land cover transitions and dynamics, land cover change studies often move on to explain the changes in terms of explanatory variables (i.e., land use change drivers) or to forecast spatial patterns of future land cover under different scenarios (i.e., land use change models) [46–51]. The success of these next stages greatly depends on the ability to carry out an accurate initial assessment of the dynamics. Moreover, apart from the need to map land cover accurately, there is also a requirement to understand the dynamics more fully. For example, a simple comparison among the land cover maps does not determine whether the observed changes derive from processes that are systematically more intensive than random processes. Over the last years, new approaches have been suggested for characterising land cover change patterns quantitatively so that any potential subsequent analyses can focus more efficiently on the important patterns and processes of change, such as the intensity analysis proposed by Aldwaik and Pontius [52]. Other studies, with a specific interest on the fragmentation of habitats for example, have focused on the calculation of landscape metrics from the initial assessment of land cover. These studies have shown that the fragmentation of forests has detrimental effects for the health of the ecosystem and the services that it is able to provide [50,53,54]. A number of indices have been created to quantify landscape structure and spatial heterogeneity based on the composition and configuration of the landscape [55–58].

To date, no study related with the assessment of land cover change in the NDR has incorporated recent analytical approaches (e.g., intensity and fragmentation analyses) and the technological and algorithmic achievements (e.g., multi-sensor data, machine learning algorithms) to improve classification accuracies and our understanding of the land cover dynamics. Therefore, there is a need for a comprehensive study of land cover change in the region. In this paper, we aim to accurately assess the land cover dynamics in the NDR over the last decades, and improve our understanding of the extent of the degradation of the delta's mangrove forest. We will do so by:

- Mapping the main land cover types of the NDR in three epochs using Landsat data, spectral-temporal metrics, and a machine learning algorithm;
- Testing the performance of the classifier when radar L-band data are added to the Landsat;
- Assessing land cover change intensity over the two periods; and
- Quantifying the mangrove forest degradation and its fragmentation using landscape metrics.

2. Study Area

The Niger Delta is a flat alluvial plain located in Nigeria on the Gulf of Guinea (Figure 1). It is the largest river delta in Africa formed primarily by sediment deposition. It has a coastline of 470 km and consists of a number of ecological zones, including mangrove swamps, freshwater swamps, forests, and lowland rain forests. The Delta has two distinct seasons (wet and dry) with an average temperature of 27 °C throughout the year and annual rainfall of 3000 to 4500 mm [13]. The Niger Delta Region covers an area of 56,000 km^2 that consists of 7 administrative states (Abia, Akwa Ibom, Anambra, Bayelsa, Delta, Imo, and Rivers) and is home to more than 33 million inhabitants (265 people per km^2; [59]). More than 70% of these people depend on the natural environment for their livelihoods.

The NDR is considered a hot spot for biodiversity in the world with 3 sites designated as Ramsar Wetlands of International Importance [60]. It is a hub for oil and gas exploration, home to 80% of the refineries in Nigeria and extensive infrastructure (e.g., c. 900 oil wells, c. 100 flow stations and gas plants, c. 1500 km trunk lines, and c. 45,000 km flow lines) [61]. Nigeria's GDP, which rose from ~292 billion USD in 2009 to over 448 billion USD in 2019 [62], is mainly generated by the oil and gas sector. Yet, the NDR remains under-developed and its inhabitants impoverished. The Nigerian Land Use Act excludes the ownership of oil minerals by the state. This is perceived by many as socially inequitable, and has resulted in continuous instability in the region [63]. Additionally, more than

220 oil spills and 17 billion cubic metres of gas flares per year, together with the impacts of the human population explosion, have led to the degradation of the Niger Delta ecosystem [9,10,19,27,64].

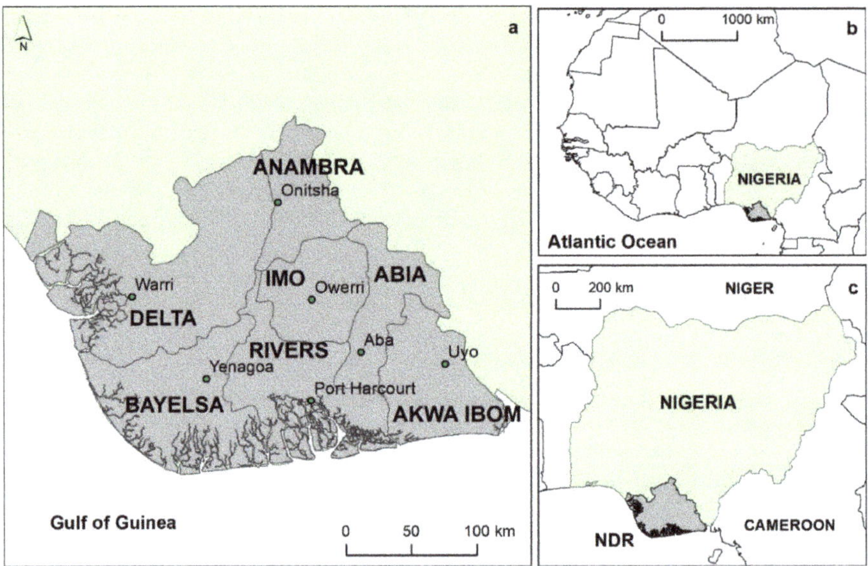

Figure 1. (**a**) Our delineation of the Niger Delta Region (comprising of the states of Abia, Akwa Ibom, Anambra, Bayelsa, Delta, Imo, and Rivers), and its location within (**b**) West Africa and (**c**) Nigeria.

3. Materials and Methods

We mapped the main land cover types in three epochs centred around 1988, 2000, and 2013, and assessed land cover change and change intensity in the two respective periods. The chosen classes were: Water, urban (i.e., built-up), woodland (i.e., lowland rainforest and freshwater forest), bareland, agricultural land, grassland, mangroves, and degraded mangroves. The choice of the classes was based on our knowledge of the area, the nomenclature used by ESA's 20 m land cover data for Africa and the 30 m-pixel Landsat-based GlobeCover30, and our desire to separate healthy mangroves from degraded ones. By definition, degraded is the land that has temporarily or permanently undergone a lowering of its capacity to deliver ecosystem services [65]. In the case of mangroves, the degraded forest has less biomass and tree cover, and is unable to provide a number of services at the same level as the healthy system, e.g., support for local livelihoods, carbon sequestration, erosion protection, provision of habitat for numerous fauna species, amongst others [66]. We also assessed the fragmentation of the mangrove forest during these two periods. Additionally, we tested the performance of the classifier when radar data are added to the optical. Figure 2 is a flowchart of our methodological framework.

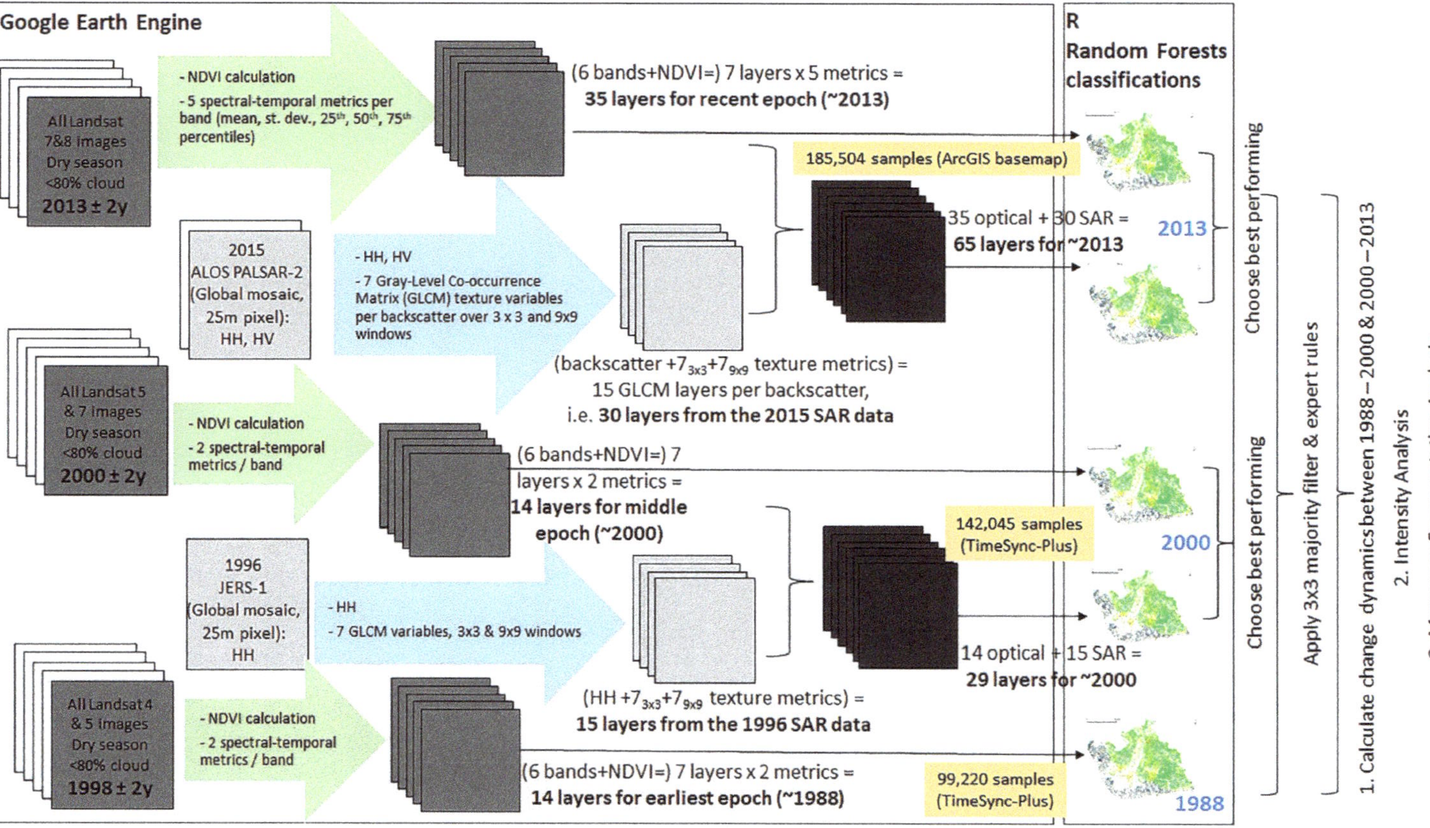

Figure 2. Methodological flowchart.

3.1. Data

3.1.1. Reference Data

Very high-resolution reference data were used for the recent epoch. This dataset is available as a MAXAR Vivid basemap within the ArcGIS software [67,68]. These cover the study area with data from November 2009 to January 2017. About 90% of the study area is covered with 46-cm-pixel data from GeoEye-1 (10 December 2010, 16 December 2011, 3 January 2013, 17 December 2013, 10 April 2014, 8 January 2015), 60-cm-pixel data from QuickBird-2 (11 February 2010, 3 October 2010, 12 June 2013), and 50-cm-pixel data from WorldView-2 (1 December 2011, 16 February 2013, 13 January 2014, 12 March 2015, 17 December 2015). Thanks to the familiarity with the study area, the broad land cover classes that were targeted in this paper were relatively easily identifiable on the very high-resolution imagery. This was also the case for the degraded mangroves, which presented the additional advantage of being spatially confined within the coastal zone, in general, and the mangrove system, in particular.

3.1.2. Landsat Data

The choice of Landsat data was driven by the need to coincide with as many other NDR studies as possible, so that comparisons could be drawn between them. Two such studies were identified: the one by Ayanlade and Drake [23] and the study by Kuenzer et al. [27]. The latter was particularly targeted, as it is the only one that has attempted to map the "degraded mangrove" class. The choice of the three epochs was also driven by the availability of the reference data and the SAR imagery.

We used all the dry season (December to February) Level 1 surface reflectance Landsat 4, 5, 7, and 8 images centred around 1988 (±2 years), 2000 (±2 years), and 2013 (±2 years) with less than 80% cloud cover from the USGS EROS Data Center for the eight WRS-2 tiles covering the study area (path 187, row57; p188, r55; p188, r56; p188, r57; p189, r55; p189, r56; p189, r57; p190, r56). Only the non-thermal bands were used, and clouds and cloud shadows were removed using F-mask [69,70]. Finally, the Normalised Difference Vegetation Index (NDVI) [71] was calculated. From the resulting 7-band image stacks (i.e., six non-thermal bands, plus the NDVI), spectral-temporal variability metrics were calculated [33,34,72,73]. For the recent epoch, five statistics for each of the seven bands were calculated: the standard deviation, the mean, and 3 percentiles (25th, 50th, and 75th). This brought the total layers for this epoch to 35. However, as data availability for the first two epochs was problematic (Figure 3), we limited the number of statistics per band to 2 (mean and st. dev.) and the total number of layers to 14.

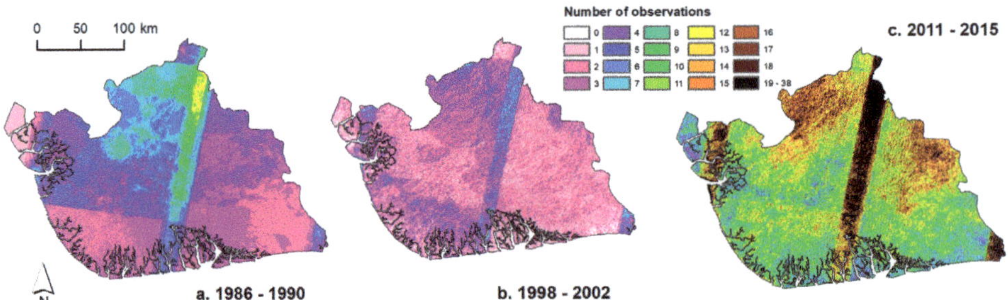

Figure 3. Number of available observations from the Landsat USGS Level 1 archive for (**a**) the first epoch; (**b**) the middle epoch, and (**c**) the more recent epoch.

3.1.3. Radar Data

Radar data were chosen for testing whether their addition to the optical metrics could improve the land cover classification. For the recent epoch, we employed the 2015 global 25 m resolution L-band Synthetic Aperture Radar data from the Advanced Land Observing Satellite-2 (ALOS-2) PALSAR-2 sensor via Google Earth Engine's API. The data are free and open access with two polarisations (HH and HV) and are currently available for 2015 to 2018. To increase the utility of the SAR data, we used Google Earth Engine to calculate a series of Gray-Level Co-Occurrence Matrix (GLCM) texture variables [72]. GLCMs are a series of localised texture metrics that quantify the statistical properties of a layer over a moving window [74]. We calculated seven GLCM layers (mean, variance, homogeneity, contrast, dissimilarity, entropy, and second moment) [75]. These statistics were calculated over both 3×3 and 9×9 windows, resulting in 15 layers per SAR backscatter (one backscatter + seven 3×3 GLCM layers + seven 9×9 GLCM layers), totalling 30 layers for the year 2015.

For the middle epoch, we acquired JAXA's 25 m resolution JERS-1 tropical region mosaics for the year 1996, the only year that such data are available over the Niger Delta Region. One polarisation is available (HH), from which we calculated 15 GLCM layers to use in the classification.

3.2. Land Cover Mapping

3.2.1. Sampling and Validation

In total, 185,504 samples were taken for the epoch centred around 2013. For the first and second epochs (i.e., 1988 and 2000), TimeSync-Plus v4.6 was used [76] to check for unchanged pixels at the 2013 sample locations. This resulted in 142,045 and 99,220 samples, respectively, for which we could confidently say that no change in the Landsat time series occurred. During classification, half of these samples were used for training and half for validation.

3.2.2. Image Classification & Post-Classification Processing

We developed the land cover classification using Random Forests classification models. Random Forests have been used successfully to classify Landsat imagery, thanks to their effective handling of correlated predictors and reduced tendency toward overfitting [77]. We used the 'RStoolbox' and 'randomForest' packages within the R statistical environment [78]. One optical only model was tested for the first epoch, while for the middle and most recent ones, we tested the performance of optical only and optical + SAR metrics (Figure 2). Based on the accuracies achieved, the outputs from the best performing models were chosen for the middle and more recent epochs. A 3×3 majority filter was applied to the outputs from all 3 epochs to get rid of the 'salt and pepper' effect of the classification. Finally, based on our knowledge of the study area, expert rules were applied to correct for some classification errors [72].

3.3. Intensity Analysis

Aldwaik and Pontius [52] devised a methodology that characterises patterns of land change quantitatively. It provides a mathematical framework that compares a uniform intensity to observed intensities of temporal changes among land cover classes (or 'categories') [79]. There are three levels of analysis, with each level exposing different types of information given the previous level of analysis. The first level, i.e., the interval level, examines how the size and speed of change vary across time intervals. The intensity of the rate of annual change is estimated using the following equations [52] (for notation, see Table S1 in the Supplementary Material):

$$S_t = \frac{area\ of\ change\ during\ interval\ [Yt\,,\ Yt+1]\,/\,area\ of\ study\ region}{duration\ of\ interval\ [Yt\,,\ Yt+1]} \times 100\%, \qquad (1)$$

$$U = \frac{area\ of\ change\ during\ all\ intervals\,/\,area\ of\ study\ region}{duration\ of\ all\ intervals} \times 100\%. \qquad (2)$$

The second level is called "category level" and it examines how the size and intensity of gross losses and gross gains in each land cover class vary across classes for each time interval. This level identifies which land cover classes are relatively dormant or active in each time interval [52]. Equations (3) and (4) provide the intensity of a class' annual gain and loss, respectively:

$$G_{tj} = \frac{area\ of\ gross\ gain\ of\ class\ j\ during\ [Yt\ ,\ Yt+1]/duration\ of\ [Yt\ ,\ Yt+1]}{area\ of\ class\ j\ at\ time\ Yt+1} \times 100\%, \quad (3)$$

$$L_{ti} = \frac{area\ of\ gross\ loss\ of\ class\ I\ during\ [Yt\ ,\ Yt+1]/duration\ of\ [Yt\ ,\ Yt+1]}{area\ of\ class\ i\ at\ time\ Yt} \times 100\%. \quad (4)$$

The third level, the "transition level", examines how the size and intensity of land cover class' transitions vary across the other classes that are available for that transition [52]. At each level, the method tests for stationarity of patterns across time intervals and identifies which land cover transitions are particularly intensive in a given period. Aldwaik and Pontius [52] provide a detailed explanation of the limitations concerning where the transition from a particular land cover class m to a class n can occur. For example, if a given land cover class n exists at a particular location at the initial time, then class n cannot gain at that place. If class n gains, then it must gain from locations that, initially, are <u>not</u> class n. If class n gains uniformly across the study area, then this class will gain from other classes, in proportion to the initial sizes of these land cover classes. Alternatively, class n might intensively avoid gaining from some particular class(es) and might intensively target gaining from some other class(es). Given the observed gross gain of class n, Equations (5) and (6) identify which other classes are intensively avoided versus targeted for gaining by class n in a given time interval:

$$R_{tin} = \frac{area\ of\ transition\ from\ i\ to\ n\ during\ [Yt\ ,\ Yt+1]/duration\ of\ [Yt\ ,\ Yt+1]}{area\ of\ class\ i\ at\ time\ Yt} \times 100\%, \quad (5)$$

$$W_{tn} = \frac{area\ of\ gross\ gain\ of\ class\ n\ during\ [Yt\ ,\ Yt+1]/duration\ of\ [Yt\ ,\ Yt+1]}{area\ that\ is\ not\ class\ n\ at\ time\ Yt} \times 100\%. \quad (6)$$

We used the *intensity.analysis* package in R to carry out the processing (https://cran.r-project.org/web/packages/intensity.analysis/vignettes/README.html).

3.4. Landscape Pattern Analysis

Post-classification comparison is most informative about changes in the composition of a landscape but gives us little—only visual—information about the spatial characteristics of these changes and the distribution of landscape elements. Landscape pattern analysis using landscape metrics provide us with additional information about the structure of changes, such as landscape fragmentation and patch aggregation or dispersion, as well as their changes in time. With the latter, we can observe changes in landscape spatial configuration through time.

We followed the approach used by Gounaridis et al. [53] and selected a number of class-level metrics [80] in order to study the changes in the spatial configuration and patterns of the 'mangrove' and 'degraded mangrove' land cover classes. We used 'Percentage of Landscape' (PLAND) as a measure of class abundance, and the 'number of patches' (NP), 'landscape patch index' (LPI), and 'patch area median' (AREA_MD) to study fragmentation of the classes of interest. With regard to patch shape analysis, we used the 'area weighted mean patch shape index' (SHAPE_AM), and for the aggregation of these classes, we used the 'area weighted mean Euclidean nearest neighbour distance index' (ENN_AM) along with its standard deviation (ENN_SD). Finally, we also used the aggregation index of 'percentage of like adjacencies' (PLADJ). Table 1 provides a listing of the selection of landscape metrics used in this study, together with a short description of their correlation with mangrove forest fragmentation. For more information, refer to McGarigal and Marks [80] who provide a full description of the metrics, including their mathematical formulas.

Table 1. Selection of landscape metrics used in this study with a short description of their relationship with mangrove forest fragmentation.

Name	Abbreviation	Description
Percentage of Landscape (%)	PLAND	Class percentage in landscape (proportional abundance)
Patch area median (ha)	AREA_MD	The median of patch areas in a class (a summary metric for the size of patches in the class, which is not influenced by very large patches)
Number of patches	NP	The number of patches in each class (simple measure of fragmentation)
Area weighted Mean Patch Shape Index	SHAPE_AM	Patch shape complexity at class level (indicative of changes at the edges)
Largest Patch Index (%)	LPI	Percentage of total landscape area occupied by the largest-sized patch (measure of dominance)
Percentage of like adjacencies (%)	PLADJ	The proportions of like adjacencies to the total number of adjacencies for the class' cells (aggregation)
Area weighted mean Euclidean nearest neighbour distance (m)	ENN_AM	Euclidean distance measured form patch edge to the closest patch edge from the same class (measures patch dispersion). Here we use the area weighted mean for the class to balance the influence of large patches.
Euclidean nearest neighbour distance Standard Deviation	ENN_SD	Measure of variation of ENN in the class (in comparison with the mean shows the form of distribution of patches in the class)

4. Results

4.1. Land Cover Mapping and Validation

Figure 4a–c are the outcomes of the classification of the metrics for the three epochs, and are accompanied by pie charts that summarise the proposition covered by each class. For the middle and latest epochs (Figure 4b,c), the combination of the optical with the SAR data produced slightly better results (Table 2) and were, therefore, the ones chosen for the subsequent analyses. The largest land cover class is by far woodland, which covers ~40% of the area (~23,000 km^2). Agricultural land is the second largest in all three time points (~12,000 km^2), while mangroves (degraded and non-degraded) and grassland occupy significant portions of the delta, too (~8000 km^2).

The classification results produced high overall accuracies of 79% (95% CI: ±3%), 83% (95% CI: ±3%), and 82% (95% CI: ±2.6%) for the three epochs, respectively (Table 2). Per-class accuracies (% correct, producer's and user's Accuracies; Table 2) were also high, with the exception of the bareland and grassland classes. The lower accuracies for these two types are attributed to the spectral confusion with the agricultural class: when fields are fallow, it gets confused with bareland, while when they are covered with vegetation, it is mostly confused with grassland (Tables S1–S5). The latter is also confused with woodland, as open woodland pixels contain a significant amount of spectral response from grasses.

Most importantly for the objective of this study, the mangrove class was mapped with high accuracy, with percentage correct and user's and producer's accuracies above 90% in all three time steps and models (Table 2). The degraded mangrove class was also mapped accurately, with producer's accuracies being consistently very high for all epochs and data combinations. However, there was some confusion between this class and the non-degraded mangroves (confusion matrices Tables S2–S6 in the Supplementary Material), resulting in lower user's accuracies, ranging from 77% to 79% for the first two time points (Table 2).

The inclusion of the SAR data in the classification of the more recent epochs generally improved the results but only slightly (Table 2). The most noteworthy improvements were achieved by the inclusion of the PALSAR-2-based metrics in the latest time point, with the user's accuracies of the water and urban classes improving by 4% (Table 2).

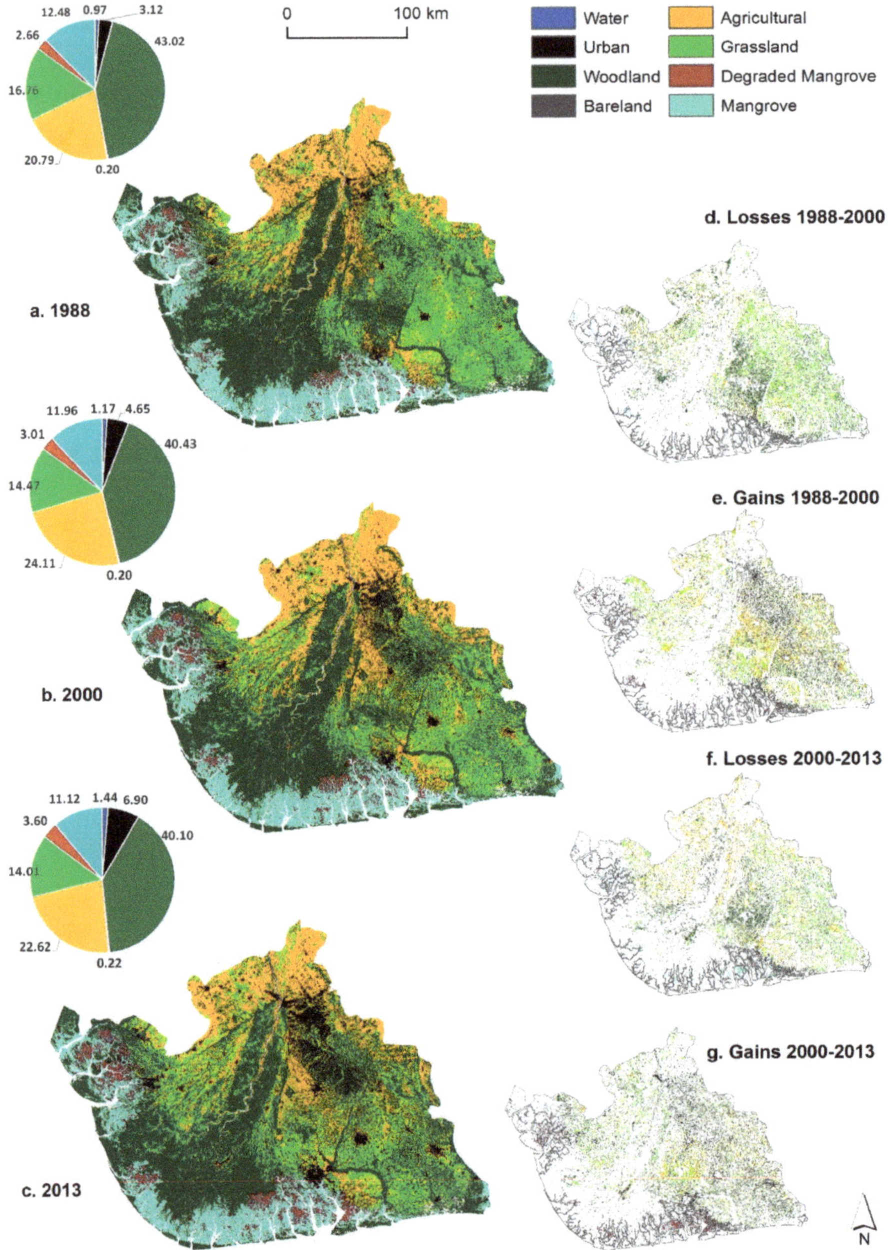

Figure 4. Land cover over the Niger Delta Region in (**a**) 1988, (**b**) 2000, and (**c**) 2013. Pie charts show the respective estimates of the area covered by each land cover type (%); scale bar corresponds to (**a**–**c**). Figures (**d**,**e**) are the losses and gains of each land cover type between 1988 and 2000; (**f**,**g**) the same for 2000–2013. The white background in (**d**–**g**) signifies persistence.

Table 2. Overall and per-class accuracy statistics for the three epochs (Wa: Water; U: Urban; Wo: Woodland; B: Bareland; A: Agricultural; G: Grassland; DM: Degraded Mangrove; M: Mangrove; CI: Confidence Interval; C = Correct; PA: Producer's Accuracy; UA = User's Accuracy).

	1988 Landsat			2000 Landsat			2000 Landsat + JERS-1			2013 Landsat			2013 Landsat + PALSAR-2		
Overall Accuracy	79.48			82.36			82.61			81.27			82.09		
95% CI	±0.003			±0.0029			±0.003			±0.0027			±0.0026		
	C	**PA**	**UA**	**C**	**PA**	**UA**	**C**	**PA**	**UA**	**C**	**PA**	**UA**	**C**	**PA**	**UA**
Wa	73	79	73	75	85	75	75	83	75	74	85	74	78	87	78
U	70	92	70	81	92	81	81	96	81	84	92	84	88	92	88
Wo	84	79	84	87	83	87	87	83	87	84	85	84	84	85	84
B	61	77	61	49	84	49	48	80	48	50	85	50	50	86	50
A	81	80	81	88	81	88	88	81	90	88	79	88	87	79	87
G	71	65	71	53	65	53	54	64	54	56	65	56	57	64	57
DM	77	82	77	78	86	78	79	85	79	86	82	86	87	82	87
M	91	90	91	90	90	90	91	90	91	90	92	90	90	93	90

4.2. Land Cover Change Dynamics

The three land cover maps were used to calculate the contingency matrix in Table 3. The matrix summarises, for the two periods, the area that has remained unchanged and the area and the type of change observed for each individual class. It also provides a summary of the area covered by each class in the beginning and in the end of each period as well as of the gains and losses they experienced. The spatial distribution of the latter is also illustrated in Figure 4d–g.

Table 3. Contingency matrix for the two periods of study representing stable (in bold) and changed areas in km^2. (a) 1988–2000; (b) 2000–2013. Wa: Water; U: Urban; Wo: Woodland; B: Bareland; A: Agricultural: G: Grassland; DM: Degraded Mangrove; M: Mangrove.

a — 2000 (km^2)

1988		Wa	U	Wo	B	A	G	DM	M	1988 total	Gross loss
	Wa	**395.70**	9.34	3.59	12.93	7.72	0.63	51.66	20.58	502.16	106.46
	U	4.30	**1444.71**	95.36	4.59	341.98	85.32	6.09	7.56	1989.91	545.20
	Wo	11.61	310.44	**193,54.71**	3.60	1655.90	2020.54	49.81	363.90	23,770.52	4415.81
	B	10.10	10.09	0.30	**72.67**	19.14	0.19	0.12	0.28	112.90	40.23
	A	20.52	543.09	647.15	39.38	**8868.25**	1439.74	7.48	5.89	11,571.48	2703.24
	G	0.55	572.51	2419.61	0.87	2883.36	**3534.56**	1.62	8.34	9421.41	5886.86
	DM	149.47	8.17	13.41	0.35	3.45	1.64	**1169.07**	454.69	1800.27	631.20
	M	40.90	26.06	536.28	0.64	8.09	6.00	535.47	**5743.70**	6897.15	1153.45
2000 Total		633.15	2924.41	230,70.43	135.03	13,787.88	7088.63	1821.33	6604.94		
Gross Gain		237.46	1479.70	3715.71	62.36	4919.64	3554.07	652.25	861.24		

b — 2013 (km^2)

2000		Wa	U	Wo	B	A	G	DM	M	2000 Total	Gross Loss
	Wa	**522.59**	2.09	4.91	10.26	4.16	0.48	76.77	13.16	634.42	111.83
	U	19.64	**2150.29**	173.38	18.35	357.87	184.36	10.72	10.19	2924.80	774.51
	Wo	10.12	371.43	**18,959.22**	21.03	1251.56	2038.85	67.20	351.52	230,70.92	4111.70
	B	58.00	5.09	1.99	**57.13**	12.33	0.38	0.09	0.09	135.10	77.97
	A	25.86	933.43	939.59	46.28	**9083.42**	2754.41	3.55	1.58	13,788.12	4704.71
	G	1.34	253.81	1784.19	5.44	1922.06	**3113.85**	5.57	2.38	7088.64	3974.79
	DM	157.76	3.64	26.83	5.21	4.52	2.07	**1158.61**	462.79	1821.42	662.81
	M	65.30	7.86	377.77	14.21	7.68	7.84	595.84	**5529.72**	6606.22	1076.50
2013 Total		860.61	3727.63	22267.88	177.91	12,643.6	8102.24	1918.34	6371.43		
Gross Gain		338.02	1577.34	3308.66	120.78	3560.18	4988.39	759.74	841.71		

4.3. Intensity Analysis

The interval level of the intensity analysis identifies the time interval in which the overall annual rate of change is faster. The total change in both intervals was found to be relatively similar: ~17% of the total area in the first period and ~15% in the second. However, the intensity of the annual area of change in the first interval is faster than in the second (1.42% and 1.16%, respectively; Figure 5). The output of Equation (2) is 1.28%, depicted as a dashed line in Figure 5. Compared to this value, the rate in the first period is considered 'fast', while in the second 'slow'.

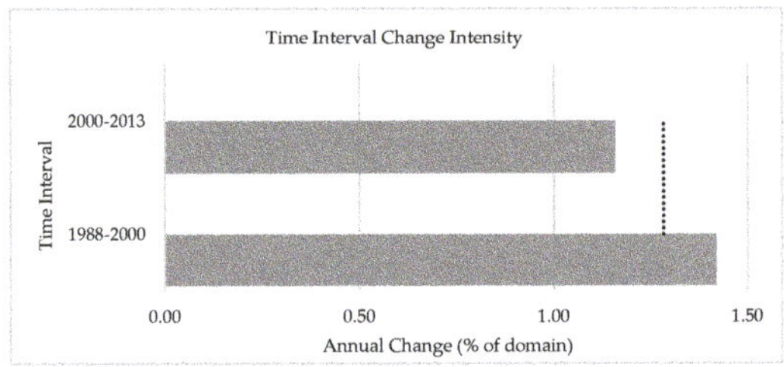

Figure 5. Intensity of the annual area of change within the two time intervals of the study. The dashed line is the uniform line (i.e., the output of Equation (2)).

Figure 6 is the graphical representation of the 'category level' of the intensity analysis. Figure 6a,c depict the size of the annual gain of loss of each land cover class in the first and the second period, respectively. Figure 6b,d show the intensity for a class' annual gain or loss, as calculated by Equations (3) and (4). The two dashed lines show the output of Equation (1) for each period, i.e., the uniform line for each period at this category intensity level [52]. When an intensity bar remains to the left of the uniform (dashed) line, then the change is relatively dormant for that land cover class and period. On the contrary, if the bar extends to the right of the dashed line, then the change is relatively active for that class and period. If, for a given land cover class, the intensity of the gains or losses remain active or dormant during all study periods, then the specific type is considered stationary.

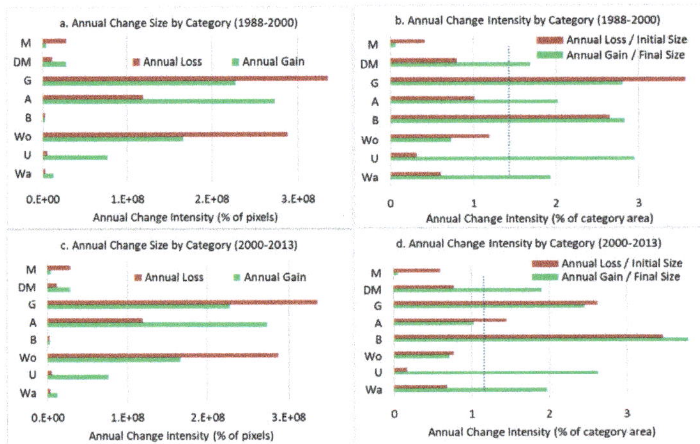

Figure 6. Category intensity analysis for the two periods. (**a,c**): gross annual area of gains and losses. (**b,d**): intensity of annual gains and losses within each land cover category. "# of elements" is the number of pixels. The dashed lines in (**b,d**) signify the uniform intensity value. Wa: Water; U: Urban; Wo: Woodland; B: Bareland; A: Agricultural: G: Grassland; DM: Degraded Mangrove; M: Mangrove.

At the transition level, the intensity analysis identifies which transitions are more intensive in a given time interval. Given the scope of the present paper and the need to keep the presentation of the results as succinct as possible, Table 4 summarises the results only for the transition from mangrove to any other class for the two periods. The outcome for all the other transitions is provided in Tables S7 and S8 of the Supplementary Material.

Table 4. Transition level intensity analysis FROM-Mangrove TO-all other classes (1988–2000 and 2000–2013). In bold and underlined: targeted classes (compared to uniform). Deg.: Degraded.

Transitions FROM	Mangrove			
Time Interval	1988–2000		2000–2013	
TO Category	Observed Annual Transition (km^2)	Transition Intensity % of 2000 Category	Observed Annual Transition (km^2)	Transition Intensity % of 2013 Category
Water	206	0.03	332	0.05
Urban	717	0.03	540	0.02
Woodland	506	0.00	1431	0.01
Bareland	40	0.04	221	**0.20**
Agricultural	485	0.00	298	0.00
Grassland	244	0.00	461	0.01
Deg. Mangrove	23,799	**1.57**	32,742	**1.80**

4.4. Landscape Pattern Analysis

Figure 7 depicts the evolution of the selected landscape metrics through time for the healthy and the degraded mangroves classes.

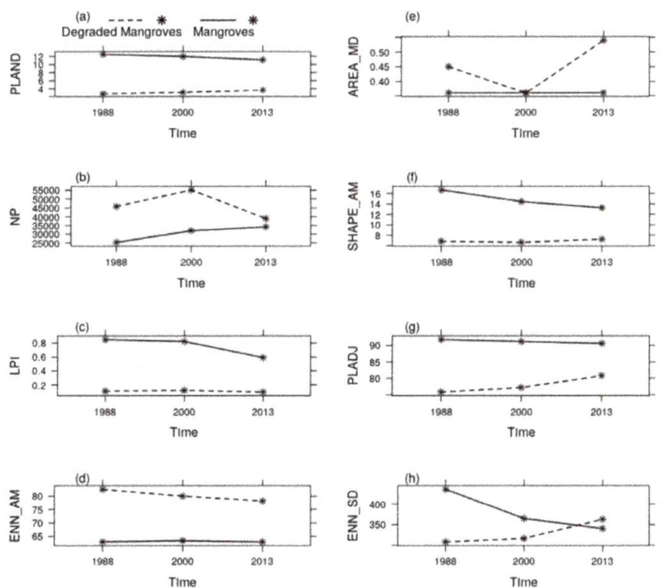

Figure 7. Landscape metrics for the mangrove and degraded mangrove classes. (**a**) Percentage of Landscape (%; PLAND); (**b**) Number of patches (NP); (**c**) Largest Patch Index (%; LPI); (**d**) Area weighted mean Euclidean nearest neighbour distance (m; ENN_AM); (**e**) Patch area median (ha; AREA_MD); (**f**) Area weighted Mean Patch Shape Index (SHAPE_AM); (**g**) Percentage of like adjacencies (%; PLADJ); (**h**) Euclidean nearest neighbour distance Standard Deviation (ENN_SD).

5. Discussion

Accurate and reliable information of land cover dynamics is essential for the sustainable management of tropical deltas and mangrove ecosystems and their capacity for ecosystem service provision. The 'traditional' remote sensing mapping approach involving the use of image mosaics of optical data from two dates, together with likelihood function maximisation image classification algorithms, is not reliable in the humid tropics due to cloud cover [29,31], data availability [27,28], and algorithm performance. This has led to conflicting land cover change estimates for the largest river delta in Africa and the failure to assess the extent of degradation of one of the most endangered ecosystems in the world [60]. Our results show that, by incorporating novel image compositing techniques, spectral-temporal metrics, and machine learning classification algorithms, a reliable assessment of the change dynamics over the Niger Delta Region can be made. Our accurate land cover estimates also allowed for a more comprehensive land change analysis that incorporates an assessment of change intensity and the fragmentation of a key component of the NDR: its mangrove forests.

5.1. Land Cover and Change Dynamics

There is an inherent difficulty in mapping land cover in tropical deltas, in general, and mangrove forests, in particular, as they are affected by seasonal and intertidal effects, with pixels often comprising of a mixture of vegetation, soil, and water due to their location between land and sea and the average tidal range in the Niger Delta being 1.5 m [9]. Nevertheless, we mapped the eight main land cover types for the entire NDR, achieving high overall accuracies in all epochs (~79% for 1988, and 82% for 2000 and 2013; Table 2) and high producer's accuracies for all classes and years. With the exception of the grassland and bareland classes, user's accuracies were also high (from 70% to 91%). Our results compare favourably with other studies in the NDR [19,23,44,45]. Regarding the mapping of degraded mangroves, one of the main objectives of this paper, our study is the first to map this accurately with user's accuracies between 77% and 87% and producer's consistently above 82%. The only other study that attempted to map degraded mangroves reported very low accuracies [27].

The results reveal some interesting dynamics:

- There is consistent net loss in mangrove and woodland types and a consistent net gain of the urban class in both periods of study
- The area covered by non-degraded mangroves was reduced by ~250 km^2 in each period (=Gross Loss – Gross Gain)
- About 10% of mangroves are degraded in each interval, and an additional 34 km^2 of mangrove were converted to urban land use in both periods
- A portion of degraded mangrove is able to bounce back into its healthier state
- The net loss for the woodland class was more than 700 km^2 in each period. A part of this class is converted to grasses (~8% and ~9%) and to agricultural land (~7% and ~5%)
- A quarter of the area mapped as grassland in the initial dates is converted to woodland by the end date
- The built-up areas increased by 47% (~900 km^2) in the first period, an area larger than the size of New York City. In the second period, the increase was smaller (~800 km^2) but still it amounted to 27% of the area covered in 2000

More specifically, according to our findings, healthy mangroves reported a net loss in both study periods: 292 km^2 in the first and 235 km^2 in the second, while degraded mangroves consistently reported a net gain (21 km^2 in the first and 97 km^2 in the second). Interestingly, our study and the studies by Kuenzer et al. [27] and James et al. [9] found a similar decrease in the overall combined (degraded and non-degraded) mangrove area. According to our results, this area was 270 km^2, while, in an almost identical period of study, Kuenzer et al. [27] found that the loss was 239 km^2. In the James et al. [9] study between 1987 to 2002, the loss was 213 km^2. However, our more accurate findings identify the total areas covered by the mangrove classes to be very different to the areas in the Kuenzer et al. [27] study: we found that mangroves and degraded mangroves occupied an area between 8697 and 8428 km^2 in the two periods, while Kuenzer et al. [27] claim that these numbers were 10,311 and 10,072 km^2, respectively. These figures differ by almost a fifth, and can play a significant role in the setting of conservation targets, management policies, and sustainability goals. Moreover, our mangrove results compare favourably with three studies that mapped mangroves as one class accurately: the study of Nwobi et al. [19], who found that mangroves occupied an area of 9115 km^2 in 2007 and 8017 km^2 in 2017; the study of Ayanlade and Drake [23] (9965 km^2 in 1987, 9255 km^2 in 2001, and 8430 km^2 in 2011); and the study by James et al. [9] (7037 km^2 in 1987 and 6824 km^2 in 2002).

While it is relatively simple to compare the results on the extent of mangroves between the different studies that mapped land cover change in the NDR, as this class is confined in the coastal belt and is always included within the study area, it is not as straightforward to compare the findings on other land cover types, as the study areas do not match. In the case of woodland, for example, the biggest land cover type in the NDR, our study found that it occupied 23,770 km^2 in 1987 and suffered net losses in both periods: ~700 km^2 in the first and ~800 km^2 in the second. The study by Ayanlade and Drake [23] also found net losses in both periods for the combined "lowland rainforest" and "freshwater forest" classes but found that these occupied 31,200 km^2 in 1987, 25,400 km^2 in 2001, and 21,470 km^2 in 2011. However, their study area far exceeds the boundaries of our delineation of the NDR. The study by Kuenzer et al. [27] also agrees that "forest" and "swamp forest" experienced net losses in both periods. They report far smaller areas than both our study and the study by Ayanlade and Drake [23]: 18,325 km^2 in 1987 and 15,408 km^2 in 2013. Finally, the Nwobi et al. [19] study also agrees that "tropical forests" were reduced but reported that these occupied 29,000 km^2 in 2007 and 25,500 km^2 in 2017. As all of these studies, including ours, reported high per-class accuracies in the mapping of forests, it is difficult to ascertain which on is closer to the true figure.

The difficulty in comparing the findings of different studies remains for the agricultural class, which we found to significantly increase in the first period (from 11,571 to 13,787 km^2) and decrease in the second (12,645 km^2 in 2013). An additional issue to the problem of relating to different study areas around the NDR is the choice of land cover nomenclature. Based on our knowledge of the region and on the classification systems of the ESA 20m African land cover data for 2016 and the GlobeLand 30 m data for 2010, we included a grassland class in our mapping efforts, which were found to decrease in the first period (from 9421 to 7089 km^2) and increase in the second (8102 km^2 in 2013). Our figures for the agricultural class are significantly lower to those in Ayanlade and Drake [23], Kuenzer et al. [27], and Nwobi et al. [19]. However, none of these studies included a separate class for grassland but, according to their spatial outputs, appear to have mapped this together with the agricultural class. We recognise that separating these classes poses difficulties, as the spectral separability between them is low: our user's accuracies for grassland are testament to that (Table 2). However, we strongly believe that it is a shortcoming to map these two classes as one, as this precludes the identification of very important land cover dynamics between either of these classes and, for example, the woodland or urban classes. If summed together, our estimates of agricultural and grassland compare favourably with those of Nwobi et al. [19], who estimated the area covered by "agricultural land" as 21,733 km^2 in 2007 and 24,179 km^2 in 2017.

An important change that occurred in both periods is the expansion of the built-up areas: from 1990 km^2 in 1988, to 2924 km^2 in 2000, to 3728 km^2 in 2013, i.e., an 87% increase. As in the previous land cover types, the difference in the extent of the study area makes comparison to the other studies difficult. For example, the Ayanlade and Drake [23] study reports much higher figures, but their study includes the city of Benin, the fourth largest Nigerian city, which lies outside of our delineation of the NDR. Similarities exist between our findings and the Nwobi et al. [19] study: their 'built-up-area' class occupied 3950 km^2 in 2007 and 5938 km^2 in 2017. Their higher estimates can be attributed to the fact that they include the city of Calabar and a number of built-up areas in the northeast of their study area that lie outside our delineation of the NDR.

According to the results of our intensity analysis (Figure 6b), in the first period of study, only mangroves and woodland demonstrated dominant gains, while all the other categories had active gains. Interestingly, only the grassland and bareland types had active annual change intensities, with the former having the largest size of losses in this period (Figure 6a). However, these two are the classes that scored lower user's accuracies and the respective intensity results need to be treated with caution. Notable results from this period are the ~5 times greater annual intensity of mangrove loss than gain and the ~10 times greater annual intensity of urban gain than loss. The intensity of agricultural expansion is also noteworthy, reporting ~2 times greater gain than loss.

In the first period, the land cover class that mangroves 'target' most intensively when they change is degraded mangroves, with a transition intensity of 1.57% of the total area of degraded mangroves in the end of the first period. This is much higher than the estimated uniform change intensity of 0.06%. An area of 535 km^2 of mangroves was degraded by the year 2000. In the second period, this change is even more intense (1.80%, higher than the uniform intensity of 0.08%) and leads to a conversion of a total of 596 km^2 of mangrove to degraded mangrove by 2013. Bareland is also found to be a targeted class for mangroves with an estimated transition intensity of 0.20% (221 km^2). Water also targets bareland, as well as mangroves and degraded mangroves, with transition intensities higher than the estimated uniform change intensity. As this is the first paper to undertake an intensity analysis in the NDR, we are unable to compare our findings to existing studies.

5.2. Fragmentation and Degradation of the Niger Delta Mangrove Forest

The Niger Delta's mangrove forest is a hub for substantial oil and gas deposits. As a consequence, it is highly vulnerable to activities of oil and gas extraction, e.g., land clearing, dredging, construction of flow stations, pipe and seismic lines, well blowouts, leakages or corrosion, equipment failure, error during operation or maintenance, accidents during transportation, sabotage, etc., as well as urbanisation, selective logging, and the proliferation of the invasive Nipa palm species (*Nypa fruticans*) that lead to the forest's destruction, fragmentation, and degradation [9,10,19,27,64].

Our land cover change and intensity analyses showed that degraded mangroves increased in both periods of study and mangroves losses were 5 times more intense than gains. To further assess the condition of the Niger Delta mangrove forest, we carried out the first ever fragmentation analysis of the area. Our fragmentation results show that the 'number of patches' (NP) for the healthy mangroves increased persistently while the 'total percentage of landscape' (PLAND) decreased (Figure 7a,b). The 'largest patch index' (LPI), a measure of dominance (Figure 7c), shows that in the second period, larger patches are on a decrease. The 'area weighted mean shape index' (SHAPE_AM; Figure 7f) is also decreasing for the healthy mangroves, in both periods: this indicates that changes are happening in the perimeter of patches, uniformly. The 'area weighted mean Euclidean nearest neighbour distance' index (ENN_AM; Figure 7d) is slightly decreasing, indicating less dispersion of the healthy mangrove patches. The standard deviation of this index (ENN_SD; Figure 7h) is decreasing but with high values compared to the mean, which indicates a more uneven distribution of patches. The high and steady values of PLADJ (Figure 7g) confirm the ENN results: the healthy mangrove patches remain relatively aggregated throughout the study period. This was expected, as mangroves are very localised within the delta and naturally only occur by the coast.

Figure 7 also shows the change in landscape metrics through time for the degraded mangroves. The size of this class (PLAND; Figure 7a) is constantly increasing but shows some fluctuation in the number of patches (NP; Figure 7b). A divergent pattern is observed in the evolution of the number of patches and the median of patch area metrics (AREA_MD; Figure 7e): NP increases in the first period and AREA_MD decreases, while in the second period, this is reversed. The latter means that this class becomes less fragmented, with more patches and lower patch size in the first period. Between 2000 and 2013, there are fewer patches and larger patch sizes, indicating that some of the first period's patches have merged to form larger ones.

A visual examination of the land cover maps and derived change maps from these revealed three areas that demonstrate higher concentrations of degraded mangrove. One such area is in the eastern part of the NDR, around the city of Port-Harcourt and the towns of Bonny, Okrika, and Degema (Figure 8a). Mangrove degradation here can be attributed to the effects of rapid urbanisation and oil extractive activities [14,17], as demonstrated by the overlap with the locations of the oil wells, the pipelines, and the oil spills in Figure 8a. At the central part of the study area, mangrove degradation is mainly due to oil spills resulting from crude oil extractive activities, notably near River Bayelsa and the towns of Nembe, Southern Ijaw, Ekeremor, Brass, and Oloibiri, where oil extraction first began as early as the 1950s (Figure 8b). The highest concentration of degraded mangroves is, however, in the western part of the NDR, in the Delta state (Figure 8c). This area shows widespread degradation, with a notable increase in the second and third date around the towns of Wari South and Wari South West. Several oil spill and gas incidents have been reported in the literature around this area and period [14,15,17,18].

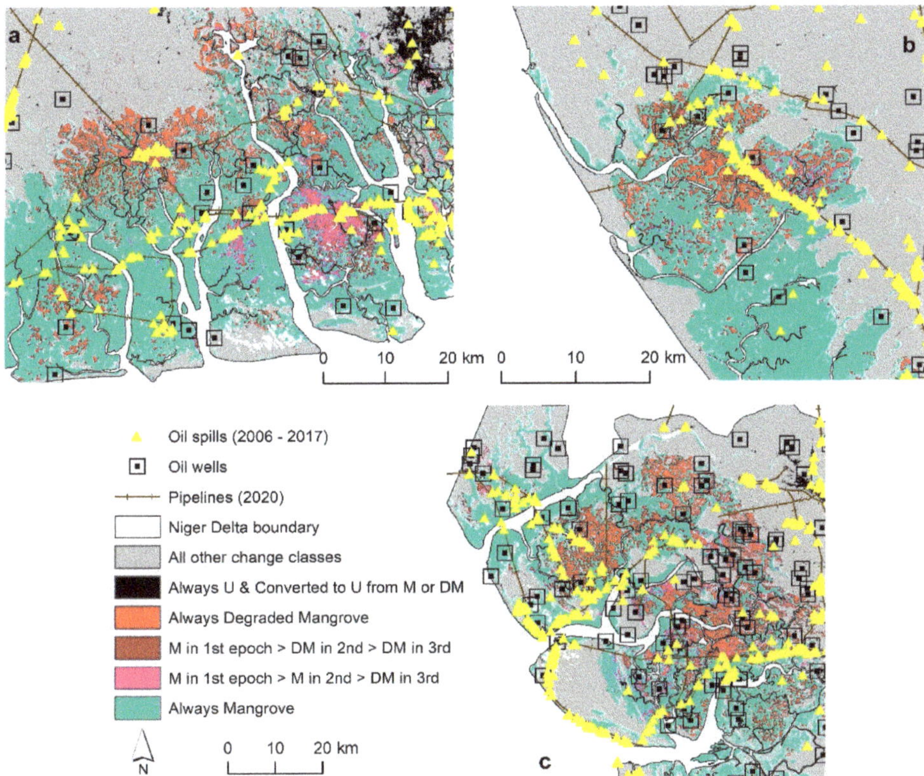

Figure 8. Oil wells, pipelines, oil spills, and mangrove degradation hotspots in three parts of the study area: (**a**) the eastern area, around the city of Port-Harcourt; (**b**) the central area, near the river Bayelsa, and (**c**) the western area around the cities of Wari South and Wari South West. U: Urban; M: Mangrove; DM: Degraded Mangrove. (Oil spill data: https://www.nosdra.gov.ng and https://oilspillmonitor.ng. Oil wells and pipeline data: https://www.shell.com.ng).

6. Conclusions

The Niger Delta Region (NDR) is an important ecosystem, providing numerous services to the millions of its human inhabitants. Despite its undisputable importance, it is under threat of degradation, mainly due to human pressure, and especially as a direct consequence of the activities related with the significant oil and gas reserves in the region. Understanding the extent of the problem requires an accurate assessment of the land cover dynamics in the region, which can only be achieved through the use of state-of-the-art remote sensing technologies and analytical techniques. Cloud contamination and gaps in the commonly employed Landsat archive makes this a fathomable task.

Here, we were able to accurately assess the land cover dynamics over a period of 25 years using the Google Earth Engine cloud computing platform to estimate spatial-temporal Landsat-based metrics in three epochs. Our results showed that mangroves, the lowland rainforests, and the freshwater forests have demonstrated a net loss, while the built-up areas have almost doubled in the period of study. By performing a land cover change intensity analysis, we were also able to demonstrate how highly intense these changes were. We also tested the ability of L-band SAR data in improving the Random Forests classifications of the main land cover types in the delta and found that these only improve the mapping of the urban and water classes, provided that more than one polarisation is available. Our results provide a valuable quantification of the land cover dynamics in the NDR and the first ever accurate assessment of the spatial extent of the degraded mangroves in the region. Such assessments are imperative for successfully addressing a number of the Sustainable Development Goals and achieving Land Degradation Neutrality by 2030, as envisaged by the United Nations LDN Target Setting Programme.

Supplementary Materials: The following are available online at http://www.mdpi.com/2072-4292/12/21/3619/s1, Table S1: Mathematical notation for Intensity Analysis [79], Table S2: Confusion matrix of the classification of the Landsat-based metrics centred around the year 1988, Table S3: Confusion matrix of the classification of the Landsat-based metrics centred around the year 2000, Table S4: Confusion matrix of the classification of the Landsat- and JERS-1-based metrics centred around the year 2000, Table S5: Confusion matrix of the classification of the Landsat-based metrics centred around the year 2013, Table S6: Confusion matrix of the classification of the Landsat-and ALOS PALSAR-2-based metrics centred around the year 2013, Table S7: Transition level intensity analysis FROM-class TO-class for 1988–2000 and 2000–2013 (all classes except Mangrove, which appears in Table 4), Table S8: Transition level intensity analysis TO-class FROM-class for 1988–2000 and 2000–2013.

Author Contributions: Conceptualization, I.I.N., E.S., S.K., G.C. and S.M.; methodology I.I.N., E.S., S.K. and T.P.H.; software: I.I.N., E.S., S.K. and T.P.H.; writing—original draft preparation: I.I.N., E.S., and S.K.; writing—review and editing: I.I.N., E.S., and T.P.H.; supervision: E.S., G.C. and S.M. All authors have read and agreed to the published version of the manuscript.

Funding: Iliya I. Nababa was funded by Petroleum Technology Development Fund (PTDF), Nigeria (grant PTDF/ED/PHD/IIN/789/15).

Acknowledgments: The authors are grateful to the USGS for the Landsat data, to JAXA for the ALOS PALSAR-2 and JERS-1 data and to Google Earth Engine for providing access to the data and the processing environment. Oil spill data were obtained from the National Oil Spill Detection and Response Agency (NOSDRA), Nigeria (https://www.nosdra.gov.ng/ and https://oilspillmonitor.ng/). Oil wells were digitised in ArcMap 10.7 [67,68] using map of oil and gas infrastructure obtained from Shell Petroleum Development Company, Nigeria (https://www.shell.com.ng/). Pipeline data were digitised from very resolution basemap imagery in ArcMap 10.7 [67,68] and an oil and gas infrastructure map obtained from Shell Petroleum Development Company, Nigeria https://www.shell.com.ng/.

Conflicts of Interest: The authors declare no conflict of interest.

References

1. Foufoula-Georgiou, E. A vision for a coordinated international effort on delta sustainability. *Deltas Landf. Ecosyst. Hum. Act.* **2013**, *358*, 3–11.
2. Ericson, J.P.; Vorosmarty, C.J.; Dingman, S.L.; Ward, L.G.; Meybeck, M. Effective sea-level rise and deltas: Causes of change and human dimension implications. *Glob. Planet. Chang.* **2006**, *50*, 63–82. [CrossRef]
3. Szabo, S.; Renaud, F.G.; Hossain, M.S.; Sebesvári, Z.; Matthews, Z.; Foufoula-Georgiou, E.; Nicholls, R.J. Sustainable development goals offer new opportunities for tropical delta regions. *Environ. Sci. Policy Sustain. Dev.* **2015**, *57*, 16–23. [CrossRef]

4. Szabo, S.; Brondizio, E.; Renaud, F.G.; Hetrick, S.; Nicholls, R.J.; Matthews, Z.; Tessler, Z.; Tejedor, A.; Sebesvari, Z.; Foufoula-Georgiou, E. Population dynamics, delta vulnerability and environmental change: Comparison of the Mekong, Ganges–Brahmaputra and Amazon delta regions. *Sustain. Sci.* **2016**, *11*, 539–554. [CrossRef] [PubMed]

5. Chow, J. Mangrove management for climate change adaptation and sustainable development in coastal zones. *J. Sustain. For.* **2018**, *37*, 139–156. [CrossRef]

6. Goudie, A.S. The drainage of Africa since the cretaceous. *Geomorphology* **2005**, *67*, 437–456. [CrossRef]

7. Spalding, M. *World Atlas of Mangroves*; Routledge: Abingdon-on-Thames, UK, 2010.

8. Zabbey, N.; Hart, A.; Erondu, E. Functional roles of mangroves of the Niger Delta to the coastal communities and national economy. In Proceedings of the 25th Annual Conference of the Fisheries Society of Nigeria (FISON), Lagos, Nigeria, 25–29 October 2010.

9. James, G.K.; Adegoke, J.O.; Saba, E.; Nwilo, P.; Akinyede, J. Satellite-based assessment of the extent and changes in the mangrove ecosystem of the Niger Delta. *Mar. Geod.* **2007**, *30*, 249–267. [CrossRef]

10. Okonkwo, C.N.P.; Kumar, L.; Taylor, S. The Niger Delta wetland ecosystem: What threatens it and why should we protect it? *Afr. J. Environ. Sci. Technol.* **2015**, *9*, 451–463.

11. Numbere, A. Impact of Hydrocarbon Pollution on the Mangrove Ecosystem of the Niger River Delta, Nigeria. Ph.D. Thesis, Saint Louis University, Saint Louis, MO, USA, 2014.

12. NDDC. *Niger Delta Regional Development Master Plan*; Niger Delta Development Commission: Port Harcourt, Nigeria, 2006; pp. 48–99.

13. World Bank. Defining an Environmental Development Strategy for the Niger Delta, Nigeria. Available online: http://documents.worldbank.org/curated/en/506921468098056629/pdf/multi-page.pdf (accessed on 18 October 2017).

14. Kadafa, A.A. Oil Exploration and Spillage in the Niger Delta of Nigeria. *Civil. Environ. Res.* **2012**, *2*, 38–51.

15. Balogun, T.F. Mapping impacts of crude oil theft and illegal refineries on mangrove of the Niger Delta of Nigeria with remote sensing technology. *Mediterr. J. Soc. Sci.* **2015**, *6*, 150. [CrossRef]

16. Onyena, A.P.; Sam, K. A review of the threat of oil exploitation to mangrove ecosystem: Insights from Niger Delta, Nigeria. *Glob. Ecol. Conserv.* **2020**, *22*, e00961. [CrossRef]

17. Duke, N.C. Oil spill impacts on mangroves: Recommendations for operational planning and action based on a global review. *Mar. Pollut. Bull.* **2016**, *109*, 700–715. [CrossRef] [PubMed]

18. Twumasi, Y.A.; Merem, E.C. GIS and remote sensing applications in the assessment of change within a coastal environment in the Niger Delta region of Nigeria. *Int. J. Environ. Res. Public Health* **2006**, *3*, 98–106. [CrossRef]

19. Nwobi, C.; Williams, M.; Mitchard, E.T.A. Rapid Mangrove Forest Loss and Nipa Palm (Nypa fruticans) Expansion in the Niger Delta, 2007–2017. *Remote Sens.* **2020**, *12*, 2344. [CrossRef]

20. Uyigue, E.; Agho, M. *Coping with Climate Change and Environmental Degradation in the Niger Delta of Southern Nigeria*; Community Research and Development Centre Nigeria (CREDC): Benin City, Nigeria, 2007.

21. Okali, D.; Eleri, E.O. *Climate Change and Nigeria: A Guide for Policy Makers*; Nigerian Environmental Study Action Team (NEST): Ibadan, Nigeria, 2004.

22. Awosika, L.F. Impacts of Global Climate Change and Sea Level Rise on Coastal Resources and Energy Development in Nigeria. In *Global Climate Change: Impact on Energy Development*; Umolu, J.C., Ed.; DAMTECH Nigeria Limited: Jos, Nigeria, 1995.

23. Ayanlade, A.; Drake, N. Forest loss in different ecological zones of the Niger Delta, Nigeria: Evidence from remote sensing. *Geojournal* **2016**, *81*, 717–735. [CrossRef]

24. Mena, C.F. Trajectories of land-use and land-cover in the northern Ecuadorian Amazon: Temporal composition, spatial configuration, and probability of change. *Photogramm. Eng. Remote Sens.* **2008**, *74*, 737–751. [CrossRef]

25. Gao, J.; Liu, Y.S. Determination of land degradation causes in Tongyu County, Northeast China via land cover change detection. *Int. J. Appl. Earth Obs. Geoinf.* **2010**, *12*, 9–16. [CrossRef]

26. Obiefuna, J.N.; Nwilo, P.C.; Atagbaza, A.O.; Okolie, C.J. Land Cover Dynamics Associated with the Spatial Changes in the Wetlands of Lagos/Lekki Lagoon System of Lagos, Nigeria. *J. Coast. Res.* **2013**, *29*, 671–679. [CrossRef]

27. Kuenzer, C.; van Beijma, S.; Gessner, U.; Dech, S. Land surface dynamics and environmental challenges of the Niger Delta, Africa: Remote sensing-based analyses spanning three decades (1986–2013). *Appl. Geogr.* **2014**, *53*, 354–368. [CrossRef]

28. Kirui, K.B.; Kairo, J.G.; Bosire, J.; Viergever, K.M.; Rudra, S.; Huxham, M.; Briers, R.A. Mapping of mangrove forest land cover change along the Kenya coastline using Landsat imagery. *Ocean Coast. Manag.* **2013**, *83*, 19–24. [CrossRef]

29. Martinuzzi, S.; Gould, W.A.; González, O.M.R. *Creating Cloud-Free Landsat ETM+ Data Sets in Tropical Landscapes: Cloud and Cloud-Shadow Removal*; Gen. Tech. Rep. IITF-32; U.S. Department of Agriculture, Forest Service, International Institute of Tropical Forestry: Washington, DC, USA, 2007.

30. Colby, J.D.; Keating, P.L. Land cover classification using Landsat TM imagery in the tropical highlands: The influence of anisotropic reflectance (vol 19, pg 1479, 2001). *Int. J. Remote Sens.* **2001**, *22*, 2655–2656.

31. Okoro, S.U.; Schickhoff, U.; Bohner, J.; Schneider, U.A. A novel approach in monitoring land-cover change in the tropics: Oil palm cultivation in the Niger Delta, Nigeria. *Erde* **2016**, *147*, 40–52. [CrossRef]

32. Frantz, D. FORCE—Landsat+ Sentinel-2 analysis ready data and beyond. *Remote Sens.* **2019**, *11*, 1124. [CrossRef]

33. Griffiths, P.; van der Linden, S.; Kuemmerle, T.; Hostert, P. Pixel-Based Landsat Compositing Algorithm for Large Area Land Cover Mapping. *IEEE J. Sel. Top. Appl. Earth Obs. Remote Sens.* **2013**, *6*, 2088–2101. [CrossRef]

34. Mueller, H.; Rufin, P.; Griffiths, P.; Siqueira, A.J.B.; Hostert, P. Mining dense Landsat time series for separating cropland and pasture in a heterogeneous Brazilian savanna landscape. *Remote Sens. Environ.* **2015**, *156*, 490–499. [CrossRef]

35. Hansen, M.C.; Egorov, A.; Roy, D.P.; Potapov, P.; Ju, J.; Turubanova, S.; Kommareddy, I.; Loveland, T.R. Continuous fields of land cover for the conterminous United States using Landsat data: First results from the Web-Enabled Landsat Data (WELD) project. *Remote Sens. Lett.* **2011**, *2*, 279–288. [CrossRef]

36. Verhulp, J.; Denner, M. The Development of the South African National Land Cover Mapping Program: Progress and Challenges. Available online: http://www.africageoproceedings.org.za/wp-content/uploads/2014/08/119_Verhulp_Denner1.pdf (accessed on 26 October 2020).

37. Basuki, T.M.; Skidmore, A.K.; Hussin, Y.A.; Van Duren, I. Estimating tropical forest biomass more accurately by integrating ALOS PALSAR and Landsat-7 ETM+ data. *Int. J. Remote Sens.* **2013**, *34*, 4871–4888. [CrossRef]

38. Nascimento, W.R.; Souza, P.W.M.; Proisy, C.; Lucas, R.M.; Rosenqvist, A. Mapping changes in the largest continuous Amazonian mangrove belt using object-based classification of multisensor satellite imagery. *Estuar. Coast. Shelf Sci.* **2013**, *117*, 83–93. [CrossRef]

39. Kamal, M.; Phinn, S.; Johansen, K. Object-Based Approach for Multi-Scale Mangrove Composition Mapping Using Multi-Resolution Image Datasets. *Remote Sens.* **2015**, *7*, 4753–4783. [CrossRef]

40. Wicaksono, P. Mangrove above-ground carbon stock mapping of multi-resolution passive remote-sensing systems. *Int. J. Remote Sens.* **2017**, *38*, 1551–1578. [CrossRef]

41. Bunting, P.; Rosenqvist, A.; Lucas, R.M.; Rebelo, L.M.; Hilarides, L.; Thomas, N.; Hardy, A.; Itoh, T.; Shimada, M.; Finlayson, C.M. The Global Mangrove WatchA New 2010 Global Baseline of Mangrove Extent. *Remote Sens.* **2018**, *10*, 1669. [CrossRef]

42. Heumann, B.W. An object-based classification of mangroves using a hybrid decision tree—Support vector machine approach. *Remote Sens.* **2011**, *3*, 2440–2460. [CrossRef]

43. Shirvani, Z.; Abdi, O.; Buchroithner, M.F. A new analysis approach for long-term variations of forest loss, fragmentation, and degradation resulting from road network expansion using Landsat time-series and object-based image analysis. *Land Degrad. Dev.* **2020**, *31*, 1462–1481. [CrossRef]

44. Onojeghuo, A.O.; Blackburn, G.A. Forest transition in an ecologically important region: Patterns and causes for landscape dynamics in the Niger Delta. *Ecol. Indic.* **2011**, *11*, 1437–1446. [CrossRef]

45. Salami, A.T.; Akinyede, J.; de Gier, A. A preliminary assessment of NigeriaSat-1 for sustainable mangrove forest monitoring. *Int. J. Appl. Earth Obs. Geoinf.* **2010**, *12*, S18–S22. [CrossRef]

46. Kamwi, J.M.; Cho, M.A.; Kaetsch, C.; Manda, S.O.; Graz, F.P.; Chirwa, P.W. Assessing the Spatial Drivers of Land Use and Land Cover Change in the Protected and Communal Areas of the Zambezi Region, Namibia. *Land* **2018**, *7*, 131. [CrossRef]

47. Geist, H.J.; Lambin, E.F. Proximate causes and underlying driving forces of tropical deforestation. *Bioscience* **2002**, *52*, 143–150. [CrossRef]

48. Quezada, M.L.; Arroyo-Rodriguez, V.; Perez-Silva, E.; Aide, T.M. Land cover changes in the Lachua region, Guatemala: Patterns, proximate causes, and underlying driving forces over the last 50 years. *Reg. Environ. Chang.* **2014**, *14*, 1139–1149. [CrossRef]

49. Campos, M.; Velazquez, A.; Verdinelli, G.B.; Skutsch, M.; Junca, M.B.; Priego-Santander, A.G. An interdisciplinary approach to depict landscape change drivers: A case study of the Ticuiz agrarian community in Michoacan, Mexico. *Appl. Geogr.* **2012**, *32*, 409–419. [CrossRef]

50. Fernandez, G.F.C.; Obermeier, W.A.; Gerique, A.; Sandoval, M.F.L.; Lehnert, L.W.; Thies, B.; Bendix, J. Land Cover Change in the Andes of Southern Ecuador-Patterns and Drivers. *Remote Sens.* **2015**, *7*, 2509–2542. [CrossRef]

51. Lei, C.G.; Wagner, P.D.; Fohrer, N. Identifying the most important spatially distributed variables for explaining land use patterns in a rural lowland catchment in Germany. *J. Geogr. Sci.* **2019**, *29*, 1788–1806. [CrossRef]

52. Aldwaik, S.Z.; Pontius, R.G. Intensity analysis to unify measurements of size and stationarity of land changes by interval, category, and transition. *Landsc. Urban Plan.* **2012**, *106*, 103–114. [CrossRef]

53. Gounaridis, D.; Zaimes, G.N.; Koukoulas, S. Quantifying spatio-temporal patterns of forest fragmentation in Hymettus Mountain, Greece. *Comput. Environ. Urban Syst.* **2014**, *46*, 35–44. [CrossRef]

54. López, S.; López-Sandoval, M.F.; Gerique, A.; Salazar, J. Landscape change in Southern Ecuador: An indicator-based and multi-temporal evaluation of land use and land cover in a mixed-use protected area. *Ecol. Indic.* **2020**, *115*, 106357. [CrossRef]

55. Coppin, P.; Jonckheere, I.; Nackaerts, K.; Muys, B.; Lambin, E. Digital change detection methods in ecosystem monitoring: A review. *Int. J. Remote Sens.* **2004**, *25*, 1565–1596. [CrossRef]

56. Liu, H.; Zhou, Q. Developing urban growth predictions from spatial indicators based on multi-temporal images. *Comput. Environ. Urban Syst.* **2005**, *29*, 580–594. [CrossRef]

57. Seto, K.C.; Fragkias, M. Mangrove conversion and aquaculture development in Vietnam: A remote sensing-based approach for evaluating the Ramsar Convention on Wetlands. *Glob. Environ. Chang.-Hum. Policy Dimens.* **2007**, *17*, 486–500. [CrossRef]

58. Chen, X.W. Using remote sensing and GIS to analyse land cover change and its impacts on regional sustainable development. *Int. J. Remote Sens.* **2002**, *23*, 107–124. [CrossRef]

59. NBS. *National Population Projection*; NBS: Abuja, Nigeria, 2018.

60. World Resources Institute. IUCN—The World Conservation Union. In *Global Biodiversity Strategy: Guidelines for Action to Save, Study, and Use Earth's Biotic Wealth Sustainably and Equitably*; World Resources Inst: Andrew Steer, DC, USA, 1992.

61. Ugochukwu, C.N.; Ertel, J. Negative impacts of oil exploration on biodiversity management in the Niger De area of Nigeria. *Impact Assess. Proj. Apprais.* **2008**, *26*, 139–147.

62. World Bank. GDP (Current US$)—Nigeria. Available online: https://data.worldbank.org/indicator/NY.GDP. MKTP.CD?end=2010&locations=NG&start=1960 (accessed on 2 August 2020).

63. Ako, R.T. Nigeria's Land Use Act: An anti-thesis to environmental justice. *J. Afr. Law* **2009**, *53*, 289–304. [CrossRef]

64. Imevbore, V.; Imevbore, A.; Gundlach, E. *Niger Delta Environmental Surveys: Vol-1-Environmental and Socio-Economic Characteristics*; Environmental Resources Managers Ltd.: Lagos, Nigeria, 1997. [CrossRef]

65. Safriel, U.; Adeel, Z. Development paths of drylands: Thresholds and sustainability. *Sustain. Sci.* **2008**, *3*, 117–123. [CrossRef]

66. Thomas, N.; Lucas, R.; Bunting, P.; Hardy, A.; Rosenqvist, A.; Simard, M. Distribution and drivers of global mangrove forest change, 1996–2010. *PLoS ONE* **2017**, *12*, e0179302. [CrossRef]

67. ESRI. ArcGIS Pro. Available online: https://www.esri.com/en-us/arcgis/products/arcgis-pro/overview (accessed on 25 September 2020).

68. Maxar; Technologies. Imagery Basemaps. Available online: https://www.maxar.com/products/imagery-basemaps (accessed on 25 September 2020).

69. Masek, J.G.; Vermote, E.F.; Saleous, N.E.; Wolfe, R.; Hall, F.G.; Huemmrich, K.F.; Gao, F.; Kutler, J.; Lim, T.K. A Landsat surface reflectance dataset for North America, 1990-2000. *IEEE Geosci. Remote Sens. Lett.* **2006**, *3*, 68–72. [CrossRef]

70. Zhu, Z.; Woodcock, C.E. Object-based cloud and cloud shadow detection in Landsat imagery. *Remote Sens. Environ.* **2012**, *118*, 83–94. [CrossRef]

71. Tucker, C.J. Red and photographic infrared linear combinations for monitoring vegetation. *Remote Sens. Environ.* **1979**, *8*, 127–150. [CrossRef]

72. Symeonakis, E.; Higginbottom, T.P.; Petroulaki, K.; Rabe, A. Optimisation of Savannah Land Cover Characterisation with Optical and SAR Data. *Remote Sens.* **2018**, *10*, 499. [CrossRef]

73. Higginbottom, T.P.; Symeonakis, E.; Meyer, H.; van der Linden, S. Mapping fractional woody cover in semi-arid savannahs using multi-seasonal composites from Landsat data. *ISPRS J. Photogramm. Remote Sens.* **2018**, *139*, 88–102. [CrossRef]

74. Haralick, R.M. Statistical and structural approaches to texture. *Proc. IEEE* **1979**, *67*, 786–804. [CrossRef]

75. Thapa, R.B.; Watanabe, M.; Motohka, T.; Shimada, M. Potential of high-resolution ALOS-PALSAR mosaic texture for aboveground forest carbon tracking in tropical region. *Remote Sens. Environ.* **2015**, *160*, 122–133. [CrossRef]

76. Cohen, W.B.; Yang, Z.G.; Kennedy, R. Detecting trends in forest disturbance and recovery using yearly Landsat time series: 2. TimeSync—Tools for calibration and validation. *Remote Sens. Environ.* **2010**, *114*, 2911–2924. [CrossRef]

77. Pal, M. Random forest classifier for remote sensing classification. *Int. J. Remote Sens.* **2005**, *26*, 217–222. [CrossRef]

78. Team, R.C. R: A Language and Environment for Statistical Computing. R Foundation for Statistical Computing, Vienna. 2017. Available online: https://www.R-Project.org (accessed on 26 October 2020).

79. Pontius, R.G.; Gao, Y.; Giner, N.M.; Kohyama, T.; Osaki, M.; Hirose, K. Design and interpretation of intensity analysis illustrated by land change in Central Kalimantan, Indonesia. *Land* **2013**, *2*, 351–369. [CrossRef]

80. McGarigal, K. *FRAGSTATS: Spatial Pattern Analysis Program for Quantifying Landscape Structure*; US Department of Agriculture, Forest Service, Pacific Northwest Research Station: Corvallis, OR, USA, 1995; Volume 351.

Publisher's Note: MDPI stays neutral with regard to jurisdictional claims in published maps and institutional affiliations.

Article

Multi-Decadal Changes in Mangrove Extent, Age and Species in the Red River Estuaries of Viet Nam

Nguyen Hong Quang [1,*], Claire H. Quinn [2], Lindsay C. Stringer [2], Rachael Carrie [2], Christopher R. Hackney [3], Le Thi Van Hue [4], Dao Van Tan [5] and Pham Thi Thanh Nga [1]

[1] Vietnam National Space Center (VNSC), Vietnam Academy of Science and Technology (VAST), 18 Hoang Quoc Viet, Hanoi 100000, Vietnam; pttnga@vnsc.org.vn

[2] Sustainability Research Institute, School of Earth and Environment, University of Leeds, Leeds LS2 9JT, UK; C.H.Quinn@leeds.ac.uk (C.H.Q.); L.Stringer@leeds.ac.uk (L.C.S.); R.H.Carrie@leeds.ac.uk (R.C.)

[3] School of Geography, Politics and Sociology, Newcastle University, Newcastle upon Tyne NE1 7RU, UK; christopher.hackney@newcastle.ac.uk

[4] Central Institute for Natural Resources and Environmental Studies, Vietnam National University (VNU), No 19 Le Thanh Tong, Hoan Kiem, Ha Noi 100000, Vietnam; huele@cres.edu.vn

[5] Faculty of Biology, Hanoi National University of Education (HNUE), 136 Xuan Thuy, Cau Giay, Ha Noi 100000, Vietnam; tandv@hnue.edu.vn

* Correspondence: nhquang@vnsc.org.vn; Tel.: +84-968-844-250

Received: 10 June 2020; Accepted: 14 July 2020; Published: 16 July 2020

Abstract: This research investigated the performance of four different machine learning supervised image classifiers: artificial neural network (ANN), decision tree (DT), random forest (RF), and support vector machine (SVM) using SPOT-7 and Sentinel-1 images to classify mangrove age and species in 2019 in a Red River estuary, typical of others found in northern Viet Nam. The four classifiers were chosen because they are considered to have high accuracy, however, their use in mangrove age and species classifications has thus far been limited. A time-series of Landsat images from 1975 to 2019 was used to map mangrove extent changes using the unsupervised classification method of iterative self-organizing data analysis technique (ISODATA) and a comparison with accuracy of K-means classification, which found that mangrove extent has increased, despite a fall in the 1980s, indicating the success of mangrove plantation and forest protection efforts by local people in the study area. To evaluate the supervised image classifiers, 183 in situ training plots were assessed, 70% of them were used to train the supervised algorithms, with 30% of them employed to validate the results. In order to improve mangrove species separations, Gram–Schmidt and principal component analysis image fusion techniques were applied to generate better quality images. All supervised and unsupervised (2019) results of mangrove age, species, and extent were mapped and accuracy was evaluated. Confusion matrices were calculated showing that the classified layers agreed with the ground-truth data where most producer and user accuracies were greater than 80%. The overall accuracy and Kappa coefficients (around 0.9) indicated that the image classifications were very good. The test showed that SVM was the most accurate, followed by DT, ANN, and RF in this case study. The changes in mangrove extent identified in this study and the methods tested for using remotely sensed data will be valuable to monitoring and evaluation assessments of mangrove plantation projects.

Keywords: mangrove development; mangrove plantation; machine learning; mangrove condition; classification; remote sensing

---

1. Introductions

Mangrove forests are one of the most biologically diverse ecosystems on Earth and deliver numerous provisioning, regulating, cultural, and supporting services that benefit the coastal and inland communities' livelihoods as well as the global atmospheric commons [1,2]. The Vietnamese government, aware of these benefits, has developed strategies for mangrove development in the Red River Delta (RRD), in order to reduce global climate change impacts and secure the livelihoods of coastal communities in the five coastal provinces in the north of the country [3]. One of the key strategies is forest plantation and restoration, involving both domestic and foreign donors (such as Denmark and Japan). Although some mangrove areas have been converted for use in aquaculture, rice, and salt farms, there has been an expansion of the mangrove forest in recent decades in some Red River estuarine areas, for example, mangrove extent increased by 538.5-ha in Thuy Truong commune between 2001 and 2016 [4]. However, there is a need for information regarding forest dynamics in terms of area cover as well as accurate methods for mapping the extent and composition of these forests with contemporary remote sensing data and methods. Compositional factors are linked to a number of variables such as size class distribution and canopy complexity, that when used in combination with others can be indicative of mangrove health or condition [5]. Therefore, understanding them can provide useful insights to support future forest management decision making.

In recent years, supervised image classification algorithms have been reported to be more accurate than unsupervised approaches [6], as the supervised outputs are trained with in situ data. Without ground-truth data, the supervised classifiers should not be applied because a minimum training dataset is required consistently [7]. However, training datasets are not always available, particularly when historic image analysis is undertaken [8]. In this paper, we applied qualified unsupervised classifiers, the iterative self-organizing data analysis technique (ISODATA) and the K-means classification, to analyze a long time series of Landsat-X data for an assessment of changes in mangrove extent. In addition, field-surveyed data were obtained to test four different learning machine image classifiers: artificial neural network (ANN), decision tree (DT), random forest (RF), and support vector machine (SVM) in order to classify mangrove age and species at a point in time of May 2019.

In the present era of digital image processing, the image fusion of multisource remote sensing data is of increasing interest and is becoming a well-established research field in the context of increasing (optical and synthetic aperture radar (SAR)) data availability [9,10]. Image fusion techniques are applied to sharpen a low resolution multispectral image using a higher resolution panchromatic layer to generate enhanced input data, resulting in new, better quality data (spectral and spatial resolution) compared to the originals [11,12]. However, sometimes it is difficult to preserve the original image spectrum. In addition, spectral distortion due to image fusion effects is a source of new information that can be used for other applications such as change detection [13]. Recently, the number of studies integrating SAR and optical images has increased as users take advantage of both of these data. Several methods of image fusion have been developed, hence the decision about which technique is the most suitable is driven by the study goals [14] and expected accuracy requirement. Here, we selected Gram–Schmidt (GS), and principal component analysis (PCA), a method which reduces the dimensionality of the information present in the original multi-band dataset, as they have been reported to be the most accurate methods [15] when fusing SPOT-7 and Sentinel-1 images, with the expectation of achieving improved mangrove species classification [16].

Remote sensing has undoubtedly been an effective tool to evaluate mangrove forests from many perspectives [17] including estimating above ground biomass [18–21], assessing mangrove health [22–24], chlorophyll [25,26], and to map changes in mangrove extent [27,28] at global [29], continental and regional [30,31], and national and smaller scales [32,33]. Most studies use remote sensing to explore the severity and consequences of mangrove loss and associated degradation. In this study, we applied three sources of remote sensing data to better understand elements of mangrove condition related to growth as well as to test the performance of different classification approaches.

2. Materials and Methods

2.1. Study Site

Mangroves and coastal wetland areas are described in [34]. Vietnam has 30 coastal provinces and cities associated with mangroves divided into four main zones: (i) Northeastern coast (Ngoc Cape to Do Son); (ii) Northern Delta (Do Son to Lach Truong River); (iii) Central coast (Lach Truong to Vung Tau); and (iv) Southern Delta (Vung Tau to Ha Tien) [4]. Our study is located in Thuy Truong commune, which is located in the Northern Delta (zone 2) at the mouth of the Thai Binh River (Figure 1A). The climate of the region is influenced by the South-East Asian tropical monsoon with four distinct seasons: spring from February to May, summer from June to August, autumn from September to November, and winter from December to January. The mean annual temperature is 23 °C and maximum and minimum monthly average temperatures are around 28 °C in July and 16 °C in January, respectively [35]. The area of Thuy Truong commune is 9.3 km^2 and home to 10,000 people. The main livelihoods are based on agriculture (rice cultivation and cash crops), aquaculture, and harvesting clams, fishes, and crabs in the nearby mangrove forests [36]. The mangrove forest has been expanded as a result of plantation efforts supported by the Danish and Japanese Red Cross programs that ended in 2006 [37]. Hence, although Vietnam's total mangrove area has declined to 62% of the original [38], in Thuy Truong, the mangrove forest has been subject to large fluctuations in extent and quality [39].

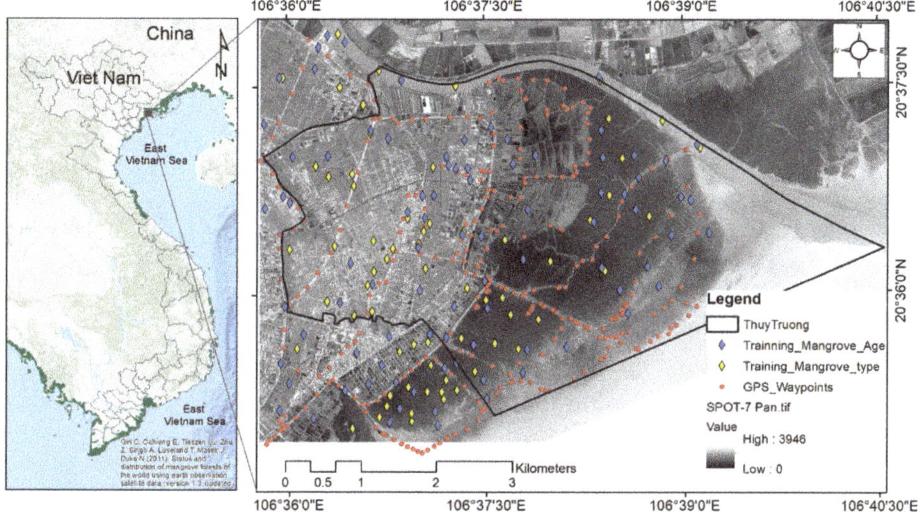

Figure 1. Location of Thuy Truong commune (central coordinates of 106°38′00E and 20°36′00N) and in situ ground-truth investigation of mangrove locations. Plots for different types (yellow diamonds) and ages (blue diamonds) were positioned with confirmation of the local people. A GPS (Garmin Montana 680) with an integrated 5 M camera was used to locate and photograph each mangrove type and age. The SPOT-7 panchromatic band with the digital number ranged from 0 to 3946 was used for the base map.

2.2. Data Collection

2.2.1. Ground-Truth Data Collection

Field investigations from 22 to 25 November 2018 were undertaken to obtain ground-truth information including 78 polygons for training mangrove species (23 polygons for accuracy assessment) and 105 polygons for training mangrove age (32 polygons for accuracy assessment). The number of samples for each class is summarized in Table 1. It was noted that later image classifications focused on the mangrove forest, however, we also needed to train other land use and land cover (LULC) layers that help to discriminate mangroves and improve the accuracy of classification algorithms. We interviewed a commune cadastral official for the mangrove plantation projects and conducted field work with five local citizens to gather mangrove age and species information and mark them on a printed map. Mangroves have been present in the study site for about 45 years, since 1975, however trees older than 10 years are all similar in terms of height, stems and color. Hence, we divided mangrove age into three categories: older than 10 years, around five years, and under three years old. Despite the original plantation projects spanning a larger area, most of the planted mangrove had been destroyed by waves, erosion, or eaten by crabs and clams.

Table 1. In situ data for supervised remote sensing image classifications and accuracy evaluation. Examples of three existing mangrove species with scientific names, local names are marked in bold for illustration.

Training for Mangrove Type	Number of Polygons	Average Area (ha)	Sum Area (ha)	Training for Mangrove Age	Number of Polygons	Average Area (ha)	Sum Area (ha)
Agriculture	8	1.6	12.4	Agriculture	10	1.9	18.8
Aquaculture	6	1.7	10.1	Aquaculture	12	2.2	26.3
Seawater	4	21.5	86.1	Seawater	5	2.5	12.4
Bare land	6	0.4	2.3	Bare land	14	0.2	3.2
Residence	6	0.9	5.1	Residence	6	1.1	6.5
River	7	3.9	27.4	River	8	1.8	14.4
Road	7	0.1	0.6	Road	13	0.4	5.0
Sonneratia caseolaris (**ban**)	9	2.1	18.9	>10 year mangrove	16	1.9	30.4
Aegiceras corniculatum (**Su**)	9	1.3	11.3	5 year mangrove	10	1.2	11.7
Kandelia obovata (**Vet**)	16	0.5	7.4	<3 year mangrove	11	1.2	13.1
Sum	78 (23)		181.6		105 (32)		141.6

2.2.2. Remote Sensing Data

Landsat-2,5 and 8, SPOT-7, and Sentinel-1 datasets were used for this research, first, to compare extracted mangrove results, and second, to fuse the optical multi-spectral and panchromatic bands with SAR backscatter bands in the Sentinel-1 data. The basic information of acquisition time, processing level, band number, and spatial resolution of the collected scenes is summarized in Table 2. The Landsat data were acquired for each 5-year period from 1975 to 2019, and with the exception of Landsat-2 data for which there were limited options, and acquisition times for the Landsat scenes were selected to minimize cloud cover (October and November). To minimize seasonal effects, the acquisition date of Landsat-8 data was chosen to be as close as possible to the acquisition date of the SPOT-7 and the Sentinel-1 images to facilitate later comparisons. The high-resolution optical remote sensing scene of SPOT-7 consisted of four multispectral bands (0–3) and one panchromatic band with 1.6 m spatial resolution. The SAR image of Sentinel-1 was processed at the ground-range detected level (10 m resolution), was pass ascending, and acquired in two polarizations (VH and VV). Sentinel-1 data were obtained from the Copernicus Open Access Hub website of European Space Agency (ESA), Landsat scenes from the United States Geological Survey (USGS), and SPOT-7 data from the Airbus group.

2.3. Methodology

Figure 2 sets out our working flow chart, which comprises three main parts: (1) The processes for the Landsat-X mangrove extent unsupervised classification; (2) the procedures to process the SPOT-7 image to classify mangrove age and fusion with the Sentinel-1 images; and (3) the processing chain including image pre-processing, speckle filtering, fusing of the VH and VV layers with the SPOT-7 image, and supervised classification of mangrove types. Minor steps such as clipping the region of interest, post-classification to convert the classified image to vector, confusion matrix (contingency matrix) calculation, band math, band conversion, etc. are not included in order to simplify the figure. Basic tasks in remote sensing image processing, like atmospheric correction [40,41], image resampling (done only for Landsat-2), SAR image pre-processing (radiometric calibration, terrain correction, and data conversion/select band to export single layer) and speckle filtering are well documented [42,43], hence they are not described in detail here. The core tasks of classifying mangrove extent, age, and species, and image fusion are explained in the following sub-sections.

Table 2. Summary of remote sensing data used (X refers to the Landsat mission of 2, 5, and 8; L1TP is data processing level 1 with precision terrain corrected; BQA stands for band quality; MSS is Multispectral Scanner Sensor; TM stands form; OLI is Operational Land Imager; Mul and Pan are short for multispectral and panchromatic bands, respectively; GPL is geometric processing level; RPL is radiometric processing level; and GRD is ground-range detected. V and H are vertical and horizontal, respectively, and coupled letters of VH and VV indicate SAR cross-polarizations).

Data	Time of Acquisition	Level	Band and Polarization	Resolution
Landsat (X)	1975/04/20 (2MSS), 1988/11/04 (5TM), 1993/11/02 (5TM), 1998/10/15 (5TM), 2003/10/10 (5TM), 2008/11/11 (5TM), 2013/10/08 (8OLI), 2018/10/06 (8OLI), 2019/05/18 (8OLI)	(2) L1TP, (5) L1TP, (5) L1TP, (5) L1TP, (5) L1TP, (5) L1TP, (8) L1TP, (8) L1TP, (8) L1TP,	(2) 4–6, (5) 1–7, BQA, (5) 1–7, BQA, (5) 1–7, BQA, (5) 1–7, BQA, (5) 1–7, BQA, (8) 1–11, BQA, (8) 1–11, BQA, (8) 1–11, BQA.	(2) 60 m (5) 30 m (5) 30 m (5) 30 m (5) 30 m (5) 30 m (5) 30 m (8) 30 m, Pan (B8)15 m (8) 30 m, Pan (B8)15 m (8) 30 m, Pan (B8)15 m
SPOT-7	2019/05/17	GPL: Sensor RPL: Basic	Band 0–3 (Mul) Pan	Mul 6 m Pan 1.5 m
Sentinel-1	2019/05/16 (ascending)	L1 GRD product	VH and VV	10 m

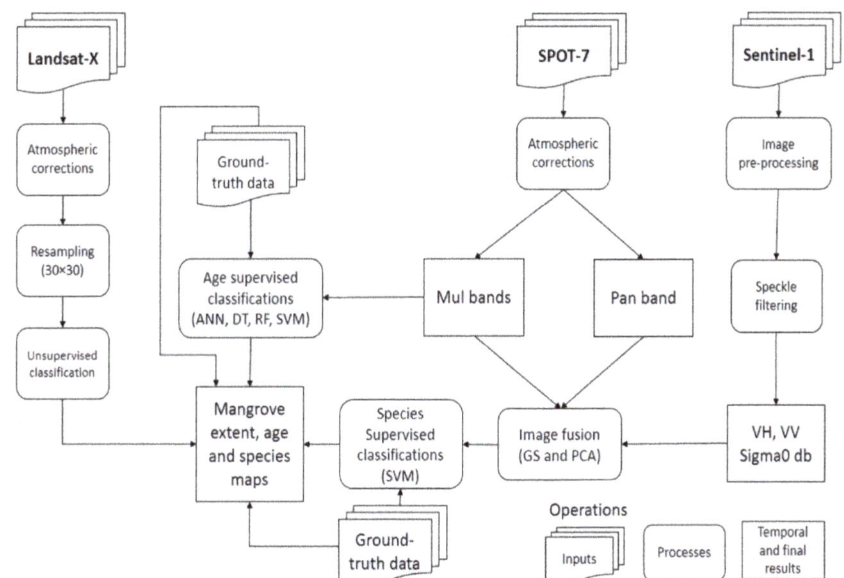

Figure 2. Flowchart of methodology used for mapping mangrove extent, age, and species. X refers to the mission number of the used Landsat images; ANN, DT, RF, SVM stand for artificial neural network, decision tree, random forest, and support vector machine, respectively; Mul and Pan are short forms of multispectral and panchromatic bands, respectively; GS and PCA indicate Gram–Schmidt and principal component analysis image fusion methods; V and H are vertical and horizontal, respectively, and coupled letters of VH and VV indicate Synthetic Aperture Radar (SAR) cross-polarizations.

2.3.1. Mangrove Age Classification

Mangrove age and growth estimations are typically quantified by means of in situ dendrometer techniques [44] and internodes [45]. However, few studies have attempted to define classifiers dealing with mangrove age estimations from remotely sensed data. We elected to use artificial neural network (ANN), decision tree (DT), random forest (RF), and support vector machine (SVM) methods from among the many available for the mangrove age estimation because (1) they are robust image supervised classification methods; (2) the advancements in machine learning (ML) approaches to model complex class signatures and accept a variety of training data [46]; and (3) because they are routinely found to have higher accuracies than the maximum likelihood method [47]. Selection of these four methods allowed us to compare results and identify the best performing method using the SPOT-7 image and the same training dataset from the field survey (Section 2.2.1).

ANN classification has been used in a wide range of applications in remote sensing. The theory and algorithm are explained in detail by Schalkoff (1992), Foody (1996), and Dreiseitl and Stephan (2020) [48–50]. Generally, ANN classification is achieved with a fundamental layered, feedforward network architecture (Figure 3) comprising a set of processing units organized in layers. Layers are connected by a weighted channel to every unit [50]. The training data are used to compute the difference (error) between the desired and actual network output; then the error is fed backward to the input layer through the network, with the weights linking the units altered in proportion to the error. The process is repeated until the error rate reaches an acceptable value of above 60% agreement between the classified and ground-truth data. Although the ANN algorithm has some advantages, in remotely sensed data classification this method has limitations when dealing with highly heterogeneous land cover types (mixed pixels) and the network can become static when the number of neurons exceeds ten [7]. In this classification, some primary parameters describing the number of neurons, maximum number of iterations, and error change are adjusted to values of 3, 300, and 0.1, respectively. The selected training method was back propagation with a weight gradient term of 0.1 and moment term of 0.5.

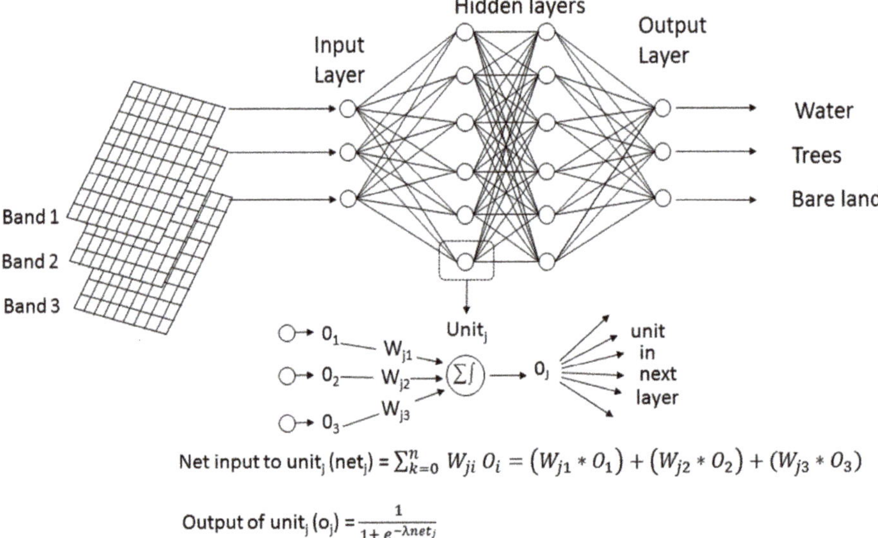

$$\text{Net input to unit}_j\ (net_j) = \sum_{k=0}^{n} W_{ji}\, O_i = \left(W_{j1} * O_1\right) + \left(W_{j2} * O_2\right) + \left(W_{j3} * O_3\right)$$

$$\text{Output of unit}_j\ (o_j) = \frac{1}{1 + e^{-\lambda net_j}}$$

Figure 3. Classification of remote sensing data by an artificial neural network (adapted from Foody 1996) where W_{ij} is the weight that connects the jth unit with its ith incoming connection; O_i and O_j are the value of the ith incoming connection and j^{th} output connection; and λ is a gain parameter, which is often set to 1.

While the conventional statistical and neural/connectionist classifiers create a single membership for each pixel at the same time, the decision tree (DT) classifier solves the problem of label assignment using a multi-stage or sequential approach [51]. The labeling process is a chain of simple decisions based on sequential test results rather than a complex decision. In terms of DT construction, there is a univariate DT, splitting features orthogonally to the axis, testing a single feature at a time while the multivariate DT splitting rule at internal nodes can differ depending on the complexity of the data and classification problem, using one or more features simultaneously. The multivariate DT is considered able to generate more accurate results than the univariate DT [52]. Two high-driven parameters of maximum tree depth and regression accuracy were set at values of 7 and 0.01.

The random forest (RF) classifier is a nonparametric and ensemble technique proposed by Breiman (2001: 5) [53], which is a "combination of tree predictors such that each tree depends on the values of a random vector sampled independently and with the same distribution for all trees in the forest". Random forests contain many decision trees, with each tree built from a random subset of training data with a random subset of predictor variables. Since the RF algorithm consists of a parametric model for prediction, it is different from traditional statistical methods [54]. Feature/feature combinations are selected using bagging, a method used to generate a training dataset by randomly drawing on replacement N examples, where N is the size of the original training set [55]. The RF approach is recommended as it has the advantage of using fully grown trees that are not pruned compared to other decision tree methods [56]. The parameters set for this method were a maximum tree depth of 10, regression accuracy of 0.01, and truncate pruned tree (yes).

Support vector machine (SVM) [57,58] is a supervised non-parametric statistical learning technique that provides good classification results from complex and noisy data [59,60]. The statistical learning theory is derived in the SVM classification system that separates the classes with a decision surface maximizing the margin between the classes. The surface is called the "optimal hyperplane" and the data points nearest the hyperplane are called "support vectors" [60]. Dealing with a large high resolution image, the SVM classifier is time-consuming to process, hence it provides a hierarchical, reduced-resolution classification process, which enables the performance to be shortened without significantly degrading the outcomes. In this study, we selected radial basic function for the Kernel type and set Gamma in a Kernel function of 0.25 and penalty parameter of 100.

2.3.2. Image Fusion

We selected Gram–Schmidt (GS) [61] and principal component analysis (PCA) [62] among many other available image fusion methods to generate higher quality (spectral and spatial resolution) MS images. These two methods presented better results compared to the modified intensity–hue–saturation (IHS) and Brovey transformation (BT) methods in a study by Quang et al. (2019) [15]. In the GS fusion technique, suitable weights assigned to the high-resolution panchromatic (PAN) layers are simulated from lower spatial multispectral bands [61,63]. Inverse GS image sharpening is then used to form the pan-sharpened spectral bands [64].

PCA is a statistical technique that identifies the key variability among variables within a dataset, reducing it to fewer dimensions or "components" of related variables that are uncorrelated with each other [14]. In this study, we fused a SPOT-7 multispectral band, a panchromatic band and a Sentinel-1 VH or VV layer once for each image fusion method to generate fused images prior to mangrove type classification.

2.3.3. Mangrove Species Categorization

Mangrove species mapping is a common application of hyperspectral remote sensing data [65,66]. Other useful information for mangrove species parcellation that can be derived from SAR RS data includes general structural information in relation to mangrove zonation [67]. Hyperspectral remote sensing data tend to provide finer-detailed information (reflectance and finer spatial resolution). An analysis of SAR backscatters on different mangrove species can help to separate mangrove species as well as provide a better understanding of the effects of different polarizations on the radar scattering for the target geographical features [68]. Hence, we employed the fused SPOT-7 with Sentinel-1 images to classify the mangrove species, applying the SVM classifier as its presentation is most accurate for mangrove age estimations. Additionally, the SPOT-7 image and the S1 VH were used separately for classifying mangrove types for comparisons with the fused images also applying the SVM classifier.

2.3.4. Mangrove Extent Classification

When it is difficult to obtain a sufficiently comprehensive set of training sites to apply a supervised classification approach, unsupervised classifications could be suitable options [28] to deliver acceptable outputs. We applied the iterative self-organizing data analysis technique (ISODATA) unsupervised classifier for nine Landsat-X datasets from 1975 to 2019 since it generated more reliable results (81.7%) than the K-means method (77.3%) in an examination by El-Rahman (2016) [69]. As typical land use and land cover (LULC) in the study site are agriculture (rice), aquaculture, residence, water bodies, and mangrove forest, we defined 10 to classify and five for maximum iterations, while other parameters were set to default. The result of ISODATA unsupervised classification of the Landsat-8 (2019) was examined for accuracy using the ground-truth data and compared with results from the SPOT-7. The results of all Landsat image classifications were used for mapping mangrove extent changes. The post-classification processes after ISODATA classification were done for all the years, an accuracy assessment for year 2019, and converting classified layers to vector files to enable the subsequent calculation of statistics.

2.3.5. Evaluation

Thirty percent of the ground-truth data was used to evaluate the classification result in terms of mangrove extent, age, and species classifications. We used descriptive and analytical statistical techniques, in which accuracies of individual categories were computed by calculating confusion matrices, user and producer accuracies, and multivariate technique of Kappa statistics for each classification [70,71]. As we conducted the field survey for only 2019, the unsupervised mangrove extents using Landsat-X was evaluated for this year. This means that we cannot assume the same accuracy for other unsupervised classifications, particularly as there is likely to be more errors in years where the spatial resolution of the RS datasets degrades (e.g., 1975, etc.). The supervised classifications were implemented separately; however, we used the same ground-truth data for each image classifier in order to compare results between them. The results of all evaluations are summarized and presented in Section 3.5.

2.3.6. Mangrove Mapping

In the post-classification process, the results of mangrove age, type, and extent were exported to vector files, allowing us to map and easily undertake statistical analyses such as zonal statistics and summary statistics in the QGIS version 3.12.0 environment. Specific layers of interest relating to mangrove age, type, and extent, etc. could be highlighted and other classes faded into the background to aid map reading and orientation. This mapping approach is consistently used in this study.

3. Results

3.1. Mangrove Age Classifications

The results showed both similarities and differences in estimated mangrove extent and age between the four methods when using the same input image and training dataset (Figure 4). Although the older mangrove extent presented similarly in DT, ANN, and SVM, large areas were classified into the five-year old mangrove category in the RF output. Various differences were shown in the youngest mangroves. RF estimated the fewest young mangroves (4.4% of total mangrove area), followed by ANN (25.8%), and SVM (29.9%), with the largest area (35.9%) in the DT map and greatest uncertainty in this class. Younger mangroves were found to be distributed further, about 2.5 km, from the coastline on the maps. This is reasonable since we know that mangroves were planted in the newer areas of sediment deposition. However, the times of mangrove planting have not been regular because it depends on the availability of funding from donor projects.

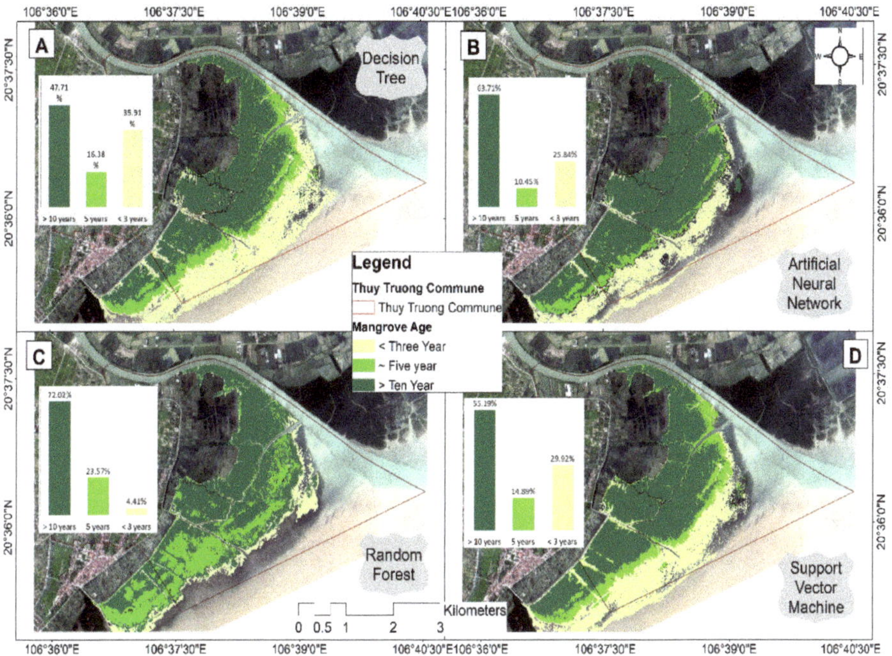

Figure 4. Maps of classifications using the multispectral data of SPOT-7 acquired on 17 May 2019 for mangroves of ages older than 10 years, around five years, and younger than three years, mapped for the four classifiers of decision tree (**A**), artificial neural network (**B**), random forest (**C**), and support vector machine (**D**). The background is the SPOT-7 true-color composition image and the red polygon shows the border of the Thuy Truong commune.

3.2. Image Fusion

Figure 5 shows the Gram–Schmidt (GS) and principal component analysis (PCA) image fusion results by color compositing of the fused bands compared with the original Sentinel-1 data. In general, there were few differences between Sentinel-1 VH and VV results using GS or PCA. In contrast, the effects of image fusion methods seemed to be greater on the color composite images. The GS enhanced the colors of the residential areas (yellow) and the mangrove (red) (C and D) more clearly than the PCA (E and F). However, these differences were in the color composites and might not affect the later mangrove classification results. The spatial resolution of the fused images (all polarization, GS, and PCA) was improved from the 10 m resolution of Sentinal-1, and 6 m resolution of the multispectral SPOT-7 to 1.5 m, as can be seen in the zoomed-in pink polygons (C and E clearly distinguish between aquaculture and mangrove, D and F distinguish between river and mangrove) compared to the same areas using original Sentinel-1 data (A and B). Another advantage of the band sharpening is the capacity to minimize cloud effects in optical images. However, in this study, we collected a cloud-free SPOT-7 scene.

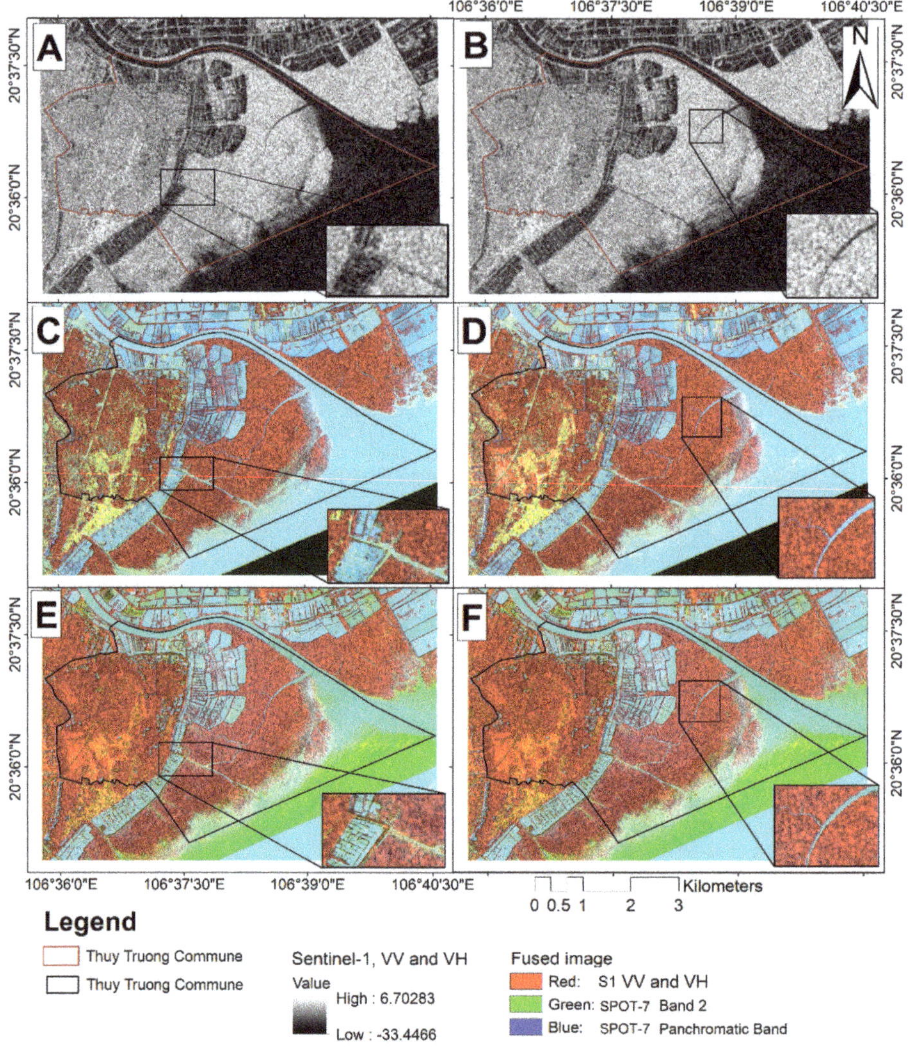

Figure 5. Demonstrations of SPOT-7 and Sentinel-1 (S1) image fusion processes where (**A**) is the original Sentinel-1 VH layer and (**B**) is Sentinel-1 VV layer (sigma0 in decibel); and (**C–F**) depict the results of the fused images using VH–GS, VV–GS, VH–PCA, and VV–PCA, respectively.

3.3. Mangrove Species Maps

Three main mangrove species were classified using the fused images and SVM classifier: the *Sonneratia caseolaris* locally called "Ban", the *Aegiceras corniculatum*, local name "Su", and *Kandelia obovata*, local name "Vet" (Figure 6). Su was present in the forest core, close to the river channel in the middle of all maps, and accounted for around 50 percent of the total mangrove area. Nonetheless, there was some mixture of Su with Ban in the core forest in the VV_GS and VV_PCA maps. Vet (26% of total mangroves) was distributed more in the southwest of the study site and it was categorized similarly in all maps. The main differences between using VH and VV S1 polarizations was the young Ban mangrove in the east of the map (C and D) was incorrectly classified as Vet in the VH_GS and VH_PCA maps according to the ground-truth investigation. In general, the use of different image fusion methods affected the mangrove species classifications less than the use of different SAR polarizations. A comparison of map A with C, and B with D, and the VV polarization fused with the SPOT-7 bands indicated a good performance for mangrove type categorization. The VH might nevertheless be more suitable for mangrove forests where the mixture of species is low. The classification of the original SPOT-7 (E) showed large areas of Vet (4.06%) were mis-classified to Su (75.83%). While percentage of Ban areas (20.11%) seemed to be similar to those of other images, the distribution was incorrect for the outer forest edge. The use of the Sentinel-1 VH layer generated mangrove species (F), and their distribution was considered accurate and agreed well with fused-image based classifications. However, the resulting resolution (10 m) was much lower than the fused images (1.5 m) and that was why the small areas of Vet and Ban were combined into the Su mangrove type.

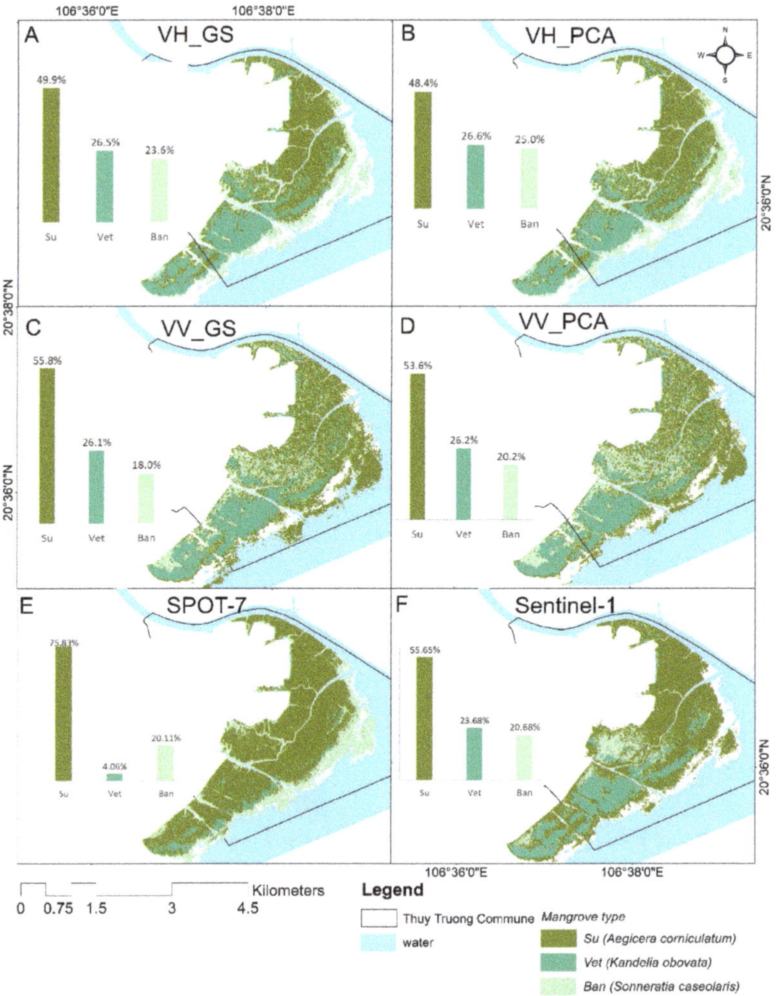

Figure 6. Maps of classified mangrove species; GS and PCA indicate Gram–Schmidt and principal component analysis image fusion methods; V and H are vertical and horizontal, respectively and coupled letters of VH and VV indicate SAR cross-polarizations. VH_GS, VH_PCA, VV_GS, and VV_PCA are combinations of fused images of SPOT-7 acquired on 17 May 2019 and Sentinel-1 polarization data (VH or VV) and the image fusion methods (GS or PCA).

3.4. Mangrove Extent Changes

Extracting mangrove extents over a long period of time (1975–2019) showed the expansion, approximately 80 m/year, of the mangrove forest to about 3.5 km seaward in Thuy Truong commune and surroundings (Figure 7). The mangrove expansion of the results of the ISODATA classifications was slower between 1975 and 1993, and slightly decreased in 1993 compared to 1988. Nonetheless, the forest has been rapidly and continuously increasing in extent for 31 years from 1998 to 2019. Based on the in situ investigation data, the mangrove forest was mostly planted when the accumulated sediment from the river was high and the base well-founded. There is a small area of mangrove fragmentation due to aquaculture ponds and the mangrove there was degraded until 2019, by which time it had mostly disappeared (see the red rectangle in years 1993 and 2019). We also tested classifying the mangrove extents using the K-means classifier. However, no significant differences between the two methods were found. Hence, we only present the ISODATA results.

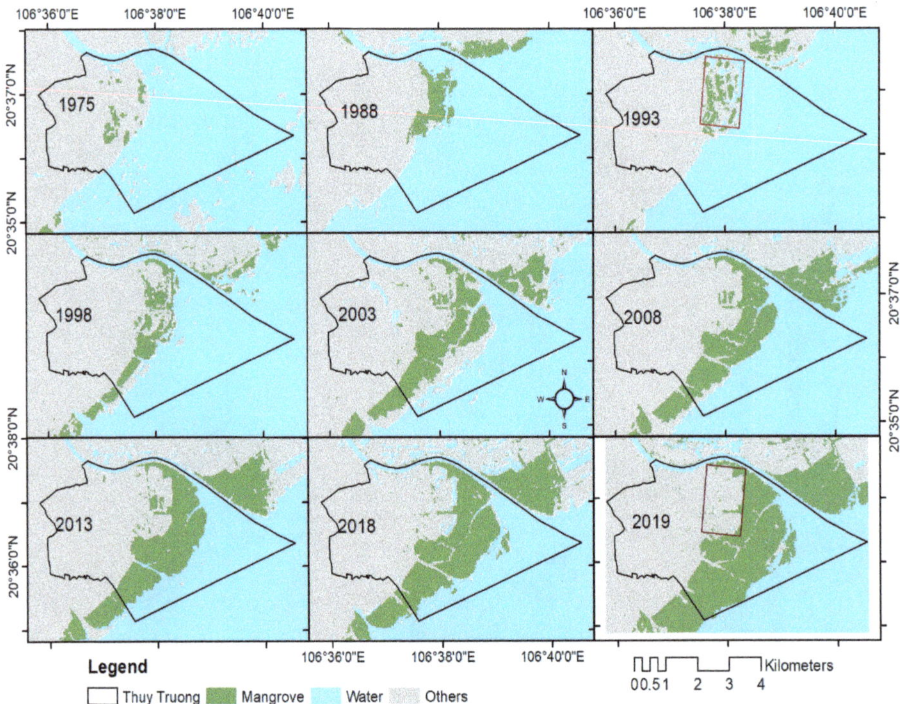

Figure 7. Changes in mangrove extent from 1975 to 2019 classified from a time series of Landsat images missions 2 to 8 (described in the Table 2) using the iterative self-organizing data analysis technique (ISODATA) classification, an unsupervised image classification approach (detailed in Section 2.3.1). The red rectangle denotes the same area of intensive aquaculture in 1993 and 2019.

Changes in mangrove extent (including all mangrove species) in ha were quantified and graphed (Figure 8). In the first 11 years, the forest expanded approximately 150 ha (1988), but had decreased by slightly over 100 ha five years later (1993). Forest extent recovered slightly by 1998, and then increased by more than 220 ha in the five years until 2003. Afterward, there was a gradually increase over the subsequent 15 years (2003–2018). A remarkable increase was found in the last year of the assessment time. It is noted that all estimates were not validated except for 2019, for which ground-truth data were available. However, the mangrove in this region grows in sediment deposits and does not mix with other vegetation, therefore unsupervised classifications are considered sufficiently accurate.

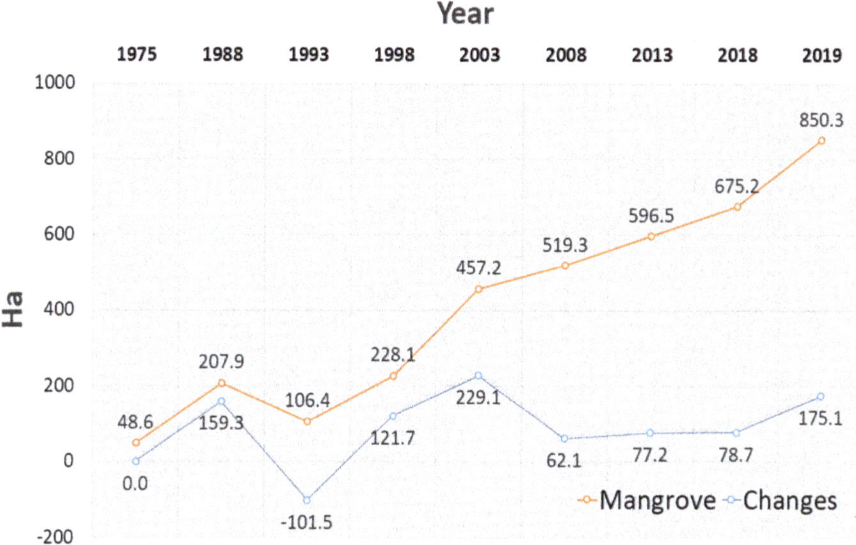

Figure 8. Total mangrove area (orange line) and changes in extent (blue line) (ha) from 1975–2019 in Thuy Truong commune.

3.5. Accuracy Assessment

3.5.1. Mangrove Age Classification

The producer and user, overall accuracy, and Kappa coefficient calculated from the four confusion matrices developed based on comparisons of the ground truth data and results of the four image classification methods (DT, ANN, FR, and SVM) for ten classes are summarized in Table 3. Although we calculated accuracy for all ten classes, we focused on three target layers of mangrove age (older than 10 years, around five years, and younger than three years). In general, the ten-year mangrove was classified at the highest accuracy (producer accuracy greater than 72.45% and user accuracy greater than 69.2%) and the five-year mangrove was the lowest (producer accuracy of 62.31% and user accuracy of 36.28%), with the exception of the low accuracy of the RF method for three-year mangroves (producer accuracy of 31.44% and user accuracy of 46.41%). With other layers, seawater was the most accurate classification, followed by river and mangrove, while road and residence were the least accurate. Comparing between the methods used (see the overall and the Kappa coefficients), the DT and SVM generated the most accurate results. RF revealed some limitations, particularly with highly mixed-pixel classes such as residence, road (narrow and long), and aquaculture.

Table 3. Accuracy indexes calculated from confusion matrices for mangrove age classification assessment. Prod. Acc. and User. Acc. are short forms of producer and user accuracy; DT, ANN, RF, and SVM indicate the image classification methods of decision tree, artificial neural network, random forest, and support vector machine, respectively.

Class	DT		ANN		RF		SVM	
	Prod. Acc. (Percent)	User Acc. (Percent)	Prod. Acc. (Percent)	User Acc. (Percent)	Prod. Acc. (Percent)	User Acc. (Percent)	Prod. Acc. (Percent)	User Acc. (Percent)
Ten_Year_Mangrove	93.86	91.14	93.31	89.77	72.45	69.20	94.67	94.62
Five_Year_Mangrove	72.33	76.74	59.59	73.70	62.31	36.28	84.39	81.36
Three_Year_Mangrove	79.11	81.01	70.36	82.92	31.44	46.41	91.09	77.34
River	97.49	99.05	93.80	99.79	97.59	48.62	98.52	98.43
Aquaculture	88.17	86.45	90.32	78.45	18.49	60.04	83.38	94.37
Residence	92.09	91.88	89.84	70.22	34.70	18.97	94.70	93.39
Road	89.31	85.28	90.48	76.17	19.24	50.68	77.65	89.86
Agriculture	98.71	98.9	98.14	97.82	52.03	78.09	99.00	99.59
Seawater	100	99.71	100.00	99.03	99.34	99.19	100.00	100.00
Bare land	68.31	80.2	1.83	18.48	46.40	41.62	84.14	64.69
	Overall Accuracy	87%	Overall Accuracy	86.72%	Overall Accuracy	55.76%	Overall Accuracy	91.96%
	Kappa Coefficient	0.89	Kappa Coefficient	0.85	Kappa Coefficient	0.50	Kappa Coefficient	0.91

3.5.2. Mangrove Species Classification

The support vector machine performances using Sentinel-1 and SPOT-7 fused images were evaluated by accuracy indexes and the Kappa coefficients (Table 4) for nine classes including mangrove species (Su, Vet, and Ban). Overall, all classes were categorized at high accuracy with around 90% overall accuracy. The Su mangrove (*A. corniculatum*) was the most accurate separation, followed by Vet (*K. obovata*) and Ban (*S. caseolaris*). Water-related layers like aquaculture, river, and seawater were the most accurate categories in contrast to bare land, which had the largest associated uncertainty. The Sentinel-1 VV polarization represented a better data source for mangrove type classification, regardless of image fusion technique, with an overall accuracy of 93% and Kappa coefficient of 0.92; compared to the use of Sentinel-1 VH polarization with overall accuracy of 89% and Kappa coefficient of 0.88. The PCA fusion method produced slightly better accuracy than the GS method, in most cases. It was interesting to look at the accuracy of the SPOT-7 and Sentinel-1 classifications with contradictory results, where the mangrove types were classified at low accuracy (around 50%) using the original SPOT-7 image, while the other classes were well separated (above 90%). UsingSentinel-1 provided the high accuracy of mangrove type and water (river and seawater) classifications (90%), in contrast, other classes such as agriculture and residence were mostly indistinguishable, with producer and user accuracies lower than 20%. This inconsistency of the producer and user accuracies of SPOT-7 and Sentinel-1 made the overall accuracies and the Kappa coefficients lower than those of the fused images for approximately 15% of SPOT-7 and 30% of Sentinel-1.

Table 4. Accuracy indexes calculated from confusion matrices for mangrove species classification assessment using SVM classifier, Prod. Acc. and User. Acc. are short forms of producer and user accuracy; VH_GS, VH_PCA, VV_GS, and VV_PCA are combinations of fused images of SPOT-7 and Sentinel-1 polarization data (VH or VV) and the image fusion methods (GS or PCA).

Class	VH_GS		VH_PCA		VV_GS		VV_PCA		SPOT-7		Sentinel-1	
	Prod. Acc. (Percent)	User Acc. (Percent)	Prod. Acc. (Percent)	User Acc. (Percent)	Prod. Acc. (Percent)	User Acc. (Percent)	Prod. Acc. (Percent)	User Acc. (Percent)	Prod. Acc. (Percent)	User Acc. (Percent)	Prod. Acc. (Percent)	User Acc. (Percent)
Su (*A. corniculatum*)	81.46	81.92	82.03	82.71	77.14	92.95	75.63	93.33	58.43	71.62	76.44	93.23
Vet (*K. obovata*)	87.34	78.65	88.14	79.39	94.13	89.60	94.45	90.50	51.22	41.03	93.87	91.12
Ban (*S. caseolaris*)	62.25	75.80	63.24	76.27	79.84	54.54	81.31	51.46	46.25	50.18	78.56	62.63
Aquaculture	90.94	94.71	91.00	94.54	98.94	93.72	98.92	93.86	95.33	65.95	45.23	38.26
Agriculture	99.21	99.50	99.25	99.52	99.88	100.00	99.88	100.00	98.60	98.48	18.95	21.65
Residence	97.52	96.89	97.23	96.53	96.59	97.29	96.58	97.31	96.07	93.33	18.45	19.34
River	99.80	99.98	99.83	99.98	97.29	99.63	97.29	99.63	97.49	98.73	89.34	91.36
Bare land	78.11	61.44	76.88	61.37	92.65	64.95	92.61	68.34	86.25	86.25	23.15	31.21
Seawater	80.31	75.63	83.21	72.14	97.62	99.97	97.62	99.97	99.96	98.97	94.12	93.67
	Overall Accuracy	89%	Overall Accuracy	90%	Overall Accuracy	93%	Overall Accuracy	93%	Overall Accuracy	78%	Overall Accuracy	60%
	Kappa Coefficient	0.88	Kappa Coefficient	0.88	Kappa Coefficient	0.92	Kappa Coefficient	0.92	Kappa Coefficient	0.74	Kappa Coefficient	0.58

3.5.3. Mangrove Extent Classification

Two confusion matrices were calculated using the ground-truth region of interest for seven classes: mangrove, aquaculture, residence, agriculture, bare land, river, and seawater. The accuracy index of producer, user accuracy, and Kappa coefficients was summarized (Table 5). With only a small number of mangrove pixels misclassified to aquaculture (compared to total pixels), the errors were low, resulting in high user accuracy of mangrove class (97.29%). In general, agriculture and aquaculture, with highly mixed pixels, suffered low accuracy. It is noted that the main task of this classification was for the mangrove layer, however, other classes would have affected the mangrove classification result. Therefore, the road, which was a reported source of error, was left out in this task, after which, the overall accuracy improved. The table showed that the accuracy of the ISODATA was slightly higher than the K-means: approximately 5% of the overall accuracy and 0.06 of the Kappa coefficient. While the seawater presented the most accurate layer, the residence in the ISODATA results (88.11%) and the aquaculture of K-means (76.32%) were classified at the lowest accuracy.

Table 5. Confusion matrices and accuracy indexes created for the unsupervised ISODATA and K-means classifiers using Landsat 8 acquired in 2019; Prod. Acc. and User. Acc. are short forms of producer and user accuracy; μ presents the averaged values.

ISODATA Classified (Pixels)	Ground-Truth (Pixels)								Summary	
	Mangrove	Aquaculture	Residence	Agriculture	Bare land	River	Seawater	Total	Prod. Acc. (Percent)	User Acc. (Percent)
Mangrove	1650	25	0	14	2	4	1	1696	99.16	97.29
Aquaculture	14	905	0	46	7	11	0	983	97.00	92.07
Residence	0	0	289	0	4	0	0	293	88.11	98.63
Agriculture	0	2	0	972	11	0	0	985	93.82	98.68
Bare land	0	0	39	4	589	0	0	632	95.93	93.20
River	0	0	0	0	0	1243	9	1252	98.81	99.28
Seawater	0	1	0	0	1	0	3382	3384	99.71	99.94
Total	1664	933	328	1036	614	1258	3392	9225	96.08μ	97.01μ
									Overall Accuracy = 97.89%	Kappa Coefficient = 0.97

K-Means Classified (Pixels)	Ground-Truth (Pixels)								Summary	
	Mangrove	Aquaculture	Residence	Agriculture	Bare land	River	Seawater	Total	Prod. Acc. (Percent)	User Acc. (Percent)
Mangrove	2754	66	1	22	8	1	0	2852	89.77	96.56
Aquaculture	261	664	0	1	0	26	0	952	76.32	69.75
Residence	0	0	255	0	9	0	0	264	95.86	96.59
Agriculture	1	0	0	744	37	0	0	782	95.38	95.14
Bare land	0	0	10	13	394	0	0	417	87.95	94.48
River	52	140	0	0	0	783	0	975	96.67	80.31
Seawater	0	0	0	0	0	0	2889	2889	100	100
Total	3068	870	266	780	448	810	2889	9131	89.77μ	96.56μ
									Overall Accuracy = 92.90%	Kappa Coefficient = 0.91

4. Discussion

Scientists and researchers have attempted to improve the accuracy of remote sensing image classification for many uses including mangrove classifications [31,72]. Consequently, dozens of image classifiers have been developed and separated into supervised and unsupervised methods [6,73]. Each classifier has its own advantages and disadvantages for a particular use, therefore choosing a single "best" classification method is a challenge. In this study, we used four machine learning algorithms: ANN, DT, RF, and SVM for mangrove age and species classification. SVM demonstrated the greatest accuracy, however it would be premature to conclude that the SVM is better than the others in remotely sensed data classification and the results found here will need to be supported by further case studies.

Image fusion is considered to improve the quality of fused images [11,74] and allows the use of different sources of data for specific applications, particularly in the context of increasing remote sensing availability. Fusing optical and SAR remote scenes is commonly undertaken to enhance cartographic object extraction and improve spatial resolution [14] as well as reducing the effects of clouds in optical images [10,15,75]. It is nevertheless difficult to say whether fused images are always better for a particular use or not. Combining more data layers can be a source of error if the added information does not support the target aim. Therefore, image pre-processing implementation is sometimes needed alongside references to previous literature. For example, we undertook backscatter analyses (Figure 9) to develop a better understanding of SAR backscatter distributions under different LULC. This allowed us to decide which data were best used for what purpose. Figure 9A's mean backscatter values for mangrove age (10, 5, and 3 years) were similar to the values for the agriculture, road, and bare land classes (both VH and VV polarizations), therefore justifying the use of Sentinel-1 data for the mangrove age classification. The mean backscatter values of mangrove species (Su, Vet, and Ban) were distinguished clearly from the mean values in other layers (Figure 9B), especially with the Sentinel-1 VV polarization. This could explain why the use of Sentinel-1 VV polarization generated the most accurate mangrove species classification.

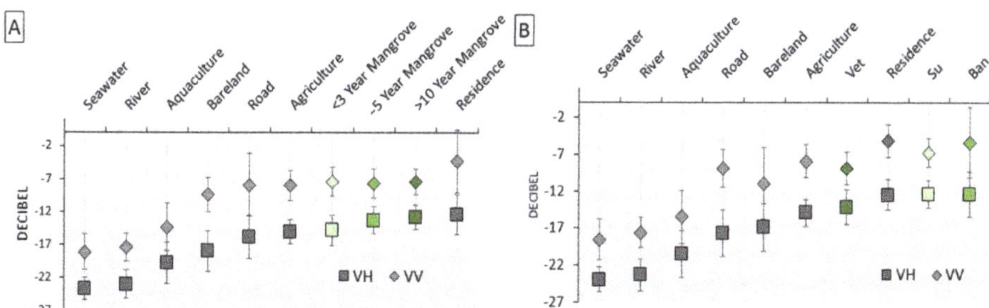

Figure 9. Sentinel-1 VH and VV backscatters on different land use and land cover; training data were used for (**A**) mangrove age and (**B**) mangrove species; V and H are vertical and horizontal, respectively, and coupled letters of VH and VV indicate SAR cross-polarizations.

Optical remote sensing, here using SPOT-7, was suitable for mangrove age and growth classification, but showed more limited capacity for classifying mangrove species for the study site of Thuy Truong. First, the received reflectance values from green vegetation surfaces (here mangrove) in optical images varied with wavelength (bands) [76]. With the SPOT-7 multispectral bands of Blue (0.455 μm–0.525 μm), Green (0.530 μm–0.590 μm), Red (0.625 μm–0.695 μm), and Near-Infrared (0.760 μm–0.890 μm), the mangrove could be discriminated from other land uses [77], even considering different ages and growth stages [78]. However, minor differences in reflectance between mangrove species could be found at the 0.760 μm–0.890 μm (NIR) [79]. In addition, the Ban mangrove (Figure 10A), Vet (Figure 10B), and Su (Figure 10C) were very similar in terms of stand, leaf, and stems based on our observations, so this also makes it difficult to discriminate between them.

Figure 10. Pictures of three mangrove species taken by the authors in Thuy Truong commune on 22 November 2018.

In terms of LULC classification accuracy in remotely sensed data processes, confusion matrices are most frequently used [71] to provide analyses of the spatial distribution of errors and a better understanding of non-stationarity in land cover errors [80]. Although these measures of accuracy are very simple [81] and widely used, it is critical that the sources of errors are not revealed. Pontius and Millones (2011) identified limitations of the Kappa indices, for example, it does not report the correct proportion, and gives information that is redundant or misleading for practical decision making [82]. With identical inputs, and comparing the accuracy of results across two or more algorithms, we could determine which method tends to generate better outputs given our specific aim. However, it is still difficult to quantify method-based errors. Uncertainty could come from the data used, perhaps as a mixed-pixel problem related to coarse spatial resolution [83], geographical distortions, atmospheric effects, or seasonal effects [84]. To prevent the effects of seasonal changes on the mangrove surface from impacting on the image classification results, we tried to collect the Landsat-X images in the same season of autumn (October–November). However, this is sometimes a challenging task.

It is useful to look at the past to understand how the present situation was reached. Mangrove extent changes have been explored by many researchers [28,31,85] to inform management practices and protect habitats. Thanks to Earth observation data archives, the ability to use remotely sensed data in these assessments is becoming more widely available and often free of charge. Most studies investigate negative aspects such as mangrove degradation, fragmentation, and conversion to other land use types [32,86–88]. Our study has found a positive outcome, with mangrove forest developing from nearly nothing (in 1975) into a large mangrove forest (in 2019), thanks to efforts of the local community, government, and philanthropic projects. Ground-truth data cannot be obtained from the past to undertake supervised mangrove classification, but the unsupervised approaches, considered less accurate than supervised algorithms [6], remain helpful. The changes in mangrove extent in Thuy Truong identified in this study and the methods for using remotely sensed data tested will be valuable to monitoring and evaluation assessments of plantation projects in the region.

5. Conclusions

We report the use of four machine learning algorithms: ANN, DT, RF, and SVM for classifying mangrove ages. The estimated ages agreed well with the ground-truth field data, and the SVM was found to be the most accurate algorithm for mangrove species classification. Sentinel-1 backscatter mechanisms for different mangrove species are basically a function of tree structure, height, and density, which, combined with multispectral bands of SPOT-7, allowed us to discriminate between the Ban mangrove, Su, and Vet at an acceptable accuracy. Our assessment of multi-decadal mangrove changes in extent used the Landsat-X image series. We found a fluctuation in the first two decades, then a constant expansion of mangrove forest during the period 1998 to 2019. For the accuracy assessment, confusion matrices, producer–user accuracy, overall accuracy, and Kappa coefficients were used to measure the extent of agreement between image-based extraction and ground-truth data. These accuracy indices showed that all the classifications were accurate, and generally greater than 75%. Further research should test SAR and optical image fusion on other mangrove species as we found supportive information of SAR backscatters for classifying different mangrove species, and gained finer resolution of the panchromatic layer of optical images.

Author Contributions: Conceptualization, N.H.Q., C.R.H., L.C.S., and L.T.V.H.; Methodology, N.H.Q., L.C.S., R.C., and C.H.Q.; Field investigation, D.V.T., R.C., C.R.H., and N.H.Q.; Data analysis, N.H.Q., P.T.T.N., C.H.Q., D.V.T., and L.T.V.H.; Writing—original draft preparation, N.H.Q.; Writing—review and editing, N.H.Q., C.R.H., C.H.Q., L.C.S., P.T.T.N, L.T.V.H., R.C., and D.V.T; Visualization, N.H.Q., R.C., and C.R.H.; Supervision, C.H.Q., L.C.S., P.T.T.N., and L.T.V.H.; Project administration, C.H.Q. and L.T.V.H. All authors have read and agreed to the published version of the manuscript.

Funding: This research was financially supported by the Newton RCUK-SEAMED project "Harnessing multiple benefits from resilient mangrove systems" funded by NAFOSTED RCUK, ESRC reference: ES/R003300/1. Publication was supported by UKRI via the University of Leeds Open Access Fund.

Acknowledgments: The authors would like to thank all the anonymous reviewers and the assistant editor Ms. Milica Kovačević for their constructive comments and suggestions on this manuscript.

Conflicts of Interest: The authors declare no conflicts of interest.

Abbreviations

List of abbreviations in this study.

No	Abbreviation	Full Name
1	ANN	Artificial Neural Network
2	DT	Decision Tree
3	GPS	Global Positioning System
4	GS	Gram–Schmidt
5	H	Horizontal
6	IHS	Intensity-Hue-Saturation
7	ISODATA	Iterative Self-organizing Data Analysis Technique
8	LULC	Land Use and Land Cover
9	MS	Multispectral
10	NIR	Near Infrared
11	PAN	Panchromatic
12	PCA	Principal Component Analysis
13	RF	Random Forest
14	RRD	Red River Delta
15	SAR	Synthetic Aperture Radar
16	SVM	Support Vector Machine
17	USGS	United States Geological Survey
18	V	Vertical

References

1. Cummings, C.A.; Todhunter, P.E.; Rundquist, B.C. Using the Hazus-MH flood model to evaluate community relocation as a flood mitigation response to terminal lake flooding: The case of Minnewaukan, North Dakota, USA. *Appl. Geogr.* **2012**, *32*, 889–895. [CrossRef]
2. Heenkenda, M.K.; Joyce, K.E.; Maier, S.W.; de Bruin, S. Quantifying mangrove chlorophyll from high spatial resolution imagery. *ISPRS J. Photogramm. Remote Sens.* **2015**, *108*, 234–244. [CrossRef]
3. VEA, The Vietnam environment administration. In *Ministry of Natural Resources and Environment*; 2016. Available online: http://vea.gov.vn/vn/truyenthong/biendoikhihau/Pages/90KH.aspx (accessed on 28 October 2019).
4. Hoa, N.H. Using Landsat imagery and vegetation indices differencing to detect mangrove change: A case in Thai Thuy District, Thai Binh Province. *J. For. Sci. Technol.* **2016**, *5*, 59–66.

5. Bakhtiyari, M.; Lee, S.Y.; Warnken, J. Seeing the forest as well as the trees: An expert opinion approach to identifying holistic condition indicators for mangrove ecosystems. *Estuar. Coast. Shelf Sci.* **2019**, *222*, 183–194. [CrossRef]

6. Hasmadi, M.; Pakhriazad, H.; Shahrin, M. Evaluating supervised and unsupervised techniques for land cover mapping using remote sensing data. *Geogr. Malays. J. Soc. Space* **2009**, *5*, 1–10.

7. Hepner, G.; Logan, T.; Ritter, N.; Bryant, N. Artificial neural network classification using a minimal training set- Comparison to conventional supervised classification. *Photogramm. Eng. Rem. Sci.* **1990**, *56*, 469–473.

8. Sohn, Y.; Rebello, N.S. Supervised and unsupervised spectral angle classifiers. *Photogramm. Eng. Rem. Sci.* **2002**, *68*, 1271–1282.

9. Amarsaikhan, D.; Douglas, T. Data fusion and multisource image classification. *Int. J. Remote Sens.* **2004**, *25*, 3529–3539. [CrossRef]

10. Belgiu, M.; Stein, A. Spatiotemporal Image Fusion in Remote Sensing. *Remote Sens.* **2019**, *11*, 818. [CrossRef]

11. Mangolini, M. Apport de la fusion d'images satellitaires multicapteurs au niveau pixel en télédétection et photo-interprétation. Ph.D. Thesis, Université de Nice Sophia-Antipolis, Nice, France, 1994.

12. Ehlers, M. Spectral characteristics preserving image fusion based on Fourier domain filtering. in Remote Sensing for Environmental Monitoring, GIS Applications, and Geology IV. In Proceedings of the International Society for Optics and Photonics (SPIE), SPIE Bellingham, WA, USA, 22 October 2004.

13. Rokni, K.; Ahmad, A.; Solaimani, K.; Hazini, S. A new approach for surface water change detection: Integration of pixel level image fusion and image classification techniques. *Int. J. Appl. Earth Obs.* **2015**, *34*, 226–234. [CrossRef]

14. Pohl, C.; van Genderen, J.L. Review article multisensor image fusion in remote sensing: Concepts, methods and applications. *Int. J. Remote Sens.* **1998**, *19*, 823–854. [CrossRef]

15. Quang, N.H.; Tuan, V.A.; Hao, N.T.P.; Hang, L.T.T.; Hung, N.M.; Anh, V.L.; Phuong, L.T.M.; Carrie, R. Synthetic aperture radar and optical remote sensing image fusion for flood monitoring in the Vietnam lower Mekong basin: A prototype application for the Vietnam Open Data Cube. *Eur. J. Remote Sens.* **2019**, *52*, 599–612.

16. Solberg, A.H.S.; Jain, A.K.; Taxt, T. Multisource classification of remotely sensed data: Fusion of Landsat TM and SAR images. *IEEE Trans. Geosci. Remote* **1994**, *32*, 768–778. [CrossRef]

17. Wang, L.; Jia, M.; Yin, D.; Tian, J. A review of remote sensing for mangrove forests: 1956–2018. *Remote Sens. Environ.* **2019**, *231*, 111223. [CrossRef]

18. Pham, T.D.; Le, N.N.; Ha, N.T.; Nguyen, L.V.; Xia, J.; Yokoya, N.; To, T.T.; Trinh, H.X.; Kieu, L.Q.; Takeuchi, W. Estimating Mangrove Above-Ground Biomass Using Extreme Gradient Boosting Decision Trees Algorithm with Fused Sentinel-2 and ALOS-2 PALSAR-2 Data in Can Gio Biosphere Reserve, Vietnam. *Remote Sens.* **2020**, *12*, 777. [CrossRef]

19. Proisy, C.; Couteron, P.; Fromard, F. Predicting and mapping mangrove biomass from canopy grain analysis using Fourier-based textural ordination of IKONOS images. *Remote Sens. Environ.* **2007**, *109*, 379–392. [CrossRef]

20. Wicaksono, P.; Danoedoro, P.; Hartono; Nehren, U. Mangrove biomass carbon stock mapping of the Karimunjawa Islands using multispectral remote sensing. *Int. J. Remote Sens.* **2016**, *37*, 26–52. [CrossRef]

21. Aslan, A.; Rahman, A.F.; Warren, M.W.; Robeson, S.M. Mapping spatial distribution and biomass of coastal wetland vegetation in Indonesian Papua by combining active and passive remotely sensed data. *Remote Sens. Environ.* **2016**, *183*, 65–81. [CrossRef]

22. Vidhya, R.; Vijayasekaran, D.; Farook, M.A.; Jai, S.; Rohini, M.; Sinduja, A.; Vi, C.; Vi, W.G. Improved classification of mangroves health status using hyperspectral remote sensing data. *Int. Arch. Photogramm. Remote Sens. Spat. Inf. Sci.* **2014**, *40*, 667. [CrossRef]

23. Chellamani, P.; Singh, C.P.; Panigrahy, S. Assessment of the health status of Indian mangrove ecosystems using multi temporal remote sensing data. *Trop. Ecol.* **2014**, *55*, 245–253.

24. Hernández-Clemente, R.; North, P.R.; Hornero, A.; Zarco-Tejada, P.J. Assessing the effects of forest health on sun-induced chlorophyll fluorescence using the FluorFLIGHT 3-D radiative transfer model to account for forest structure. *Remote Sens. Environ.* **2017**, *193*, 165–179. [CrossRef]

25. Mohammed, G.H.; Colombo, R.; Middleton, E.M.; Rascher, U.; van der Tol, C.; Nedbal, L.; Goulas, Y.; Pérez-Priego, O.; Damm, A.; Meroni, M.; et al. Remote sensing of solar-induced chlorophyll fluorescence (SIF) in vegetation: 50 years of progress. *Remote Sens. Environ.* **2019**, *231*, 111177. [CrossRef]

26. Pastor-Guzman, J.; Atkinson, P.M.; Dash, J.; Rioja-Nieto, R. Spatiotemporal variation in mangrove chlorophyll concentration using Landsat 8. *Remote Sens.* **2015**, *7*, 14530–14558. [CrossRef]

27. Vaiphasa, C. Remote sensing techniques for mangrove mapping. Ph.D. Thesis, International Institute for Geo-information Science and Earth Observation (ITC), Enschede and Wageningen University, Enschede, The Netherlands, 2006.

28. Green, E.P.; Clark, C.D.; Mumby, P.J.; Edwards, A.J.; Ellis, A.C. Remote sensing techniques for mangrove mapping. *Int. J. Remote Sens.* **1998**, *19*, 935–956. [CrossRef]

29. Blasco, F.; Gauquelin, T.; Rasolofoharinoro, M.; Denis, J.; Aizpuru, M.; Caldairou, V. Recent advances in mangrove studies using remote sensing data. *Mar. Freshwater Res.* **1998**, *49*, 287–296. [CrossRef]

30. Hoan, N.T.; Duong, N.D.; Tateishi, R. Combination of ADEOS II–GLI and MODIS 250m Data for Land Cover Mapping of Indochina Peninsula. In Proceedings of the 26th Asian Conference on Remote Sensing and 2nd Asian Space Conference, Hanoi, Vietnam, 7–11 November 2005.

31. Lymburner, L.; Bunting, P.; Lucas, R.; Scarth, P.; Alam, I.; Phillips, C.; Ticehurst, C.; Held, A. Mapping the multi-decadal mangrove dynamics of the Australian coastline. *Remote Sens. Environ.* **2020**, *238*, 111185. [CrossRef]

32. Long, J.B.; Giri, C. Mapping the Philippines' mangrove forests using Landsat imagery. *Sensors* **2011**, *11*, 2972–2981. [CrossRef]

33. Tong, P.H.S.; Auda, Y.; Populus, J.; Aizpuru, M.; Habshi, A.A.; Blasco, F. Assessment from space of mangroves evolution in the Mekong Delta, in relation to extensive shrimp farming. *Int. J. Remote Sens.* **2004**, *25*, 4795–4812. [CrossRef]

34. Tri, N.H.; Adger, W.; Kelly, P. Natural resource management in mitigating climate impacts: The example of mangrove restoration in Vietnam. *Glob. Environ. Chang.* **1998**, *8*, 49–61.

35. Tri, N.H.; Adger, W.N.; Kelly, P.M. Social vulnerability to climate change and extremes in coastal Vietnam. *World Dev.* **1999**, *27*, 249–269.

36. Pham, Q.T. An analysis of soil characteristics for agricultural land use orientation in Thai Thuy District, Thai Binh Province. *VNU J. Sci. Earth Sci.* **2007**, *23*, 105–109.

37. Powell, N.; Osbeck, M.; Tan, S.B.; Toan, V.C. Mangrove restoration and rehabilitation for climate change adaptation in Vietnam. World Resources Report, Washington DC. *World Resour. Rep.* **2011**, 1–22.

38. Macintosh, D.J.; Ashton, E.C. *A Review of Mangrove Biodiversity Conservation and Management*; Centre for Tropical Ecosystems Research: Aarhus, Denmark, 2002.

39. Giang, H.; Manh, D.; Huy, N. Use of Salt-Marsh Site Classification for Mangrove Forest Development and Reforestation in the Coastal Area of Thai Binh Province in the Context of Climate Change. In *International Conference on Asian and Pacific Coasts*; Springer: Berlin/Heidelberg, Germany, 2019.

40. Ouaidrari, H.; Vermote, E.F. Operational atmospheric correction of Landsat TM data. *Remote Sens. Environ.* **1999**, *70*, 4–15. [CrossRef]

41. Lu, D.; Mausel, P.; Brondizio, E.; Moran, E. Assessment of atmospheric correction methods for Landsat TM data applicable to Amazon basin LBA research. *Int. J. Remote Sens.* **2002**, *23*, 2651–2671. [CrossRef]

42. Rahman, M.R.; Thakur, P.K. Detecting, mapping and analysing of flood water propagation using synthetic aperture radar (SAR) satellite data and GIS: A case study from the Kendrapara District of Orissa State of India. *Egypt. J. Remote Sens. Space Sci.* **2017**, *21*, S37–S41. [CrossRef]

43. Cian, F.; Marconcini, M.; Ceccato, P. Normalized Difference Flood Index for rapid flood mapping: Taking advantage of EO big data. *Remote Sens. Environ.* **2018**, *209*, 712–730. [CrossRef]

44. Nazim, K.; Ahmed, M.; Shaukat, S.S.; Khan, M.U.; Ali, Q.M. Age and growth rate estimation of grey mangrove Avicennia marina (Forsk). *Vierh Pakistan. Pak. J. Bot.* **2013**, *45*, 535–542.

45. Duarte, C.M.; Thampanya, U.; Terrados, J.; Geertz-Hansen, O.; Fortes, M.D. The determination of the age and growth of SE Asian mangrove seedlings from internodal counts. *Mangroves Salt Marshes.* **1999**, *3*, 251–257. [CrossRef]

46. Maxwell, A.E.; Warner, T.A.; Fang, F. Implementation of machine-learning classification in remote sensing: An applied review. *Int. J. Remote Sens.* **2018**, *39*, 2784–2817. [CrossRef]

47. Yu, L.; Liang, L.; Wang, J.; Zhao, Y.; Cheng, Q.; Hu, L.; Liu, S.; Yu, L.; Wang, X.; Zhu, P.; et al. Meta-discoveries from a synthesis of satellite-based land-cover mapping research. *Int. J. Remote Sens.* **2014**, *35*, 4573–4588. [CrossRef]

48. Foody, G.M. Relating the land-cover composition of mixed pixels to artificial neural network classification output. *Photogramm. Eng. Rem. Sci.* **1996**, *62*, 491–498.

49. Dreiseitl, S.; Ohno-Machado, L. Logistic regression and artificial neural network classification models: A methodology review. *J. Biomed. Inform.* **2002**, *35*, 352–359. [CrossRef]

50. Schalkoff, R. *Pattern Recognition: Statistical, Structural and Neural Approaches*; John Wiley: Toronto, ON, Canada, 1992.

51. Pal, M.; Mather, P.M. An assessment of the effectiveness of decision tree methods for land cover classification. *Remote Sens. Environ.* **2003**, *86*, 554–565. [CrossRef]

52. Brodley, C.E.; Utgoff, P.E. *Multivariate Versus Univariate Decision Trees*; Department of Computer and Information Science, University of Massachusetts: Amherst, MA, USA, 1992.

53. Breiman, L. Random forests. *Mach. Learn.* **2001**, *45*, 5–32. [CrossRef]

54. Noi, P.T.; Degener, J.; Kappas, M. Comparison of multiple linear regression, cubist regression, and random forest algorithms to estimate daily air surface temperature from dynamic combinations of MODIS LST data. *Remote Sens.* **2017**, *9*, 398. [CrossRef]

55. Breiman, L. Bagging predictors. *Mach. Learn.* **1996**, *24*, 123–140. [CrossRef]

56. Quinlan, J.R. *C4. 5: Programs for Machine Learning*; Elsevier: Amsterdam, The Netherlands, 2014.

57. Ben-Hur, A.; Horn, D.; Siegelmann, H.T.; Vapnik, V. Support vector clustering. *J. Mach. Learn. Res.* **2001**, *2*, 125–137. [CrossRef]

58. Vapnik, V.N. *The Nature of Statistical Learning Theory*; Springer: New York, NY, USA, 1995.

59. Wu, T.F.; Lin, C.J.; Weng, R.C. Probability estimates for multi-class classification by pairwise coupling. *J. Mach. Learn. Res.* **2004**, *5*, 975–1005.

60. Hsu, C.W.; Chang, C.C.; Lin, C.J. *A Practical Guide to Support Vector Classification*; National Taiwan University: Taipei, Taiwan, 2003; pp. 1396–1400.

61. Laben, C.A.; Brower, B.V. Process for Enhancing the Spatial Resolution of Multispectral Imagery Using Pan-sharpening. U.S. Patent No. 6,011,875, 4 January 2000.

62. Abdi, H.; Williams, L.J. Principal component analysis. *Wiley Interdiscip. Rev. Comput. Stat.* **2010**, *4*, 433–459. [CrossRef]

63. Aiazzi, B.; Alparone, L.; Baronti, S.; Selva, M. MS+ Pan image fusion by an enhanced Gram–Schmidt spectral sharpening. In Proceedings of the 26th EARSeL symposium, Warsaw, Poland, 29 May–1 June 2006; Millpress: Rotterdam, The Netherlands, 2007.

64. Kumar, U.; Mukhopadhyay, C.; Ramachandra, T. Pixel based fusion using IKONOS imagery. *Int. J. Recent Trends Eng.* **2009**, *1*, 173.

65. Kamal, M.; Phinn, S. Hyperspectral data for mangrove species mapping: A comparison of pixel-based and object-based approach. *Remote Sens.* **2011**, *3*, 2222–2242. [CrossRef]

66. Wang, L.; Sousa, W.P.; Gong, P.; Biging, G.S. Comparison of IKONOS and QuickBird images for mapping mangrove species on the Caribbean coast of Panama. *Remote Sens. Environ.* **2004**, *91*, 432–440. [CrossRef]

67. Held, A.; Ticehurst, C.; Lymburner, L.; Williams, N. High resolution mapping of tropical mangrove ecosystems using hyperspectral and radar remote sensing. *Int. J. Remote Sens.* **2003**, *24*, 2739–2759. [CrossRef]

68. Quang, N.H.; Tuan, V.A.; Le Hang, T.T.; Manh Hung, N.; Thi Dieu, D.; Duc Anh, N.; Hackney, C.R. Hydrological/Hydraulic Modeling-Based Thresholding of Multi SAR Remote Sensing Data for Flood Monitoring in Regions of the Vietnamese Lower Mekong River Basin. *Water* **2020**, *12*, 71. [CrossRef]

69. El-Rahman, S.A. Hyperspectral image classification using unsupervised algorithms. *IJACSA Int. J. Adv. Comput. Sci. Appl.* **2016**, *7*, 198–205.

70. Zhuang, X.; Engel, B.A.; Xiong, X.; Johannsen, C.J. Analysis of classification results of remotely sensed data and evaluation of classification algorithms. *Photogramm. Eng. Remote Sci.* **1995**, *61*, 427–432.

71. Congalton, R.G. A review of assessing the accuracy of classifications of remotely sensed data. *Remote Sens. Environ.* **1991**, *37*, 35–46. [CrossRef]

72. Kuenzer, C.; Bluemel, A.; Gebhardt, S.; Quoc, T.V.; Dech, S. Remote sensing of mangrove ecosystems: A review. *Remote Sens.* **2011**, *3*, 878–928. [CrossRef]

73. Thakur, S.; Mondal, I.; Ghosh, P.B.; Das, P.; De, T.K. A review of the application of multispectral remote sensing in the study of mangrove ecosystems with special emphasis on image processing techniques. *Spat. Inf. Res.* **2020**, *28*, 39–51. [CrossRef]

74. Ehlers, M. Multisensor image fusion techniques in remote sensing. *ISPRS J Photogramm.* **1991**, *46*, 19–30. [CrossRef]

75. Lu, Z.; Dzurisin, D.; Jung, H.S.; Zhang, J.; Zhang, Y. Radar image and data fusion for natural hazards characterisation. *Int. J. Image Data Fusion* **2010**, *1*, 217–242. [CrossRef]

76. Vaiphasa, C.; Ongsomwang, S.; Vaiphasa, T.; Skidmore, A.K. Tropical mangrove species discrimination using hyperspectral data: A laboratory study. *Estuar. Coast. Shelf Sci.* **2005**, *65*, 371–379. [CrossRef]

77. Rahimizadeh, N.; Kafaky, S.B.; Sahebi, M.R.; Mataji, A. Forest structure parameter extraction using SPOT-7 satellite data by object-and pixel-based classification methods. *Environ. Monit. Assess.* **2020**, *192*, 43. [CrossRef] [PubMed]

78. Rasolofoharinoro, M.; Blasco, F.; Bellan, M.F.; Aizpuru, M.; Gauquelin, T.; Denis, J. A remote sensing based methodology for mangrove studies in Madagascar. *Int. J. Remote Sens.* **1998**, *19*, 1873–1886. [CrossRef]

79. Wan, L.; Lin, Y.; Zhang, H.; Wang, F.; Liu, M.; Lin, H. GF-5 Hyperspectral Data for Species Mapping of Mangrove in Mai Po, Hong Kong. *Remote Sens.* **2020**, *12*, 656. [CrossRef]

80. Comber, A.; Fisher, P.; Brunsdon, C.; Khmag, A. Spatial analysis of remote sensing image classification accuracy. *Remote Sens. Environ.* **2012**, *127*, 237–246. [CrossRef]

81. Story, M.; Congalton, R.G. Accuracy assessment: A user's perspective. *Photogramm. Eng. Remote Sci.* **1986**, *52*, 397–399.

82. Pontius, R.G.; Millones, M. Death to Kappa: Birth of quantity disagreement and allocation disagreement for accuracy assessment. *Int. J. Remote Sens.* **2011**, *32*, 4407–4429. [CrossRef]

83. Hsieh, P.-F.; Lee, L.C.; Chen, N.-Y. Effect of spatial resolution on classification errors of pure and mixed pixels in remote sensing. *IEEE Trans. Geosci. Remote* **2001**, *39*, 2657–2663. [CrossRef]

84. Hashiba, H.; Kameda, K.; Sugimura, T.; Takasaki, K. Analysis of landuse change in periphery of Tokyo during last twenty years using the same seasonal landsat data. *Adv. Space Res.* **1998**, *22*, 681–684. [CrossRef]

85. Manson, F.; Loneragan, N.R.; McLeod, I.M.; Kenyon, R.A. Assessing techniques for estimating the extent of mangroves: Topographic maps, aerial photographs and Landsat TM images. *Mar. Freshw. Res.* **2001**, *52*, 787–792. [CrossRef]

86. Vo, Q.T.; Oppelt, N.; Leinenkugel, P.; Kuenzer, C. Remote sensing in mapping mangrove ecosystems—An object-based approach. *Remote Sens.* **2013**, *5*, 183–201. [CrossRef]

87. Kirui, K.B.; Kairo, J.G.; Bosire, J.; Viergever, K.M.; Rudra, S.; Huxham, M.; Briers, R.A. Mapping of mangrove forest land cover change along the Kenya coastline using Landsat imagery. *Ocean Coast. Manag.* **2013**, *83*, 19–24. [CrossRef]

88. Terchunian, A.; Klemas, V.; Segovia, A.; Alvarez, A.; Vasconez, B.; Guerrero, L. Mangrove mapping in Ecuador: The impact of shrimp pond construction. *Environ. Manag.* **1986**, *10*, 345–350. [CrossRef]

 *remote sensing*

Article

Remote Sensing of Mangroves and Estuarine Communities in Central Queensland, Australia

Debbie Chamberlain [1,2,*], **Stuart Phinn** [2] **and Hugh Possingham** [1,3]

[1] Centre for Biodiversity and Conservation Science, School of Biological Sciences, The University of Queensland, St. Lucia, QLD 4072, Australia; h.possingham@uq.edu.au

[2] Remote Sensing Research Centre, School of Earth and Environmental Sciences, The University of Queensland, St. Lucia, QLD 4072, Australia; s.phinn@uq.edu.au

[3] The Nature Conservancy, 4245 Fairfax Drive, Arlington, VA 22203, USA

[*] Correspondence: d.chamberlain@uq.edu.au; Tel.: +61-4-0182-8535

Received: 21 November 2019; Accepted: 3 January 2020; Published: 6 January 2020

Abstract: Great Barrier Reef catchments are under pressure from the effects of climate change, landscape modifications, and hydrology alterations. With the use of remote sensing datasets covering large areas, conventional methods of change detection can expose broad transitions, whereas workflows that excerpt data for time-series trends divulge more subtle transformations of land cover modification. Here, we combine both these approaches to investigate change and trends in a large estuarine region of Central Queensland, Australia, that encompasses a national park and is adjacent to the Great Barrier Reef World Heritage site. Nine information classes were compiled in a maximum likelihood post classification change analysis in 2004–2017. Mangroves decreased (1146 hectares), as was the case with estuarine wetland (1495 hectares), and saltmarsh grass (1546 hectares). The overall classification accuracies and Kappa coefficient for 2004, 2006, 2009, 2013, 2015, and 2017 land cover maps were 85%, 88%, 88%, 89%, 81%, and 92%, respectively. The cumulative area of open forest, estuarine wetland, and saltmarsh grass (1628 hectares) was converted to pasture in a thematic change analysis showing the "from–to" change. We generated linear regression relationships to examine trends in pixel values across the time series. Our findings from a trend analysis showed a decreasing trend (p value range = 0.001–0.099) in the vegetation extent of open forest, fringing mangroves, estuarine wetlands, saltmarsh grass, and grazing areas, but this was inconsistent across the study site. Similar to reports from tropical regions elsewhere, saltmarsh grass is poorly represented in the national park. A severe tropical cyclone preceding the capture of the 2017 Landsat 8 Operational Land Imager (OLI) image was likely the main driver for reduced areas of shoreline and stream vegetation. Our research contributes to the body of knowledge on coastal ecosystem dynamics to enable planning to achieve more effective conservation outcomes.

Keywords: Landsat; estuary; protected area; land use; land cover; change detection; time series; Great Barrier Reef

---

1. Introduction

Coastal marine ecosystems are among the most diverse and productive in the world, and they provide critical habitats for a wide variety of plants, fish, shellfish, and other wildlife [1–3]. Coastal and near-shore marine ecosystems are facing unprecedented pressures from land use modification. Many studies have analysed change dynamics in wetland ecosystems due to the utilisation of remote sensing techniques [4–6] resulting from a combination of two factors: (1) greater open access to longer time series of image archives and their derived products and (2) more easily accessed tools for using remote sensing data and their products to monitor change from local to global scales. The Landsat satellite archive has provided new opportunities for assessing historical changes in landscapes [7], including coastal ecosystems.

Estuarine wetlands are located at the interface of land and sea and are essential support mechanisms in the marine and terrestrial systems. The inter-realm connectivity of coastal wetlands features strongly in integrated conservation planning approaches [8]. An improved understanding of land–sea connectivity dynamics is crucial to the health of coastal fisheries species' populations [9]. However, connected ecosystems are traditionally studied as separate entities, despite the potential for interactions between them to have consequences for their health and functioning [10]. Estuary-dependent fisheries species are important because they contribute 75% of the total value of Australia's commercial fisheries catches and 90% by numbers of Australia's recreational fish catch [11]. Coastal wetlands that support fisheries are a diverse assemblage of marshes, mangroves, forested wetlands, and estuaries. Mangroves are coastal forests with unique adaptations to saline conditions, and they form a characteristic vegetation zone along sheltered bays, tidal inlets and estuaries in the tropics and subtropics, globally [12,13]. These wetland types fulfil critical roles in ecosystem functions, and they provide many highly valued ecosystem services: raw materials and food, coastal protection, erosion control, water purification, the maintenance of fisheries, carbon sequestration, tourism, recreation, education, and research [14]. Now evident is a drastic decline in ecosystem services on which human society depends from changes in land use and land cover within coastal wetlands [15]. Coastal land use and land cover (LULC) change is illustrated by clearing and modifying coastal habitats and artificial barriers to flow. For example, one of the highest risks to the Great Barrier Reef that has been identified by the Australian Government is the degradation to coastal habitats and connectivity impairment as a result of land use changes affecting the region's ecosystems [16].

The major threats to coastal wetlands are climate change, clearing (through urban areas, ports, and industry development), dieback, changes in hydrology (e.g., the restriction or alteration of flows), and pollution [17–19]. Additionally, overfishing, cattle grazing, pest animals, the use of recreational vehicles, and fire have had impacts on some components of wetland systems [20]. Pressures can be subtle but may result in considerable changes in ecosystem functioning. These threats are often related to, for example, hydrological change (including the development of ponded pasture) may significantly alter water quality, and heavy and sustained grazing pressure of marine grasslands can dramatically alter ground cover. Thus, the modification of ecosystems affects both habitat value and the filtration and retention capacity of those areas [21]. On Queensland's east coast, agriculture and the urban development of infrastructure with berms, ponded pasture, dams, seawalls, and roads on coastal plains impose threats to the resilience of mangroves and associated wetlands [22]. For example, in the Mackay region of Queensland, Pioneer River mangroves have been reclaimed on average by 5 ha each year over the last 50 years [23], with a total loss of 26% since European settlement [24]. Mangrove–fishery links are well-recognised [25], but, to expedite conservation efforts, it is necessary to quantify the spatio–temporal scales of change in mangrove habitats (e.g., disturbance, loss, and regrowth) [26,27]. Indeed, the array of benefits that are offered by wetlands makes it critical that they are monitored, maintained, and restored where and whenever possible [28]. Paradoxically, within Great Barrier Reef coastal provinces, ecosystem effects and cumulative impacts on fishery resources are poorly understood [29]. Moreover, disparate jurisdictional responsibilities hinder assessment efforts. With the ongoing loss of these systems, Australia's commercial and recreational fisheries are becoming depleted nation-wide [30].

Quantifying LULC changes is not only crucial for the evaluation of services but also the protection of coastal wetland ecosystems, and remote sensing technology provides one of the most useful ways to monitor wetland dynamics [6]. Protected areas such as national parks often depend on the landscapes surrounding them and their hydrologic connections to maintain flows of organisms, water, nutrients, and energy. Park managers have little authority over the surrounding landscape, although land use changes and hydrology alterations can have major impacts on the integrity of a protected area [31,32]. In Queensland, public and private lands are surveyed by the Statewide Landcover and Trees Study

(SLATS), which uses satellite imagery to monitor woody vegetation clearing in native vegetation including mangroves and estuarine regions [33], but no studies have focused on using remote sensing to map biome variability and change dynamics in Great Barrier Reef catchments on a landscape scale.

The fundamental broad objectives of this research are to assess the regional drivers of wetland degradation in order to assist in maintaining the values that underpin estuarine ecosystem integrity within and outside the boundary of a protected area. The research presented here can aid the prediction of responses under future change scenarios (e.g., climate shifts/disturbance). The change detection process identifies differences in the state of an object or phenomenon by observation at different times [34]. Possible classification inaccuracies and a lack of consensus in regional-scale LULC approaches necessitates the employment of more than one method for comparative purposes and to aid validation, particularly for change detection in complex landscapes such as coastal wetlands [35].

Our study focuses on change dynamics by using four methods of change analysis: post-classification change analysis with a supervised classification technique, visual interpretation, thematic change dynamics, and trend analysis. There has been an increased use of supervised classification techniques in comparison with unsupervised techniques in the last decade [36]. The supervised pixel-based maximum likelihood ML classification is the most common method in remote sensing image data analysis, and it is often applied as a benchmarking algorithm [37–39]. Supervised change analyses for wetland mapping in remote sensing studies have previously focused on coast line dynamics or reclamation activities [40,41]; few studies have examined wetland change dynamics within and surrounding the border of a protected area and adjacent to a World Heritage Site. We provide a description of observed changes through maps generated for a fourteen-year time period (2004–2017) on a two/three-yearly basis. The research contributes to the fields of land cover characterisation, landscape dynamics, and conservation planning.

The objectives of this study are: (1) to quantify how the coastal landscape (mangroves and associated communities) has spatially and temporally changed in a period of 14 years (2004–2017) within a region that is subjected to intense commercial and recreational fishing; (2) to assess the implications of landscape change to biodiversity within and outside the boundary of a national park; (3) to inform regional land planning, conservation efforts, and policy-makers. In summary, the study addresses the important question: has significant, human-induced change occurred in the coastal landscape, resulting in altered ecosystem function that could have possible repercussions for the fishery resource?

2. Materials and Methods

2.1. Study Area and Data Sources

Our study area is located within the northeast coast drainage division of the Central Queensland coast, specifically the Plane Creek Basin catchment of the Mackay Whitsunday Natural Resource Region, Central Queensland, Australia. Rocky Dam Creek and Cape Palmerston National park are positioned in the Ince Bay Receiving Waters adjacent to the World Heritage listed Great Barrier Reef. The primary intensive land use is the cultivation of sugar cane, making up 18% of the catchment area, with Mackay being the largest sugar-producing region in Australia [42]. Grazing is also an important land use, accounting for 42% [21]. The region's estuaries directly support several commercial fisheries, e.g., East Coast Inshore Fin Fish Fishery, East Coast Otter Trawl Fishery, and Coral Reef Fin Fish Fishery [43]. Additionally, recreational fishing is a considerable activity in the region, with 24.8% of the population participating in fishing for recreation, far greater than the state average of 15.1% [44]. The Mackay Whitsunday Natural Resource Region supports extensive areas of estuarine and mangrove wetlands, these being dominant features of the coastal landscape [45]. Mangroves and associated communities cover 62,094 ha of tidal land in the region, with nine wetland areas recognised as nationally important [46]. The total area of the Rocky Dam Creek sub-catchment is 53,697.5 hectares. Cape Palmerston National Park is listed as a category II protected area on the International Union for Conservation of Nature (IUCN) World Database on Protected Areas [47] and covers 7200 hectares (Figure 1). Ten ecosystem types listed as endangered in the IUCN Red List are present (Table 1 and Figure 2). The areal extent of the sub-catchment and coastal zone used for land cover classification has 53,302.05 hectares of a variety of land cover types. The study area is located between latitude 21°27′–21°37′S and longitude 149°17′–149°26′E.

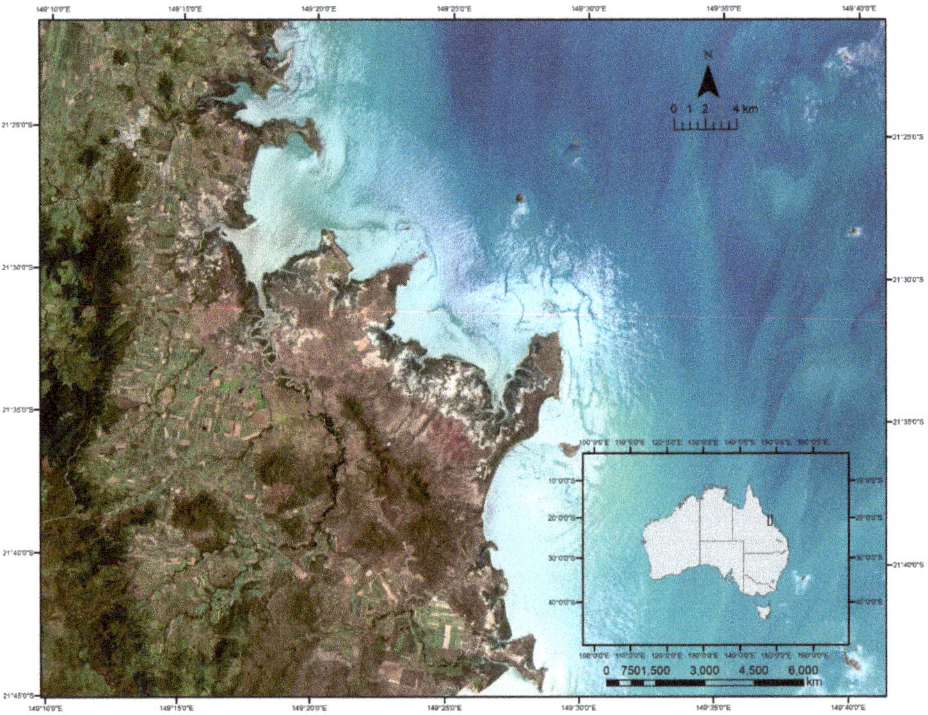

Figure 1. Study site—Rocky Dam Creek and Cape Palmerston National Park Central Queensland—Sentinel-2B composite image visualised by using the red, green, and blue wavelength bands, captured 31 January 2018 at 00:22:57 provided by United States Geological Survey (USGS).

Table 1. (IUCN)-listed endangered ecosystems occurring at the study site [20]—Rocky Dam Creek/Cape Palmerston National Park.

Regional Ecosystem	Extent in Reserves	Description	Structure
8.1.4	Low	*Schoenoplectus subulatus* and/or *Eleocharis dulcis* sedgeland or *Paspalum vaginatum* tussock grassland	Sedgeland
8.1.5	Low	*Melaleuca* spp and/or *Eucalyptus tereticornis* and/or *Corymbia tessellaris* with a ground stratum of salt tolerant grasses and sedges, in a narrow zone adjoining tidal ecosystems	Woodland
8.2.2	Low	Semi-evergreen microphyll vine thicket to vine forest on coastal dunes	Closed forest
8.3.1	Low	Semi-deciduous to evergreen notophyll to mesophyll vine forest and/or sclerophyll emergent forest, fringing streams or in the vicinity of water courses	Closed forest, riverine wetland or fringing riverine wetland

Table 1. *Cont.*

Regional Ecosystem	Extent in Reserves	Description	Structure
8.3.2	Low	*Melaleuca viridiflora* var. *viridiflora* on seasonally inundated alluvial plains with impeded drainage	Woodland on floodplain
8.3.4	Low	Freshwater wetlands with permanent water and aquatic vegetation	Forbland, palustrine wetland (e.g., vegetated swamp)
8.3.5	Low	*Eucalyptus platyphylla* and/or *Lophostemon suaveolens* and/or *Corymbia clarksoniana* woodland on alluvial plains	Open Forest
8.5.3	Low	*Eucalyptus drepanophylla* and/or *Corymbia clarksoniana* and/or *E. platyphylla* and/or *C. dallachiana* and/or *Melaleuca viridiflora* woodland on broad low rises and gently sloping tertiary sand planes	Woodland
8.12.26	Low	*Corymbia tessellaris* and/or *Eucalyptus tereticornis* open forest on hill slopes of islands and coastal areas on Mesozoic-to-Proterozoic igneous rocks, as well as tertiary acid-to-intermediate volcanic rocks; habitat for the Proserpine Rock Wallaby	Open Forest
8.12.27	Low	*Corymbia tessellaris* and/or *Eucalyptus tereticornis* and/or *C. intermedia* and/or *C. clarksoniana* open forest with a secondary tree layer of *Livistona decora* on low hills on Mesozoic-to-Proterozoic igneous rocks	Open Forest

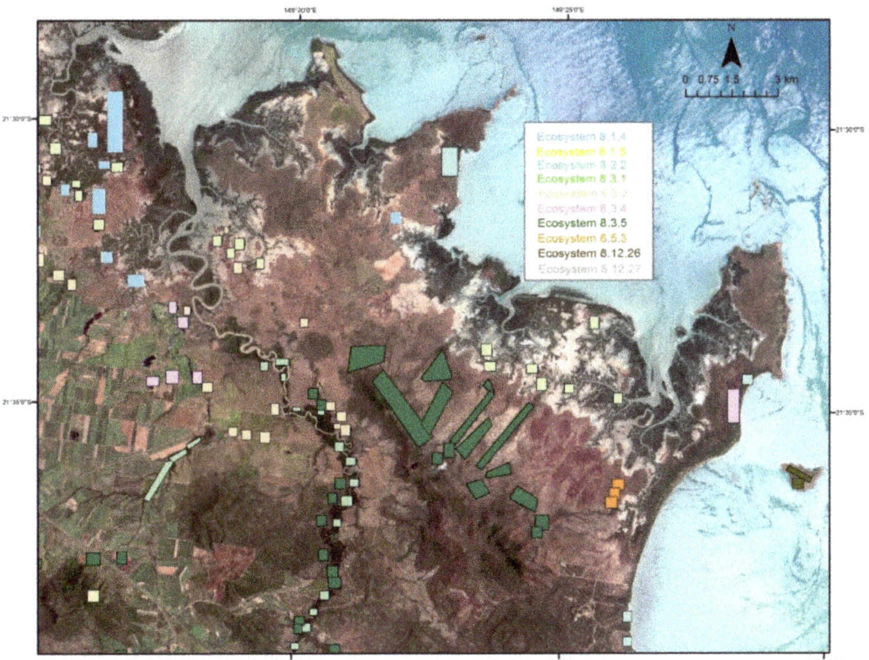

Figure 2. Sentinel-2B image captured 31 January 2018 with IUCN-endangered ecosystems [20]—Rocky Dam Creek/Cape Palmerston National Park.

Seven broadly recognised mangrove communities occur throughout the region. Within the high rainfall areas of the Central Queensland coast bioregion, estuarine wetlands are about equally dominated by saltpan and samphire flats along the high intertidal area; yellow and orange mangroves (*Ceriops tagal* and *Bruguiera* spp.) dominate along the mid-intertidal area; and the stilted mangrove (*Rhizophora stylosa*) dominates in the lower intertidal area [45]. Twenty-three tree and shrub species of mangroves are present [48]. Landscape elevation ranges from 238 m to sea level; therefore, the study site is not solely within the legislative constraints of the defined coastal area of the Queensland Government (i.e., 5 km from the coastline or where land reaches the height of 10 m; Australian Height Datum [29]).

We used the Sentinel-2B image captured on 31 January 2018 (spatial resolution of 10 m) to overlay the major surface water features of the Plane Creek Basin catchment from the Australian Hydrological Geospatial Fabric (Geofabric) [49] (Figure 3). The map provides a hydrological visualisation of topographically consistent spatial surface water features and stream connectivity. Geospatial stream data are useful for natural resource managers, as streams can be traced upstream and downstream to identify drainage networks and water movement within the catchment area. Here, the map is included to provide information on water connections and hydrology throughout the sub-catchment.

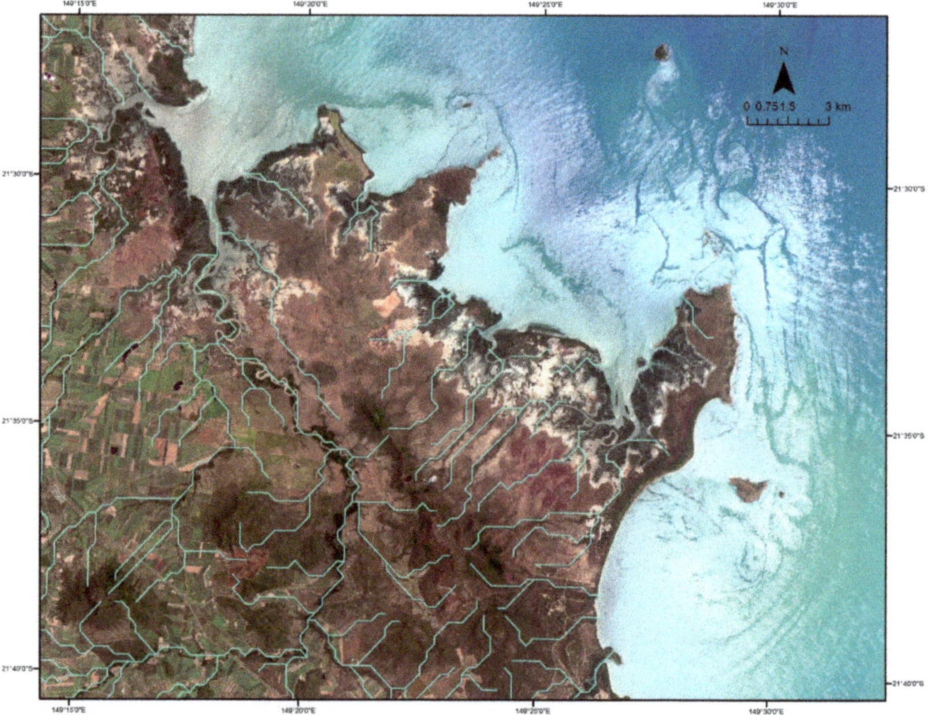

Figure 3. Sentinel-2B image captured 31 January 2018 overlaid with the Australian Hydrological Geospatial Fabric (Geofabric) [49] showing the drainage networks and hydrological connections of Rocky Dam Creek/Cape Palmerston National Park.

For change detection, we used the Landsat satellite archive images captured in April, August, and September for the years 2004, 2006, 2009, 2013, 2015 and 2017 (Table 2). We acquired images from the United States Geological Survey (USGS) Earth Explorer Landsat Archive at level 1T, (except for 2009) which has systematic radiometric and geometric correction applied to the data by incorporating ground control points and topographic accuracy by utilizing a digital elevation model. The 2009 image was derived from USGS Collection 1 and processed at Tier 1. The 2004, 2006 and 2009 images were taken from the Landsat 5 Thematic Mapper (TM). The 2013, 2015 and 2017 images were taken from Landsat 8 OLI (Operational Land Imager) and TIRS (Thermal Infrared Sensor). Data from Landsat satellites are spatially and geometrically consistent, and they comply with UTM projection [50]. We derived maps from imagery acquired in winter and early spring, as cloud cover inhibits image availability in warmer months [51]. Tidal information corresponding to each image date and time are from the Bureau of Meteorology [52] (Table 2). Medium-resolution satellite imagery is suitable for mapping mangrove and wetland areas on a regional scale [53]. There are two reasons for selecting Landsat imagery: (1) It is acquired at regular intervals, and (2) it is freely available from USGS. We acquired data from two independent sources for use as ground truth data: (1) Queensland Herbarium from 2003, 2006, 2009, 2011, 2013, 2015, and 2017 [54] and (2) Google Earth images from 2005, 2009, 2013, and 2016. Local expert knowledge was included for validation, as this has become an important component of mapping methodology [55].

Table 2. Image dates and observed sea levels for Hay Point tidal gauge, Central Queensland.

Image Date	Identifier	Observed Sea Level (m)
13 August 2004	Path 92 Row 75	4.361
3 August 2006	Path 92 Row 75	2.871
28 September 2009	Path 92 Row 75	4.582
7 August 2013	Path 92 Row 75	4.736
12 August 2015	Path 92 Row 75	4.572
27 April 2017	Path 92 Row 75	6.66

2.2. Image Analysis

All Landsat and Sentinel images were acquired from Earth Explorer and georeferenced to UTM WGS 1984. Data collection included satellite images and ancillary data with local expert knowledge. Prior to the classification of the Landsat images, we stacked the total number of bands in each satellite image by producing a composite. The Mask tool was used to extract the study area and exclude urban coastal regions, the inland region, and the open ocean [56]. Expert knowledge informed the land class designation. Thematic maps produced from the supervised classification were used for change detection in the analyses of thematic change dynamics and the time-series (Figure 4). All image analyses were performed in ArcGIS Version 10.6.1.

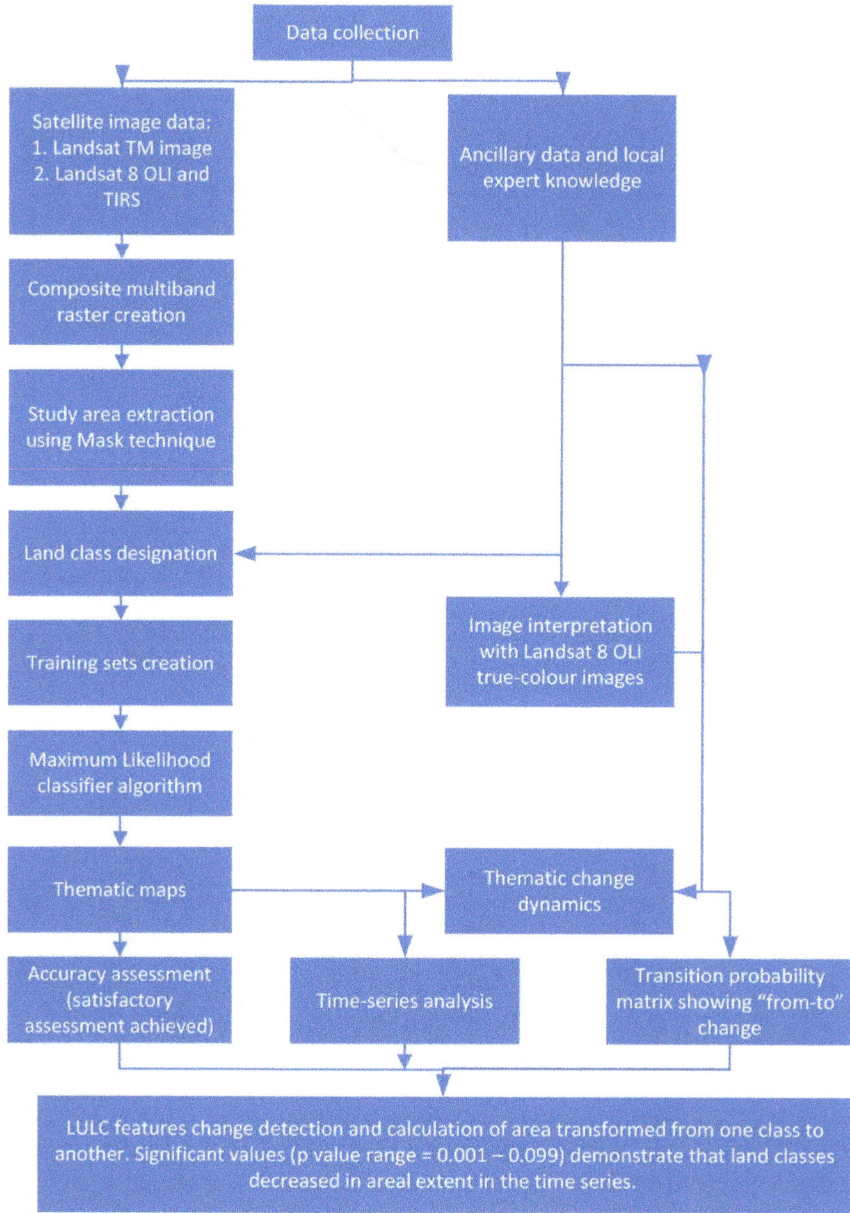

Figure 4. Methodological workflow used for land use and land cover (LULC) change detection.

2.3. Image Classification

We used a supervised classification method and the ML clustering algorithm with composite images of all Landsat bands. Supervised classification has been the most frequent method by which the remotely sensed data of mangrove areas have been classified, and the ML algorithm has been found to be a robust technique that is capable of repeated refinement and reclassification [14]. With supervised training, it is important that the training area be a homogeneous sample of the respective class but at the same time includes the range of variability for the class [57]. Therefore, more than one training area per class was used. An accurate classification depends on the extent of overlap between class signatures. The ML classifier minimizes the total error in the classification if the estimate of the underlying probability distribution is correct [58]. Based on Bayes' theorem, the ML algorithm uses a discrimination function to assign pixels to the class with the highest likelihood [59]. Images were classified by using spectral signatures that were obtained from training samples. Training sample polygons represent distinct sample areas of various land cover types to be classified. Distinguishable classes represented by the training samples were examined from the spectral band characteristics [60]. By using the statistical tools in ArcGIS, we determined the samples to be distinguishable by their histograms and distinct scatter-plots [59]. Between 7 and 140 training polygons were generated for each feature class. We classified images into nine information (land) classes: cropping/grazing, oceanic, sand beach, open forest, mangrove forest, estuarine wetland, saltpan, bare mudflat, and saltmarsh grass. Wetland land-use classes include emergent vegetation, riparian vegetation, and riverine and palustrine wetlands (e.g., vegetated swamps) (Table 3).

Table 3. Land classes used in the classification analysis.

Class	Class Type	Class Description
1	Cropping/Grazing	Lands covered with temporary crops followed by harvest and a bare soil period/pasture land used for grazing cattle
2	Oceanic	Coastal seawater occurring along the coastline and seaward, including the estuaries and mouths of rivers and streams
3	Sand beach	Smooth, sloping accumulations of sand and gravel along the shoreline
4	Estuarine wetland	Coastal tree swamps that are non-tidal, wooded wetlands and are covered or saturated by water for all or part of the year; includes emergent vegetation, riparian vegetations, and riverine and palustrine wetlands (e.g., vegetated swamps). Covers the habitat types of *Melaleuca* spp. and *Eucalyptus* spp,
5	Open forest	Grades from woodland species 18–30 m tall to open forest up to 50 m tall, e.g., *Eucalyptus tereticornis*, *Eucalyptus platyphylla* on parallel dunes, alluvial plains, undulating low hills, lowlands and foothills, frequently with a shrub layer of *Acacia* spp.
6	Mangrove forest	Closed forest to open shrubland of mangrove species, the seaward edge and fringe of waterways dominated by *Rhizophora stylosa*, with *Ceriops tagel* and *Bruguiera* spp in the lower intertidal. Situated on marine clay plains and estuaries
7	Saltpan	Samphire open forbland on saltpans and plains adjacent to mangroves
8	Bare mudflat	Tidal flats of coastal wetland areas where sediments have been deposited by tides and rivers/streams, composed of estuarine silts, clays, and marine animal detritus
9	Saltmarsh grass	*Sporobolus virginicus* tussock grassland and other ground layer species on marine sediments; usually forms a narrow belt between mangroves and alluvial communities in the upper coastal intertidal zone between land and open saltwater or brackish water that is regularly flooded by the tides

2.4. Accuracy Assessment

We visually compared the spectral classes with reference data derived from high spatial resolution true-colour aerial and satellite images in Google Earth and the Queensland Herbarium at corresponding dates to the Landsat images in order to verify land cover classification accuracy. A stratified random sampling design was appropriate for the accuracy assessment. The random point application in

ArcGIS generated approximately 500 unbiased random points, and the validation points were projected onto the classified maps and compared to produce the error matrix. Subsequently, we determined the corresponding change detection thematic map's user's accuracies, producer's accuracies, overall accuracy, and Kappa coefficient [61] (Table 4).

Table 4. Accuracy assessment of Landsat images captured in 2004 and 2017 at Rocky Dam Creek/Cape Palmerston National Park.

Land Class Name	Producer's Accuracy (%)		User's Accuracy (%)	
	2004	2017	2004	2017
Cropping_Grazing	0.84	0.84	0.83	0.91
Oceanic	0.92	0.97	1.00	1.00
Sand beach	1.00	0.83	1.00	0.50
Open forest	0.86	0.95	0.98	1.00
Mangrove forest	0.81	0.87	0.95	0.96
Estuarine wetland	0.72	0.90	0.72	0.69
Saltpan	1.00	0.96	1.00	0.86
Bare mudflat	1.00	0.93	0.51	1.00
Saltmarsh grass	0.92	1.00	0.39	0.67
Producer's accuracy			0.89	0.94
Overall accuracy	0.95%	100%		
Kappa coefficient	0.85	0.92		

2.5. Change Detection

Change detection procedures estimate that a change in the reflectance of the study area results from a corresponding change in surface cover or surface material [62]. We characterised changes by using a suite of analytics. The first two methods of change detection were post-classification change analysis, and visual interpretation of images and comparison with Google Earth images from similar dates. Using local expert knowledge, we applied the on-screen digitisation of Landsat 2017 true-colour images to show areas of mangroves, saltpan/saltmarsh grass and estuarine wetland that have been altered from 2004 to the oceanic information class in 2017. The third method of change detection was the use of thematic change dynamics by using a remote sensing software tool to portray the dynamics of land cover change that occurred at Rocky Dam Creek/Cape Palmerston National Park. This tool measures the transition dynamics of a land cover class to another class at a given extent. Firstly, an initial state image was input as a time 1 image (2004), and then a final state image was input as time 2 (2017). Only changed areas were taken to visualize the overall dynamics. The output was a transition probability matrix signifying the "from–to" change that exemplified the past and present state of different land cover classes. The fourth and final method of change detection was a raster-based trend analysis where a time-series stack of thematic maps that was constructed into a 3D array and indexed via row and column was built to get a time-series vector. The occurrence of each class in each pixel across the time-series (i.e., the maximum spatial extent for each class) was used to create the output map. For example, the maximum extent map for open forest would show pixels that contained at least one occurrence of that class across the time-series array. We fitted a linear regression model to the data array with a slope value related to areal cover change per year in the time-series. The trend analysis measured the net change between pixels through the time series data, integrating six raster datasets with the same spatial extent, the output being a spatial map of slope and trend analysis. Information classes were combined to give a broad statistical appraisal of the region's LULC change dynamics. The map showing slope of the regression line is displayed as ranked data that are representations of the data's spatial attributes; this was created with the Jenks optimization method, a data clustering method that determines the best arrangement of values into different classes. The method seeks to reduce the variance within classes and maximize the variance between classes. If the values increment in time, they have a positive slope (red area) and, in the case of a decreasing regression line, a negative slope (green area). Temporal relationships were evaluated among the years by using the Pearson's correlation coefficient r value. Values represent the direction and magnitude of land cover change through the time-series. Post classification change analysis, visual interpretation, and thematic change were quantified by using Erdas Imagine Version 16.5.1 and the Spatial Analyst Toolset in ArcGIS Version 10.6.1. ArcGIS and Python routines were used for the time series analysis.

3. Results

The aim of this research is to quantify how a large, tropical, coastal region with estuarine-dependent fisheries has spatially and temporally changed in a period of 14 years (2004–2017). We used four methods of change detection in the analysis.

3.1. Post Classification Change Analysis

Our post-classification change detection analysis for the Rocky Dam Creek/Cape Palmerston National Park region with images from 2004 and 2017 quantified the amount of change in each land cover type (Figure 5). Information classes with a positive change demonstrated an increase in percentage area, and those with a negative change described a decrease. A positive change was apparent for four of the information classes: cropping/grazing (0.45%, 1329 hectares); oceanic (14%, 13,280 hectares), the large variability for the oceanic class can be explained by the time of day the two images were taken (the 2004 image was captured at low tide and the 2017 image was captured at high tide) (Table 1); saltpan (1.69%, 1565 hectares); and sand flat (0.12%, 126 hectares). A negative change occurred for five of the information classes: open forest (−0.05%, −1851 hectares), mangrove forest (−3.31%, −1147 hectares), estuarine wetland (−3.88%, −1496 hectares), bare mudflat (−5.49%, −2627 hectares), and saltmarsh grass (−3.65%, −1551 hectares) (Table 5). Notably, we found a low occurrence of saltmarsh grass within the park boundary. The overall classification accuracy and Kappa coefficient for 2004, 2006, 2009, 2013 2015, and 2017 land cover maps were 85%, 88%, 88%, 89% 81% and 92%, respectively, which were acceptable accuracy levels (Table 4; Tables S1 and S2 in Supplementary Materials). These values represent the general precision level that can be expected in mapping LULC when using the supervised classification technique.

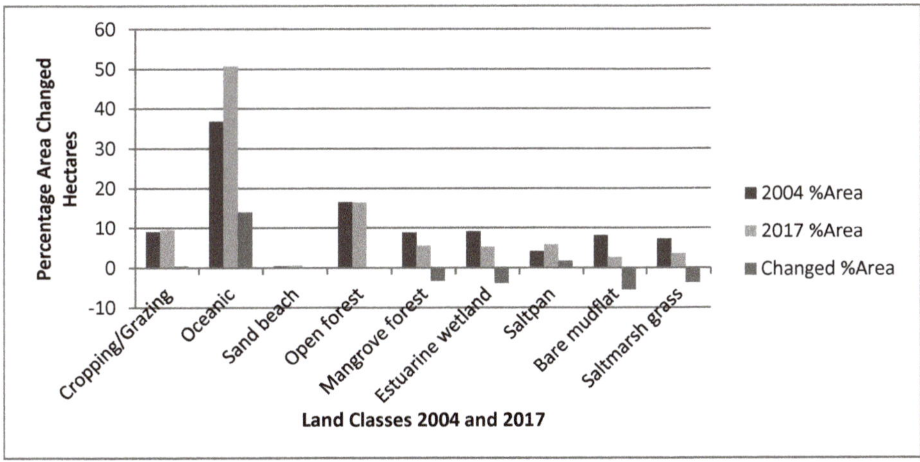

Figure 5. Land use differences between 2004 and 2017 at Rocky Dam Creek/Cape Palmerston National Park.

Table 5. Post-classification change statistics for five land use classes: mangrove forest, estuarine wetland, saltmarsh grass, bare mudflat, and saltpan at Rocky Dam Creek/Cape Palmerston National Park.

Land Use Class	Pixel Count 2017	Area in 2017 (Ha)	Percent Area in 2017 (%)	Percent Area Increase (Decrease) from 2004 to 2017 (%)	Increase (Decrease) from 2004 to 2017 (Ha)
Mangrove forest	40,018	3601	5.53	−3.31	−1147
Estuarine wetland	37,643	3388	5.21	−3.88	−1496
Saltmarsh grass	25,860	2327	3.57	−3.65	−1551
Bare mudflat	18,925	1703	2.6	−5.49	−2627
Saltpan	41,980	3778	5.8	1.69	1565

3.2. Image Interpretation

At the high tide, marine and estuarine waters flood the bays, intertidal flats, and channels of the region which, during wet season events, are diluted to brackish levels in some areas by freshwater flooding and stream flow from the catchment. Freshwater areas are shallow and receive water from stream flow and floodout. Water is otherwise saline throughout and less than 6 m deep. The tidal range is 7 m. The site is a good example of a diverse, hydrologically related aggregation of marine, estuarine, and freshwater wetlands within the Central Queensland Coast bioregion [46]. The Landsat 2017 true-colour composite image exhibits the delineated polygonal outlines of the mangrove (Figure 6), saltpan/saltmarsh grass (Figure 7) and estuarine wetland (Figure 8) classes from the 2004 Landsat image that changed to the oceanic information class. Inundation occurred over a substantial area of three information classes: mangrove forest (87 hectares), saltpan/saltmarsh grass (49 hectares), and estuarine wetland (17 hectares).

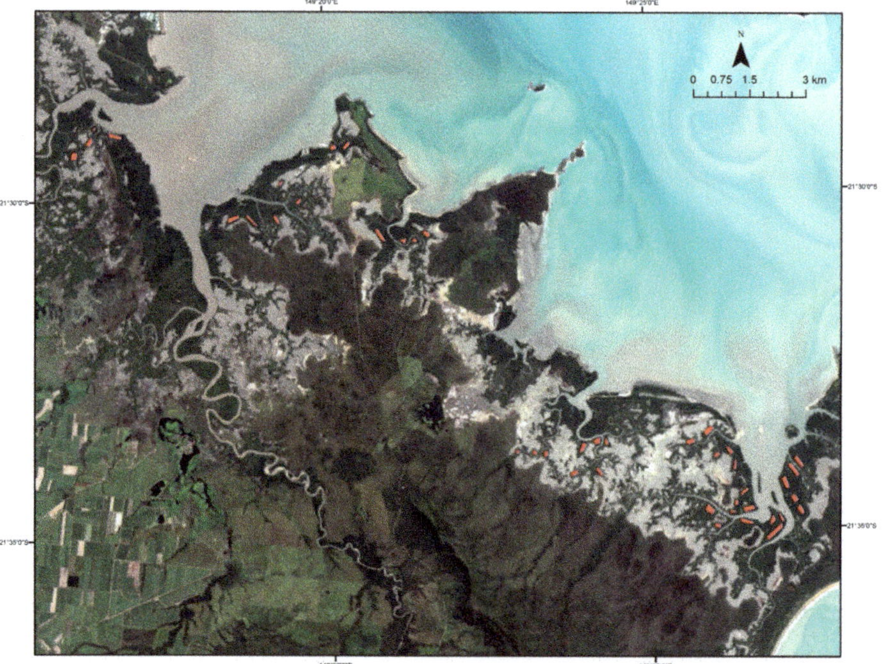

Figure 6. Landsat 8 Operational Land Imager (OLI) true-colour image captured 27 April 2017 that highlights areas of mangroves that changed to the oceanic information class (shown in red) at Rocky Dam Creek/Cape Palmerston National Park.

Figure 7. Landsat 8 OLI true-colour image captured 27 April 2017 that highlights the areas of saltpan/saltmarsh grass that changed to the oceanic information class (shown in orange) at Rocky Dam Creek/Cape Palmerston National Park.

Figure 8. Landsat 8 OLI true-colour image captured 27 April 2017 that highlights areas of estuarine wetland that changed to the oceanic information class (shown in pink) at Rocky Dam Creek/Cape Palmerston National Park.

3.3. Thematic Change Dynamics

The thematic change summary matrix 2004–2017 shows the number of pixels in each of the nine land classes per zone (Table S3 in the Supplementary Materials) and the net percentage loss of each of the nine land cover classes per zone (Table S4 in the Supplementary Materials). The mangrove land class transitioned to three land classes: (1) open forest (net loss = 5.08%, 597 hectares), (2) estuarine wetland (net loss = 8.51%, 273 hectares), and (3) saltmarsh grass (net loss = 6.26%, 145 hectares). Three information classes transitioned to the cropping/grazing land class over the 14-year time period: the open forest (net loss = 3.3%, 75 hectares), the estuarine wetland (net loss = 4.95%, 552 hectares), and the saltmarsh grass, which deteriorated considerably (net loss = 4.26%, 1001 hectares). A total of 1628 hectares of coastal vegetation transformed into pasture. A proportion of the cropping/grazing land class changed to saltpan (net loss = 5.38%, 192 hectares). The open forest land class claimed a proportion of the estuarine wetland (net loss = 2.16%, 1122 hectares). The bare mudflat information class transitioned to three land classes: (1) saltpan (net loss = 14.44%, 591 hectares), (2) saltmarsh grass (net loss = 2.38%, 66 hectares), and (3) sand flat (net loss = 66.15%, 192 hectares).

3.4. Time Series Analysis

In a spatio–temporal analysis, we used the single explanatory variable of time to explore how the predictor variable, LULC, changed over the landscape 2004–2017. The map of the regression slope shows positive and negative values across the study site (Figure 9). The forested area (mangrove forest and open forest) was the dominant region that occupied the mid-way point in the data range, indicating that this area generally had little inter-annual variability and no decrease on average; this is the yellow area (average r value = 0.24) (Figure 9). A predominantly negative slope of the regression line indicates vegetation thinning and was pronounced in the land classes of cropping/grazing, estuarine wetland, and saltmarsh grass—these are represented by the dark green area (Pearson correlation coefficient r value range = −0.84–0.2) (Figure 10). The total vegetation decline in the cropping/grazing class was 1896 hectares. Strong positive slope values show a high exposure in the system through time but small inter-annual variability. High exposure areas include the saltmarsh grass adjacent to Rocky Dam Creek, saltpan, and bare mudflat land classes across the site extent—these are shown as the orange/red area (Pearson correlation coefficient r value range = 0.37–0.97) (Figure 10). The total change in the combined classes was 9375 hectares. Many areas throughout the study region displayed significant values (*p* value range = 0.001–0.099), demonstrating that land classes decreased in areal extent in the time series, e.g., the saltmarsh grass south of the saltpan zone, the estuarine wetland south of the saltmarsh grass zone, the bare mudflat in the stream inlets, scattered grazing sites, the open forest at Glendower Point (coastal mid-way point of the site) and the north-east boundary of the park, and the fringing mangroves along the coastline and stream channels.

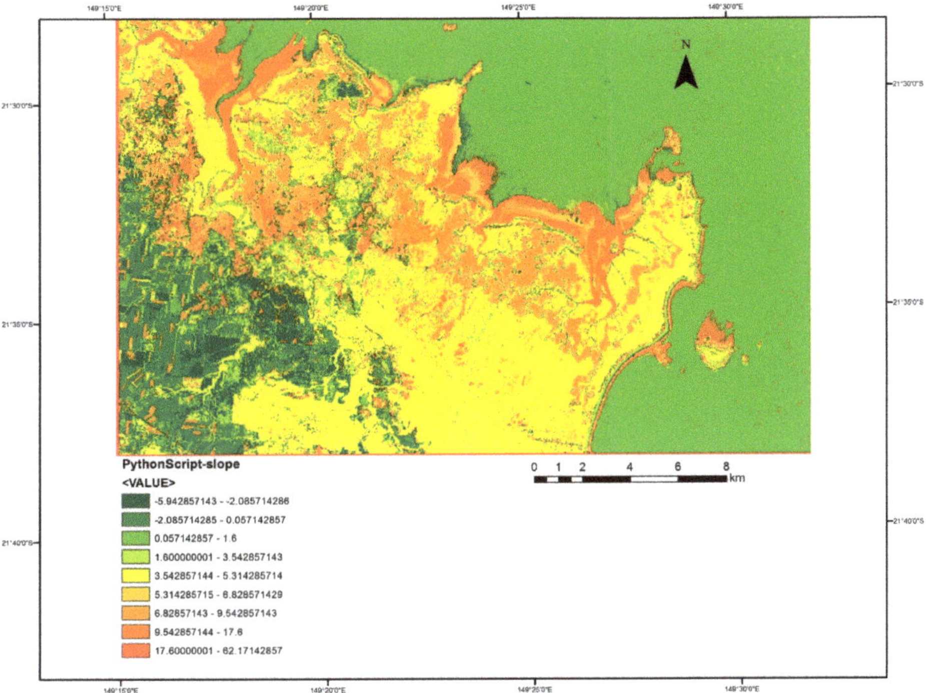

Figure 9. Map of regression slope that highlights positive and negative values across the study site 2004–2017. High positive slope values show an increase in areal extent through time and are most pronounced in the saltpan and bare mudflat land classes across the site extent (shown in red); high negative slope values show a decrease in areal extent through time and are most pronounced in the land classes: cropping/grazing, estuarine wetland, and saltmarsh grass, as well as the fringing mangroves along the coastline and stream channels (shown in green) at Rocky Dam Creek/Cape Palmerston National Park.

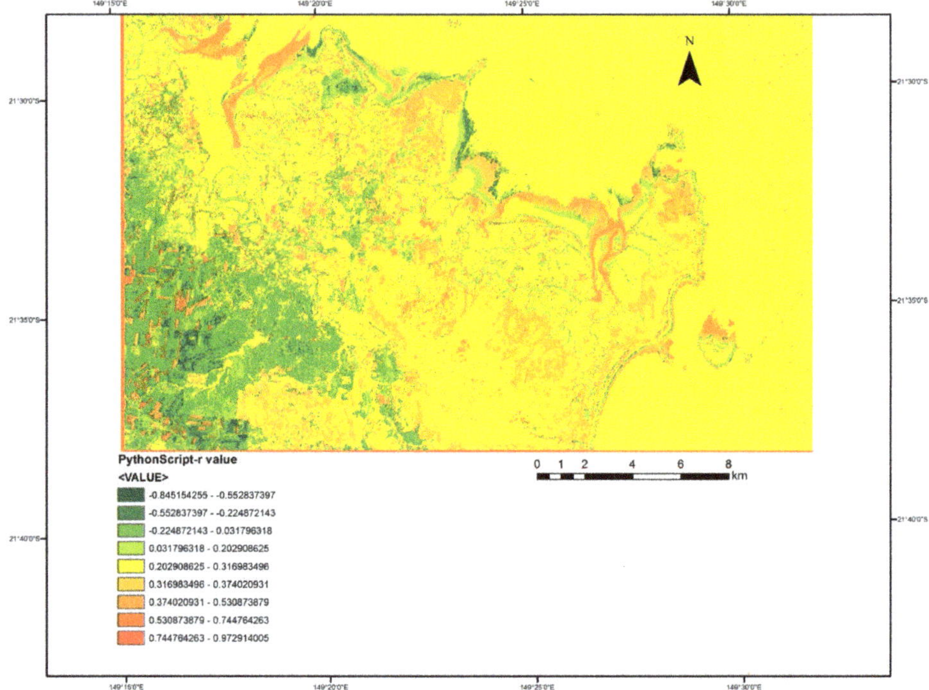

Figure 10. Map of Pearson correlation coefficient r values across the study site 2004–2017. A strong positive correlation is highlighted in the land classes: saltpan and bare mud flat (shown in red); a strong negative correlation is highlighted in the land classes: cropping/grazing, estuarine wetland, and saltmarsh grass (shown in green). Significant values (p value range = 0.001–0.099) demonstrate that the following land classes decreased in areal extent in the time series: saltmarsh grass, estuarine wetland, bare mudflat in the stream inlets, scattered grazing sites, open forest at Glendower Point (coastal mid-way point of the site) and the north-east boundary of the park, and fringing mangroves along the coastline and stream channels at Rocky Dam Creek/Cape Palmerston National Park.

4. Discussion

4.1. Change Dynamics in Estuarine Ecosystems

Old world tropical mangroves found in the Indo-Pacific, including tropical Australia, possess notable attributes of species diversity, richness, abundance, and succession, and they are therefore considered to be the most dominant and important mangroves globally [63,64]. We examined changes in their vegetation structure and connectivity within a spatially extensive estuarine region of Central Queensland, Australia by using four methods of change detection. This work has relevance to the maintenance of biodiversity and ecological processes because it explores: (1) the distribution of critical wetland habitats in relation to their proximity to threats from human development; (2) temporal change in the distribution and abundance of wetland habitats correlated (spatially) with temporal change in human activities of varying types (fishing, coastal development, agriculture, erosion, and hydrology modification); and (3) interactions that occur at scales larger than a protected area's boundary that affect the maintenance of biodiversity values.

An analysis of classified maps revealed that gradual ecosystem change occurred across large areas and various habitats. Other studies such as that by Kanniah et al. [65] in the Southern Peninsula, Malaysia, and that of Tran and Fischer [66] in Ca Mao Province, Vietnam, confirm that protected area status does not guarantee the encroachment control of long-term anthropogenic influence, and the downsizing of mangrove communities continues. Competing demands for available resources, especially in coastal provinces, drives change in hydrology and land use outside a protected areas' administrative boundary, thus affecting ecological processes within such as movement of organisms, water availability, and connectivity functions [31]. Mangroves are well recognised as fragile ecosystems that play an important role in linking marine and terrestrial systems [67]. Nevertheless, it is apparent from our results and most previously published research that mangrove ecosystems are in decline [68–70]. The mangroves in our study region are located in the mid and lower intertidal zone and are constrained not by land surface temperature, as in semi-arid regions [71,72], but by air/seawater temperature, freshwater levels, and other geospatial properties. Whilst air/seawater temperature was not measured in this study, other studies have shown a linkage between mangrove deforestation and anthropogenic climate change [73]. The frequency and intensity of cyclones and storms has increased as a consequence of greater sea-surface temperature, with further escalation predicted [74,75]. The Landsat 2017 image was captured in April immediately following the impact of a severe tropical cyclone [16]. We suggest that the mangrove decline in our study (1147 hectares) was a result of change in sediment profiles, defoliation, and inundation from a coastal cyclone [74,76]. Furthermore, because effects outside the boundary of a protected area manifest themselves within that boundary, alterations in hydrology for pasture and direct trampling may be linked to the decline in mangroves, saltmarsh grass, and estuarine wetlands. Similarly, Al-Hamdan et al. [77], by using Landsat satellite data in Tanzania during the period of 2000–2010, found a net deforestation of mangroves with a net agriculture expansion. Our result for mangrove transition (1015 hectares) from the thematic change analysis is comparable to Chen et al. [78], who observed the transition of mangroves to other land uses during the period of 1985–2013 in the Honduras.

River and stream flow regulation is pervasive in Australia. River regulation affects riverine vegetation by fundamentally altering the flow regime, thus changing the hydrology and flow across a range of different spatial and temporal scales [79]. Though there is recognition by government agencies of the alienation of flow-dependent ecosystems that are attributed to anthropogenic barriers [22], regulating structures such as culverts, pipes, road crossings, weirs, and ponded pasture still potentially cause connectivity disruption in our study region [80]. Further, the inclusion of regulating structures interrupts stream flow regimes, floodplain–wetland connections, biotic responses, channel formation, and sediment transfer [81]. The fragmentation of riverine vegetation with corresponding environmental degradation from flow regulation has been observed with Landsat imagery in other studies, such as those by Das and Pal [82] in India and Antwi et al. [83] in Ghana. We suggest that the reduction in estuarine wetland in our study (1496 hectares) including endangered ecosystems, in combination with a significant declining trend in vegetation extent (thinning) (Table 1, Figure 2), was primarily due to the alteration of natural flow regimes through stream regulation, which affected the processes that sustain riparian vegetation communities.

Saltmarsh grass is recognized as providing climate benefits through carbon sequestration as well as other ecosystem benefits, e.g., storm surge erosion protection and ontogenetic habitat for fisheries species [9,84]. Tropical saltmarsh grass is poorly represented in protected areas and crudely acknowledged for its ecosystem services when confronted with alternative land uses. In a recent review, Wegscheidl et al. [85] identified a lack of quantitative information needed to substantiate the value of Australia's saltmarshes, both locally and regionally. Likewise, we found a low occurrence of saltmarsh grass in Cape Palmerston National park, and there was an apparent decreasing trend in the vegetation extent (thinning) of saltmarsh grass throughout the study site. Saltpan and bare mudflat areas exhibited inconsistency across the study site. There were high levels of bare mudflat accretion in the main channels and areas of coastline; however, stream inlets and drainage networks showed inter-annual variability (long-term trend increases and decreases) due to tidal fluctuations and a decrease through time. Similar to those reported by previous studies in tropical regions [86], our results suggest that the saltmarsh–mudflat system in the landward region of the Cape Palmerston

National Park shows instability and is degrading over time, possibly due to climatic factors such as recent cyclonic activity, sea level variability, and prolonged inundation.

The need for scientifically-based regional-scale land use planning around protected areas is integral in human-dominated landscapes to balance conservation goals with livelihood needs for crops, pasture, and other ecosystem services [31]. The decline in wetland ecosystems in our study could be attributed to both direct and indirect effects. Direct effects could include altered vegetation composition and structure from trampling by grazing animals and the modification of ground morphology. An indirect effect could include the draining and hydrological disturbances that convert wetlands to agricultural and grazing land, resulting in tidal disruption and vegetation fragmentation. Across the sub-catchment, the cumulative area of open forest, estuarine wetland, and saltmarsh grass (1628 hectares) was converted to pasture. Riverine landscapes are highly valued in Australia for grazing and are often preferred by livestock because of their vegetation, shade, and water [87]. Though the Sarina Inlet–Ince Bay Aggregation is a designated important wetland under Australian federal biodiversity conservation policy [46], implementation is lacking [88]. The land classes that are open forest and estuarine wetland transition to cropping/grazing is a similar result to that obtained by Haque and Basak [37], who found that forested land transitioned to either shallow water or settlement in Bangladesh during 1980–2010. The result by Toure et al. [89] in Senegal with Landsat imagery and ML classification highlighted the unexpected transition of agriculture to saltpan, as was the case for areas of cropping/grazing in our study (192 hectares).

The significant declining trend observed for open forest, fringing mangroves, estuarine wetlands, and vegetation levels in scattered grazing sites was inconsistent across the study area. This inconsistency illustrates how multiple forms of change can co-occur within relatively close proximity. We suggest that the decline in shoreline vegetation cover was the direct result of a severe tropical cyclone that impacted the coast in March 2017, and we also suggest that grazing-induced, ubiquitous vegetation degradation contributed to and will continue to exacerbate the loss of resilience in these systems.

4.2. Comparison with SLATS

The SLATS program was initiated by the Queensland Government to provide factual information on land cover and trends in land clearing, tree growth, and regrowth on public and private lands [90]. The SLATS data are based on the supervised classification of multiple Landsat satellite images and digital terrain models at a resolution of approximately 30 m, with maps on woody vegetation clearing (and replacement land cover) that are the result of the anthropogenic removal of vegetation [91]. SLATS has clear differences with our study in that SLATS does not include any vegetation loss caused by natural tree death or natural disasters (e.g., cyclones) when calculating woody vegetation clearing rates. Further, SLATS applies radiometric standardisation to the Landsat images. Finally, topographic corrections are used to increase accuracy in areas of high slope [92]. However, as our study area is generally of low, flat elevation, we deemed the correction unnecessary. An inspection of SLATS maps from 2004 to 2017 in ArcGIS displayed similarities with our results with many sites cleared of woody vegetation and converted to pasture, particularly along the boundary of the national park, in the north-east, north-west, south, and central areas. According to SLATS, the total converted vegetation in the Plane Creek catchment is 3536 hectares, and, by digitizing the Rocky Dam Creek sub-catchment pasture polygons in ArcGIS, we found a total of 1100 hectares. Though the total SLATS pasture profile is smaller than our results for the thematic change (1628 hectares) and time series (1896 hectares), our results nevertheless reflect a variable but significant impact on the coastal region that was likely caused by an intense climatic event.

4.3. Limitations of the Study

Remote sensing data and tools are fundamental methods for measuring LULC, but there are critical drawbacks in the change detection of wetlands. The first drawback is that classification errors from the individual-date images can affect the final change detection accuracy, and, although ground-truth data engender the development of accurate LULC classification and accuracy assessments, errors can still occur [93]. Foody [94] argued that accuracy values cannot be appropriately interpreted by readers or users unless a detailed account of the approach to accuracy assessment is provided. The lack of robust validation could have serious implications for some users and may lessen their confidence in remote sensing as a source of land cover data. Therefore, the validation methods that detail the user's and producer's accuracies of change with Kappa coefficient and which include the confusion matrix for the 2004 and 2017 images (Table 4) have been given to allow for replication. The second drawback is that during high tides, there can be a sharp decline in the spectral reflectance of mangroves, especially in the NIR and SWIR regions [95,96]. Our study used a combined binary change detection and time series analysis approach, illustrating that it is beneficial to use multiple images in change detection research since apparent changes between any two images could be due to irrelevant causes such as tide, sea surface state, and water constituents. The third drawback is that, ideally, change detection requires precise image alignment, which is difficult to achieve, and the fourth drawback is that post classification comparison-based binary approaches that are used for hard classifications, i.e., comparatively broad scale classifications, may not detect subtle transformations in land cover modification in which the land cover type may have been altered but not changed (e.g., a thinned forest or saltmarsh degradation), ensuing an inappropriate representation [97].

4.4. Implications for the Conservation of Estuarine Ecosystems

Quantifying the level of coastal wetland fragmentation and landscape connectivity is an essential component of contemporary strategies that are aimed at biodiversity conservation and fishery sustainability [98]. The results presented here are noteworthy from two viewpoints. The first is nationally—in Australia, there is no nationally consistent approach to quantify the area of mangrove or saltmarshes, and historical benchmarks are scarce [99]. Our results inform the Australian inventory of spatio–temporal distribution, as they show important changes in the representation of coastal vegetation classes, particularly mangroves and saltmarsh grasses, in the tropical catchments of the Eastern seaboard. The second viewpoint is regionally—natural resource management is hampered by complex management arrangements that provide challenges to achieving environmental sustainability and are additional to increasing pressures from natural and anthropogenic forces [100]. Our findings raise concerns that lands surrounding the Cape Palmerston National Park are under threat, and, because interactions occur at scales larger than a protected areas' boundary, repercussions arise for the environmental stability of the entire region. Watson et al. [101] argue that the occurrence of threatened species is widespread outside protected areas, and plants are one of the most poorly represented taxonomic groups. Furthermore, protected areas are not exempt from anthropogenic impacts; for example, Jones et al. [102] identified an increase of human pressure of 1.5% on IUCN listed protected areas categories I and II between 1993 and 2009. Particularly evident in our study was the decline of estuarine wetlands, which include endangered ecosystems: the broad leaf tea-tree *Malaleuca viridiflora* and semi-evergreen microphyll vine thicket-to-vine forest [103] (Figure 2).

There is a need for a more comprehensive understanding of the ecosystem value assigned to Australia's coastal landscapes. This information is a high priority and needed to support evidence-based decision-making and conservation actions that attribute socio–economic value, warranting ecosystem protection and repair [85]. For example, the Australian Government listed subtropical and temperate coastal saltmarsh as a vulnerable ecological community under the Environment Protection and Biodiversity Conservation Act 1999 (EPBC) in 2013 [104]. Carbon sequestration pathways designate saltmarshes (and other coastal wetlands) as disproportionality valuable in sequestering carbon dioxide compared to terrestrial ecosystems [105]. Therefore, it is an opportune time to apply protection to these communities. We propose that the vulnerable listing be extended to tropical saltmarsh regions. The 1998–2003 historical occurrence of mangrove dieback in local estuaries, which affected >30 km^2 of remnant mangrove cover [106,107], failed to conclusively identify the causative agent (agricultural herbicides and flooding were implicated in the event). However, recent northern Australian mangrove dieback has been linked to climate change as the most likely cause [108]. The declining trend in fringing mangroves found in our study is concomitant to a loss of ecosystem services that are provided by the coastal habitat–fishery linkage, as the service value of mangroves has been observed to be higher at the seaward edge [109]. Notwithstanding the Australian government's efforts to provide protection to Great Barrier Reef catchments [16], ecosystem degradation is ongoing.

Two key factors determine the extent to which coastal habitats can recover and the associated fauna rejuvenate from a major acute (pulse-like) disturbance such as a cyclone: (1) the time window until the next major acute disturbance [110] and (2) the extent and intensity of chronic (press-like) disturbances, such as disruption in sediment/water profiles [74] and elevated mean seawater temperatures suppressing recovery rates [64] during that window. Predictions that tropical cyclones will increase in frequency and intensity in Australia in the coming decades [111] have been accompanied by projections of an escalation in storm surges and extreme sea-levels under future climate change [112]. Improving the resilience of Great Barrier Reef coastal ecosystems requires active landscape protection and restoration approaches to maintain as many biodiversity and ecosystem functions as possible [113].

5. Conclusions

In this paper, we have used a stratification approach to examine different types of change by analysing land cover types. Ancillary data and local expert knowledge were necessary to expose long-term trends and formulate explanations in a region that surrounds and includes a national park that has, until now, been largely devoid of significant direct anthropogenic impact. Whereas the reasons for such changes could generally be explained with detailed field-based data sets, such information does not exist at the requisite spatial and temporal scales. Remote sensing datasets, e.g., Landsat imagery, provided the only feasible method to enumerate the trends in LULC in the spatially extensive study area. Areal reduction in threatened and endangered ecosystems (e.g., mangroves and estuarine wetlands) occurred within Cape Palmerston National Park and its surroundings. We found a decreasing trend in the vegetation extent of estuarine wetlands, saltmarsh grass, and grazing areas. Significant declining values were observed in open forest, fringing mangroves, estuarine wetlands, and saltmarsh grass, albeit on localized scales, with a mosaic of ensemble change across the study site. The occurrence of a severe tropical cyclone immediately preceding capture of the 2017 Landsat image was likely the main agent in the declining trend for shoreline and stream vegetation. Long-term grazing pressure contributed to vegetation degradation and loss of resilience on a landscape scale. SLATS maps confirm that many sites in the sub-catchment were cleared of woody vegetation and converted to pasture during our time period. To maintain ecosystem services and encourage habitat–fishery linkages, effective monitoring action is crucial to understand recovery and set in place adaptive management approaches. Historical occurrence of mangrove dieback in the region, coupled with the recent calls for the increased monitoring of northern Australian mangrove ecosystems due to dieback connections to climate change, could be extended to Great Barrier Reef catchments. Future studies by the authors will create virtual constellation synergies by integrating optical land imaging systems with similar characteristics, e.g., Landsat and Sentinel-2, firstly to assess post-cyclone recovery and secondly to explore ecosystem variability forced by climate controls in Great Barrier Reef catchments. Our research contributes to the body of knowledge on coastal ecosystem dynamics to enable planning to achieve more effective conservation outcomes.

Supplementary Materials: The following are available online at http://www.mdpi.com/2072-4292/12/1/197/s1, Table S1: Accuracy assessment of Landsat image captured in 2004, Rocky Dam Creek/Cape Palmerston National Park, Table S2: Accuracy assessment of Landsat image captured in 2017, Rocky Dam Creek/Cape Palmerston National Park, Table S3: Thematic Change Summary Matrix 2004–2017 with number of pixels in each land class per zone Rocky Dam Creek/Cape Palmerston National Park, Table S4: Thematic Change Summary Matrix 2004–2017 with percentage of land classes occurring in each zone Rocky Dam Creek/Cape Palmerston National Park.

Author Contributions: Conceptualization, S.P.; formal analysis, D.C.; methodology, D.C.; resources, H.P.; supervision, S.P. and H.P.; writing—original draft, D.C.; writing—review and editing, S.P. and H.P. All authors have read and agreed to the published version of the manuscript.

Funding: This research received no external funding.

Acknowledgments: Technical support was gratefully received from Steve Bryant of Mackay State High School (Mackay, QLD 4740, Australia).

Conflicts of Interest: The authors declare no conflict of interest.

References

1. Allgeier, J.E.; Layman, C.A.; Mumby, P.J.; Rosemond, A.D. Biogeochemical implications of biodiversity and community structure across multiple coastal ecosystems. *Ecol. Monogr.* **2015**, *85*, 117–132. [CrossRef]
2. Cloern, J.E.; Abreu, P.C.; Carstensen, J.; Chauvaud, L.; Elmgren, R.; Grall, J.; Greening, H.; Johansson, J.O.; Kahru, M.; Sherwood, E.T.; et al. Human activities and climate variability drive fast-paced change across the world's estuarine-coastal ecosystems. *Glob. Chang. Biol.* **2016**, *22*, 513–529. [CrossRef] [PubMed]
3. Costanza, R.; de Groot, R.; Sutton, P.; van der Ploeg, S.; Anderson, S.J.; Kubiszewski, I.; Farber, S.; Turner, R.K. Changes in the global value of ecosystem services. *Glob. Environ. Chang.* **2014**, *26*, 152–158. [CrossRef]
4. Al-Nasrawi, A.; Hopley, C.; Hamylton, S.; Jones, B. A spatio-temporal assessment of landcover and coastal changes at Wandandian Delta System, Southeastern Australia. *J. Mar. Sci. Eng.* **2017**, *5*, 55. [CrossRef]
5. Ballanti, L.; Byrd, K.; Woo, I.; Ellings, C. Remote sensing for wetland mapping and historical change detection at the Nisqually River Delta. *Sustainability* **2017**, *9*, 191. [CrossRef]
6. Li, D.; Lu, D.; Wu, M.; Shao, X.; Wei, J. Examining land cover and greenness dynamics in Hangzhou Bay in 1985–2016 using Landsat time-series data. *Remote Sens.* **2017**, *10*, 32. [CrossRef]
7. Schmidt, M.; Lucas, R.; Bunting, P.; Verbesselt, J.; Armston, J. Multi-resolution time series imagery for forest disturbance and regrowth monitoring in Queensland, Australia. *Remote Sens. Environ.* **2015**, *158*, 156–168. [CrossRef]
8. Reuter, K.E.; Juhn, D.; Grantham, H.S. Integrated land-sea management: Recommendations for planning, implementation and management. *Environ. Conserv.* **2016**, *43*, 181–198. [CrossRef]
9. Sheaves, M.; Barnett, A.; Bradley, M.; Abrantes, K.G.; Brians, M.; James Cook University. *Life History Specific Habitat Utilisation of Tropical Fisheries Species*; FRDC Project No 2013-046; James Cook University: Townsville, Australia, 2016.
10. Peirson, W.; Davey, E.; Jones, A.; Hadwen, W.; Bishop, K.; Beger, M.; Capon, S.; Fairweather, P.; Creese, B.; Smith, T.F.; et al. Opportunistic management of estuaries under climate change: A new adaptive decision-making framework and its practical application. *J. Environ. Manag.* **2015**, *163*, 214–223. [CrossRef]
11. Creighton, C.; Boon, P.I.; Brookes, J.D.; Sheaves, M. Repairing Australia's estuaries for improved fisheries production—What benefits, at what cost? *Mar. Freshw. Res.* **2015**, *66*, 493–507. [CrossRef]
12. Barbier, E.B.; Hacker, S.D.; Kennedy, C.; Koch, E.W.; Stier, A.C.; Silliman, B.C. The value of estuarine & coastal ecosystem services. *Ecol. Monogr.* **2011**, *81*, 169–193.
13. van der Stocken, T.; Carroll, D.; Menemenlis, D.; Simard, M.; Koedam, N. Global-scale dispersal and connectivity in mangroves. *Proc. Natl. Acad. Sci. USA* **2019**, *116*, 915–922. [CrossRef] [PubMed]
14. Green, K.; Congalton, R.; Tukman, M. *Imagery and GIS: Best Practices for Extracting Information from Imagery*; Esri Press: Redlands, CA, USA, 2017.
15. Ferro-Azcona, H.; Espinoza-Tenorio, A.; Calderón-Contreras, R.; Ramenzoni, V.C.; Gómez País, M.; Mesa-Jurado, M.A. Adaptive capacity and social-ecological resilience of coastal areas: A systematic review. *Ocean Coast. Manag.* **2019**, *173*, 36–51. [CrossRef]
16. Commonwealth of Australia. *Reef 2050 Long-Term Sustainability Plan-July 2018*; Department of the Environment, Commonwealth of Australia: Canberra, Australia, 2018; pp. 9–13, 92–93.
17. Fang, X.; Hou, X.; Li, X.; Hou, W.; Nakaoka, M.; Yu, X. Ecological connectivity between land and sea: A review. *Ecol. Res.* **2018**, *33*, 51–61. [CrossRef]
18. Schaffelke, B.; Mellors, J.; Duke, N.C. Water quality in the Great Barrier Reef region: Responses of mangrove, seagrass and macroalgal communities. *Mar. Pollut. Bull.* **2005**, *51*, 279–296. [CrossRef] [PubMed]
19. Spalding, M.D.; Ruffo, S.; Lacambra, C.; Meliane, I.; Hale, L.Z.; Shepard, C.C.; Beck, M.W. The role of ecosystems in coastal protection: Adapting to climate change and coastal hazards. *Ocean Coast. Manag.* **2014**, *90*, 50–57. [CrossRef]
20. Queensland Herbarium. Regional Ecosystem Description Database (REDD) Version 11 (December 2018). Available online: https://apps.des.qld.gov.au/regional-ecosystems/ (accessed on 1 March 2019).

21. Reef Catchments Limited. *Natural Resource Management Plan, Mackay Whitsunday Isaac 2014–2024*; Reef Catchments Limited: Mackay, Australia, 2014; pp. 1–29, 60–87.

22. Moore, M. *Mackay Whitsunday Fish Barrier Prioritisation Report*; Catchment Solutions: Mackay, Australia, 2015.

23. Duke, N.C.; Roelfsema, C.; Tracey, D.; Godson, L. *Preliminary Investigation into the Dieback of Mangroves in the Mackay Region: Initial Assessment and Primary Causes*; University of Queensland: Brisbane, Australia, 2001.

24. Sheaves, M.; Brookes, J.; Coles, R.; Freckelton, M.; Groves, P.; Johnston, R.; Winberg, P. Repair and revitalisation of Australia's tropical estuaries and coastal wetlands: Opportunities and constraints for the reinstatement of lost function and productivity. *Mar. Policy* **2014**, *47*, 23–38. [CrossRef]

25. Olds, A.D.; Connolly, R.M.; Pitt, K.A.; Pittman, S.J.; Maxwell, P.S.; Huijbers, C.M.; Moore, B.R.; Albert, S.; Rissik, D.; Babcock, R.C.; et al. Quantifying the conservation value of seascape connectivity: A global synthesis. *Glob. Ecol. Biogeogr.* **2015**, *25*, 3–15. [CrossRef]

26. Carrasquilla-Henao, M.; Juanes, F. Mangroves enhance local fisheries catches: A global meta-analysis. *Fish Fish.* **2016**, 1–15. [CrossRef]

27. Manson, F.J.; Loneragan, N.R.; Skilleter, G.A.; Phinn, S.R. An evaluation of the evidence for linkages between mangroves and fisheries. *Oceanogr. Mar. Biol.* **2005**, *43*, 485–515.

28. Adam, P. Salt Marsh Restoration. In *Coastal Wetlands*, 2nd ed.; Perillo, G.M.E., Wolanski, E., Eds.; Elsevier: Amsterdam, The Netherlands, 2019; pp. 817–861. [CrossRef]

29. Great Barrier Reef Marine Park Authority. *Great Barrier Reef Region Strategic Assessment: Strategic Assessment Report*; GBRMPA: Townsville, Australia, 2014; pp. 197–281, 325–327, 441–476.

30. Creighton, C. *Revitalising Australia's Estuaries*; Final Report—2012-036-DLD; Fisheries Research and Development Corporation: Canberra, Australia, 2013; pp. 1–38, 145–165.

31. DeFries, R.; Karanth, K.K.; Pareeth, S. Interactions between protected areas and their surroundings in human-dominated tropical landscapes. *Biol. Conserv.* **2010**, *143*, 2870–2880. [CrossRef]

32. Jones, D.A.; Hansen, A.J.; Bly, K.; Doherty, K.; Verschuyl, J.P.; Paugh, J.I.; Carle, R.; Story, S.J. Monitoring land use and cover around parks: A conceptual approach. *Remote Sens. Environ.* **2009**, *113*, 1346–1356. [CrossRef]

33. Macintosh, A. *The National Greenhouse Accounts & Land Clearing: Do the Numbers Stack Up?* Research Paper No. 38; Australia Institute: Canberra, Australia, 2007.

34. Singh, A. Digital change detection techniques using remotely-sensed data. *Int. J. Remote Sens.* **1989**, *10*, 989–1003. [CrossRef]

35. Lu, D.; Li, G.; Moran, E. Current situation and needs of change detection techniques. *Int. J. Image Data Fusion* **2014**, *5*, 13–38. [CrossRef]

36. Gómez, C.; White, J.C.; Wulder, M.A. Optical remotely sensed time series data for land cover classification: A review. *ISPRS J. Photogramm. Remote Sens.* **2016**, *116*, 55–72. [CrossRef]

37. Haque, M.I.; Basak, R. Land cover change detection using GIS and remote sensing techniques: A spatio-temporal study on Tanguar Haor, Sunamganj, Bangladesh. *Egypt. J. Rem. Sens. Space Sci.* **2017**, *20*, 251–263. [CrossRef]

38. Jones, T.; Glass, L.; Gandhi, S.; Ravaoarinorotsihoarana, L.; Carro, A.; Benson, L.; Ratsimba, H.; Giri, C.; Randriamanatena, D.; Cripps, G. Madagascar's mangroves: Quantifying nation-wide and ecosystem specific dynamics, and detailed contemporary mapping of distinct ecosystems. *Remote Sens.* **2016**, *8*, 106. [CrossRef]

39. Richards, J.A.; Jia, X. *Remote Sensing Digital Image Analysis: An Introduction*, 4th ed.; Springer: Berlin/Heidelberg, Germany, 2006; pp. 1–78, 137–160, 193–238, 295–328, 389–413, 423–428.

40. Barik, K.K.; Mitra, D.; Annadurai, R.; Tripathy, J.K.; Nanda, S. Geospatial analysis of coastal environment: A case study on Bhitarkanika mangroves, East coast of India. *Indian J. Geo-Mar. Sci.* **2016**, *45*, 492–498.

41. Xu, C.; Pu, L.; Zhu, M.; Li, J.; Chen, X.; Wang, X.; Xie, X. Ecological security and ecosystem services in response to land use change in the coastal area of Jiangsu, China. *Sustainability* **2016**, *8*, 816. [CrossRef]

42. Folkers, A.; Rohde, K.; Delaney, K.; Flett, I. *Mackay Whitsunday Water Quality Improvement Plan 2014–2021*; Reef Catchments (Mackay Whitsunday Isaac) Limited: Mackay, Australia, 2014; pp. 11–33, 43–63, 107.

43. Pascoe, S.; Innes, J.; Tobin, R.; Stoeckle, N.; Paredes, S.; Dauth, K. *Beyond GVP: The Value of Inshore Commercial Fisheries to Fishers and Consumers in Regional Communities on Queensland's East Coast*; FRDC Project No 2013-301; Fisheries Research and Development Corporation: Canberra, Australia, 2016; pp. 1–79.

44. Webley, J.; McInnes, K.; Teixeira, D.; Lawson, A.; Quinn, R. *Statewide Recreational Fishing Survey 2013-14*; Department of Agriculture and Fisheries: Brisbane, Australia, 2015; pp. 82–83.

45. Reef Catchments. *State of the Region Report, Mackay Whitsunday Isaac*; Reef Catchments: Mackay, Australia, 2013.

46. Department of the Environment. Directory of Important Wetlands in Australia, Sarina Inlet-Ince Bay Aggregation-QLD053. Available online: https://www.environment.gov.au/cgi-bin/wetlands/report.pl (accessed on 1 March 2019).

47. IUCN. World Database on Protected Areas Cape Palmerston-Rocky Dam in Australia. Available online: https://protectedplanet.net/23846 (accessed on 1 March 2019).

48. Duke, N.C.; Larkum, A.W.D. Mangroves and Seagrasses. In *The Great Barrier Reef: Biology, Environment and Management*; Hutchings, P., Kingsford, M., Eds.; CSIRO Publishing: Collingwood, Australia, 2008; pp. 1–18.

49. Bureau of Meteorology. *Australian Hydrological Geospatial Fabric (Geofabric) Product Guide Version 3.0*; Bureau of Meteorology: Canberra, Australia, 2015.

50. USGS. *Landsat Collection 1 Level 1 Product Definition Version 1.0*; United States Geological Survey: South Dakota, USA, 2017.

51. Lyons, M.B.; Phinn, S.R.; Roelfsema, C.M. Long term land cover and seagrass mapping using Landsat and object-based image analysis from 1972 to 2010 in the coastal environment of South East Queensland, Australia. *ISPRS J. Photogramm. Remote Sens.* **2012**, *71*, 34–46. [CrossRef]

52. Bureau of Meteorology. Tide Predictions for Australia, South Pacific and Antarctica. Available online: http://www.bom.gov.au/australia/tides/ (accessed on 1 March 2019).

53. Kuenzer, C.; Bluemel, A.; Gebhardt, S.; Quoc, T.V.; Dech, S. Remote sensing of mangrove ecosystems: A review. *Remote Sens.* **2011**, *3*, 878–928. [CrossRef]

54. Neldner, V.J.; Niehus, R.E.; Wilson, B.A.; McDonald, W.J.F.; Ford, A.J.; Accad, A. *The Vegetation of Queensland: Descriptions of Broad Vegetation Groups. Version 4.0*; Queensland Herbarium, Department of Environment and Science: Brisbane, Australia, 2019.

55. Australian Bureau of Agricultural and Resource Economics and Sciences. *Guidelines for Land Use Mapping in Australia: Principles, Procedures and Definitions*, 4th ed.; Australian Bureau of Agricultural and Resource Economics and Sciences: Canberra, Australia, 2011.

56. Long, J.B.; Giri, C. Mapping the Philippines' mangrove forests using Landsat imagery. *Sensors* **2011**, *11*, 2972–2981. [CrossRef] [PubMed]

57. Soto-Berelov, M.; Hislop, S. *Approaches Used for Pixel Based Time Series Analysis of Landsat Data: Literature Review*; RMIT University: Melbourne, Australia, 2016.

58. Vogelmann, J.E.; Gallant, A.L.; Shi, H.; Zhu, Z. Perspectives on monitoring gradual change across the continuity of Landsat sensors using time-series data. *Remote Sens. Environ.* **2016**, *185*, 258–270. [CrossRef]

59. Abdul Aziz, A.; Phinn, S.; Dargusch, P.; Omar, H.; Arjasakusuma, S. Assessing the potential applications of Landsat image archive in the ecological monitoring and management of a production mangrove forest in Malaysia. *Wetl. Ecol. Manag.* **2015**, *23*, 1049–1066. [CrossRef]

60. Nguyen, H.-H.; McAlpine, C.; Pullar, D.; Johansen, K.; Duke, N.C. The relationship of spatial–temporal changes in fringe mangrove extent and adjacent land-use: Case study of Kien Giang coast, Vietnam. *Ocean Coast. Manag.* **2013**, *76*, 12–22. [CrossRef]

61. Jensen, J.R. *Introductory Digital Image Processing: A Remote Sensing Perspective*, 4th ed.; Pearson Education: Glenview, IL, USA, 2015; pp. 501–582.

62. El-Hattab, M.M. Applying post classification change detection technique to monitor an Egyptian coastal zone (Abu Qir Bay). *Egypt. J. Rem. Sens. Space Sci.* **2016**, *19*, 23–36. [CrossRef]

63. Ajai; Bahuguna, A.; Chauhan, H.B.; Sen Sarma, K.; Bhattacharya, S.; Ashutosh, S.; Pandey, C.N.; Thangaradjou, T.; Gnanppazham, L.; Selvam, V.; et al. Mangrove inventory of India at community level. *Natl. Acad. Sci. Lett.* **2013**, *36*, 67–77. [CrossRef]

64. Saintilan, N.; Rogers, K.; Kelleway, J.J.; Ens, E.; Sloane, D.R. Climate change impacts on the coastal wetlands of Australia. *Wetlands* **2018**, *38*, 1–10. [CrossRef]

65. Kanniah, K.; Sheikhi, A.; Cracknell, A.; Goh, H.; Tan, K.; Ho, C.; Rasli, F. Satellite images for monitoring mangrove cover changes in a fast growing economic region in Southern Peninsular Malaysia. *Remote Sens.* **2015**, *7*, 14360–14385. [CrossRef]

66. Tran, L.X.; Fischer, A. Spatiotemporal changes and fragmentation of mangroves and its effects on fish diversity in Ca Mau Province (Vietnam). *J. Coast Conserv.* **2017**, *21*, 355–368. [CrossRef]

67. Olds, A.D.; Albert, S.; Maxwell, P.S.; Pitt, K.A.; Connolly, R.M. Mangrove-reef connectivity promotes the effectiveness of marine reserves across the western Pacific. *Glob. Ecol. Biogeogr.* **2013**, *22*, 1040–1049. [CrossRef]

68. Eyoh, A.; Ebom, O. Spatio-temporal analysis of land use/land cover change trend of Akwa Ibom State, Nigeria from 1986–2016 using remote sensing and GIS. *Intl. J. Sci. Res.* **2015**, *5*, 1805–1809. [CrossRef]

69. Jones, T.; Ratsimba, H.; Ravaoarinorotsihoarana, L.; Glass, L.; Benson, L.; Teoh, M.; Carro, A.; Cripps, G.; Giri, C.; Gandhi, S.; et al. The dynamics, ecological variability and estimated carbon stocks of mangroves in Mahajamba Bay, Madagascar. *J. Mar. Sci. Eng.* **2015**, *3*, 793–820. [CrossRef]

70. Richards, D.R.; Friess, D.A. Rates and drivers of mangrove deforestation in Southeast Asia, 2000–2012. *Proc. Natl. Acad. Sci. USA* **2016**, *113*, 344–349. [CrossRef] [PubMed]

71. Duke, N.C.; Kovacs, J.M.; Griffiths, A.D.; Preece, L.; Hill, D.J.E.; van Oosterzee, P.; Mackenzie, J.; Morning, H.S.; Burrows, D. Large-scale dieback of mangroves in Australia. *Mar. Freshw. Res.* **2017**, *68*, 1816–1829. [CrossRef]

72. Hickey, S.M.; Phinn, S.R.; Callow, N.J.; Van Niel, K.P.; Hansen, J.E.; Duarte, C.M. Is climate change shifting the poleward limit of mangroves? *Estuaies Coasts* **2017**, *40*, 1215–1226. [CrossRef]

73. Rodriguez, W.; Feller, I.C.; Cavanaugh, K.C. Spatio-temporal changes of a mangrove–saltmarsh ecotone in the northeastern coast of Florida, USA. *Glob. Ecol. Conserv.* **2016**, *7*, 245–261. [CrossRef]

74. Asbridge, A.; Lucas, R.; Accad, A.; Dowling, R. Mangrove response to environmental changes predicted under varying climates: Case studies from Australia. *Curr. For. Rep.* **2015**, *1*, 178–194. [CrossRef]

75. Hoque, M.A.-A.; Phinn, S.; Roelfsema, C. A systematic review of tropical cyclone disaster management research using remote sensing and spatial analysis. *Ocean Coast. Manag.* **2017**, *146*, 109–120. [CrossRef]

76. McInnes, K. *Wet Tropics Cluster Report, Climate Change in Australia: Projections for Australia's Natural Resource Management Regions*; CSIRO and Bureau of Meteorology: Canberra, Australia, 2015.

77. Al-Hamdan, M.Z.; Oduor, P.; Flores, A.I.; Kotikot, S.M.; Mugo, R.; Ababu, J.; Farah, H. Evaluating land cover changes in Eastern and Southern Africa from 2000 to 2010 using validated Landsat and MODIS data. *Int. J. Appl. Earth Obs. Geoinf.* **2017**, *62*, 8–26. [CrossRef]

78. Chen, C.-F.; Son, N.-T.; Chang, N.-B.; Chen, C.-R.; Chang, L.-Y.; Valdez, M.; Centeno, G.; Thompson, C.; Aceituno, J. Multi-decadal mangrove forest change detection and prediction in Honduras, Central America, with Landsat Imagery and a markov chain model. *Remote Sens.* **2013**, *5*, 6408–6426. [CrossRef]

79. Kingsford, R. Flow alteration and its effect on Australia's riverine vegetation. In *Vegetation of Australian Riverine Landscapes: Biology, Ecology and Management*; Capon, S., James, C., Eds.; CSIRO Publishing: Clayton South, Australia, 2016; pp. 261–268.

80. Greet, J.; Cousens, R.D.; Webb, J.A. More exotic and fewer native plant ppecies: Riverine vegetation patterns pssociated with altered seasonal flow patterns. *River Res. Appl.* **2013**, *29*, 686–706. [CrossRef]

81. Greet, J.O.E.; Webb, J.A.; Cousens, R.D. The importance of seasonal flow timing for riparian vegetation dynamics: A systematic review using causal criteria analysis. *Freshw. Biol.* **2011**, *56*, 1231–1247. [CrossRef]

82. Das, R.T.; Pal, S. Exploring geospatial changes of wetland in different hydrological paradigms using water presence frequency approach in Barind Tract of West Bengal. *Spat. Inf. Res.* **2017**, *25*, 467–479. [CrossRef]

83. Antwi, E.K.; Boakye-Danquah, J.; Asabere, S.B.; Yiran, G.A.B.; Loh, S.K.; Awere, K.G.; Abagale, F.K.; Asubonteng, K.O.; Attua, E.M.; Owusu, A.B. Land use and landscape structural changes in the ecoregions of Ghana. *J. Disaster Res.* **2014**, *9*, 452–467. [CrossRef]

84. Gonneea, M.E.; Maio, C.V.; Kroeger, K.D.; Hawkes, A.D.; Mora, J.; Sullivan, R.; Madsen, S.; Buzard, R.M.; Cahill, N.; Donnelly, J.P. Salt marsh ecosystem restructuring enhances elevation resilience and carbon storage during accelerating relative sea-level rise. *Estuar. Coast. Shelf Sci.* **2019**, *217*, 56–68. [CrossRef]

85. Wegscheidl, C.J.; Sheaves, M.; McLeod, I.M.; Hedge, P.T.; Gillies, C.L.; Creighton, C. Sustainable management of Australia's coastal seascapes: A case for collecting and communicating quantitative evidence to inform decision-making. *Wetl. Ecol. Manag.* **2016**, *25*, 3–22. [CrossRef]

86. Ward, D.P.; Petty, A.; Setterfield, S.A.; Douglas, M.M.; Ferdinands, K.; Hamilton, S.K.; Phinn, S. Floodplain inundation and vegetation dynamics in the Alligator Rivers region (Kakadu) of northern Australia assessed using optical and radar remote sensing. *Remote Sens. Environ.* **2014**, *147*, 43–55. [CrossRef]

87. Jones, C.; Vesk, P. Grazing. In *Vegetation of Australian Riverine Landscapes: Biology, Ecology and Management*; Capon, S., James, C., Eds.; CSIRO Publishing: Clayton South, Australia, 2016; Chapter 17; pp. 307–323.

88. Brock, M.A. Australian wetland plants and wetlands in the landscape: Conservation of diversity and future management. *Aquat. Ecosyst. Health Manag.* **2003**, *6*, 29–40. [CrossRef]

89. Toure, M.A.; Ndiaye, M.L.; Traore, V.B.; Faye, G.; Cisse, B.; Ndiaye, A.; Wade, C.T. Using of Landsat images for land use changes detection in the ecosystem: A case study of the Senegal River Delta. *Int. J. Environ. Agric. Biotechnol.* **2016**, *1*, 200–209.

90. Department of Environment and Science. *Land Cover Change in Queensland 2016–17-and-2017–18: A Statewide Landcover and Trees Study (SLATS) Summary Report*; Department of Environment and Science: Brisbane, Australia, 2018.

91. Simmons, B.A.; Law, E.A.; Marcos-Martinez, R.; Bryan, B.A.; McAlpine, C.; Wilson, K.A. Spatial and temporal patterns of land clearing during policy change. *Land Use Policy* **2018**, *75*, 399–410. [CrossRef]

92. Department of Environment and Science. *Statewide Landcover and Trees Study (SLATS): Overview of Method*; Department of Environment and Science: Brisbane, Australia, 2018.

93. Lu, D.; Weng, Q. A survey of image classification methods and techniques for improving classification performance. *Int. J. Remote Sens.* **2007**, *28*, 823–870. [CrossRef]

94. Foody, G.M. Assessing the accuracy of land cover change with imperfect ground reference data. *Remote Sens. Environ.* **2010**, *114*, 2271–2285. [CrossRef]

95. Younes Cárdenas, N.; Joyce, K.E.; Maier, S.W. Monitoring mangrove forests: Are we taking full advantage of technology? *Int. J. Appl. Earth Obs. Geoinf.* **2017**, *63*, 1–14. [CrossRef]

96. Zhang, X.; Tian, Q. A mangrove recognition index for remote sensing of mangrove forest from space. *Curr. Sci.* **2013**, *105*, 1149–1155.

97. Foody, G.M. Status of land cover classification accuracy assessment. *Remote Sens. Environ.* **2002**, *80*, 185–201. [CrossRef]

98. Lucas, R.; Lule, A.V.; Rodríguez, M.T.; Kamal, M.; Thomas, N.; Asbridge, E.; Kuenzer, C. Spatial ecology of mangrove forests: A remote sensing perspective. In *Mangrove Ecosystems: A Global Biogeographic Perspective*; Rivera-Monroy, V.H., Lee, S.Y., Eds.; Springer International Publishing: Cham, Switzerland, 2017; pp. 87–112.

99. Rogers, K.; Boon, P.I.; Branigan, S.; Duke, N.C.; Field, C.D.; Fitzsimons, J.A.; Kirkman, H.; Mackenzie, J.R.; Saintilan, N. The state of legislation and policy protecting Australia's mangrove and salt marsh and their ecosystem services. *Mar. Policy* **2016**, *72*, 139–155. [CrossRef]

100. Phinn, S.; Joyce, K.; Scarth, P.; Roelfsema, C. The role of integrated information acquisition and management in the analysis of coastal ecosystem change. In *Remote Sensing of Aquatic Coastal Ecosystem Processes: Science and Management Applications*; Richardson, L.L., LeDrew, E.F., Eds.; Springer: Dordrecht, The Netherlands, 2006; Volume 9, pp. 217–249.

101. Watson, J.E.; Evans, M.C.; Carwardine, J.; Fuller, R.A.; Joseph, L.N.; Segan, D.B.; Taylor, M.F.; Fensham, R.J.; Possingham, H.P. The capacity of Australia's protected-area system to represent threatened species. *Conserv. Biol.* **2011**, *25*, 324–332. [CrossRef]

102. Jones, K.R.; Venter, O.; Fuller, R.A.; Allan, J.R.; Maxwell, S.L.; Negret, P.J.; Watson, J.E.M. One-third of global protected land is under intense human pressure. *Science* **2018**, *360*, 788–791. [CrossRef]

103. Department of the Environment. Broad Leaf Tea-Tree (*Melaleuca Viridiflora*) Woodlands in High Rainfall Coastal North Queensland. In *Community and Species Profile and Threats Database*. Available online: http://www.environment.gov.au/sprat (accessed on 1 March 2019).

104. Department of the Environment. Subtropical and temperate coastal saltmarsh. In *Community and Species Profile and Threats Database*. Available online: http://www.environment.gov.au/sprat (accessed on 1 March 2019).

105. McLeod, E.; Chmura, G.L.; Bouillon, S.; Salm, R.; Björk, M.; Duarte, C.M.; Lovelock, C.E.; Schlesinger, W.H.; Silliman, B.R. A blueprint for blue carbon: Toward an improved understanding of the role of vegetated coastal habitats in sequestering CO_2. *Front. Ecol. Environ.* **2011**, *9*, 552–560. [CrossRef]

106. Abbot, J.; Marohasy, J. Has the herbicide diuron caused mangrove dieback? A re-examination of the evidence. *Hum. Ecol. Risk Assess. Int. J.* **2011**, *17*, 1077–1094. [CrossRef]

107. Duke, N.C.; Bell, A.M.; Pederson, D.K.; Roelfsema, C.M.; Bengtson Nash, S. Herbicides implicated as the cause of severe mangrove dieback in the Mackay region, NE Australia: Consequences for marine plant habitats of the GBR World Heritage Area. *Mar. Pollut. Bull.* **2005**, *51*, 308–324. [CrossRef] [PubMed]

108. Lucas, R.; Finlayson, C.M.; Bartolo, R.; Rogers, K.; Mitchell, A.; Woodroffe, C.D.; Asbridge, E.; Ens, E. Historical perspectives on the mangroves of Kakadu National Park. *Mar. Freshw. Res.* **2018**, *69*, 1047–1063. [CrossRef]

109. Barbier, E.B. The value of coastal wetland ecosystem services. In *Coastal Wetlands: An Integrated Ecosystem Approach*, 2nd ed.; Perillo, G.M.E., Wolanski, E., Eds.; Elsevier: Amsterdam, The Netherlands, 2019; pp. 947–964.

110. Sippo, J.Z.; Lovelock, C.E.; Santos, I.R.; Sanders, C.J.; Maher, D.T. Mangrove mortality in a changing climate: An overview. *Estuar. Coast. Shelf Sci.* **2018**, *215*, 241–249. [CrossRef]

111. Emanuel, K.A. Downscaling CMIP5 climate models shows increased tropical cyclone activity over the 21st century. *Proc. Natl. Acad. Sci. USA* **2013**, *110*, 12219–12224. [CrossRef]

112. CSIRO; Bureau of Meteorology. *Climate Change in Australia: Projections for Australia's Natural Resource Management Regions*; CSIRO and Bureau of Meteorology: Canberra, Australia, 2015.

113. Waterhouse, J.; Schaffelke, B.; Bartley, R.; Eberhard, R.; Brodie, J.; Star, M.; Thorburn, P.; Rolfe, J.; Ronan, M.; Taylor, B.; et al. *2017 Scientific Consensus Statement: Land Use Impacts on Great Barrier Reef Water Quality and Ecosystem Condition*; Queensland Government: Brisbane, Australia, 2017.

 *remote sensing*

Article

A Novel Approach to Modelling Mangrove Phenology from Satellite Images: A Case Study from Northern Australia

Nicolas Younes [1,2,*,†], Tobin D. Northfield [1,2,3], Karen E. Joyce [1,2], Stefan W. Maier [1,2,4], Norman C. Duke [5] and Leo Lymburner [6]

[1] Centre for Tropical Environmental and Sustainability Science, Cairns, QLD 4878, Australia; tnorthfield@wsu.edu (T.D.N.); karen.joyce@jcu.edu.au (K.E.J.); stefan.maier@maitec.com.au (S.W.M.)

[2] College of Science and Engineering, James Cook University, Townsville, QLD 4811, Australia

[3] Department of Entomology, Tree Fruit Research and Extension Center, Washington State University, 110 N Western Ave., Wenatchee, WA 98801, USA

[4] Maitec, P.O. Box U19, Charles Darwin University, Darwin, NT 0815, Australia

[5] Centre for Tropical Water and Aquatic Ecosystem Research (TropWATER), James Cook University, Townsville, QLD 4811, Australia; norman.duke@jcu.edu.au

[6] Geoscience Australia, CNR Jerrabomberra Ave and Hindmarsh Drive, Symonston, ACT 2609, Australia; Leo.Lymburner@ga.gov.au

* Correspondence: nicolas.younescardenas@my.jcu.edu.au

† Current address: Fenner School of Environment and Society, Australian National University, Linnaeus Way, Acton ACT 2601, Australia.

Received: 22 September 2020; Accepted: 30 November 2020; Published: 8 December 2020

Abstract: Around the world, the effects of changing plant phenology are evident in many ways: from earlier and longer growing seasons to altering the relationships between plants and their natural pollinators. Plant phenology is often monitored using satellite images and parametric methods. Parametric methods assume that ecosystems have unimodal phenologies and that the phenology model is invariant through space and time. In evergreen ecosystems such as mangrove forests, these assumptions may not hold true. Here we present a novel, data-driven approach to extract plant phenology from Landsat imagery using Generalized Additive Models (GAMs). Using GAMs, we created models for six different mangrove forests across Australia. In contrast to parametric methods, GAMs let the data define the shape of the phenological curve, hence showing the unique characteristics of each study site. We found that the Enhanced Vegetation Index (EVI) model is related to leaf production rate (from in situ data), leaf gain and net leaf production (from the published literature). We also found that EVI does not respond immediately to leaf gain in most cases, but has a two- to three-month lag. We also identified the start of season and peak growing season dates at our field site. The former occurs between September and October and the latter May and July. The GAMs allowed us to identify dual phenology events in our study sites, indicated by two instances of high EVI and two instances of low EVI values throughout the year. We contribute to a better understanding of mangrove phenology by presenting a data-driven method that allows us to link physical changes of mangrove forests with satellite imagery. In the future, we will use GAMs to (1) relate phenology to environmental variables (e.g., temperature and rainfall) and (2) predict phenological changes.

Keywords: GAMs; Generalized Additive Models; EVI; Landsat; mangrove forests; phenology; time series analysis

1. Introduction

Around the world, the effects of changing plant phenology are evident in many ways: from earlier and longer growing seasons to altering the relationships between plants and their natural pollinators [1–3]. Remote sensing techniques allow us to detect subtle changes in plant phenology, and here we present a novel approach to describe phenological cycles of mangrove ecosystems. We contribute to a better understanding of mangrove phenology by investigating physical changes of mangrove ecosystems and how the evidence of change is captured by satellite images. Accurately modelling and predicting mangrove phenology will help us understand not only the seasonal variations but also the long-term trends in the natural cycles of these forests. New models, such as the one presented here, will advance our understanding of how drought, heatwaves and other extreme weather events affect mangrove health and growth. Similar to using sea temperature to predict coral bleaching events, we could use phenology to predict mangrove dieback events akin to those of 2015 and 2016 in the Gulf of Carpentaria in northern Australia.

Phenology is related to the life cycle events of plants and animals and their relationship to climatic and other abiotic factors [3,4]. Plant phenology also plays an important role in the carbon cycle in the form of sequestration and storage. Phenological cycles of plants ensure that leafing, flowering and fruiting events occur during the most appropriate season to achieve maximum growth and reproductive success. Mangrove phenology is often described at the species level by relating the time of year when trees flower, fruit or defoliate with suspected drivers like temperature and rainfall [5,6]. For example, [7] described the phenology and distribution of *Avicennia marina* mangroves along the Australian coastline, and [8] described the flowering and leafing phenologies of mangroves in the Darwin region. While these descriptions provide a very valuable baseline for comparison, they often lack the spatial extent and frequency needed for phenological studies [9].

We can monitor mangrove phenology using remote sensing or we can collect in situ data. The main advantage of in situ monitoring is that it provides information at the tree and species level, where observations can be very detailed over a wide range of variables. However, in situ monitoring is challenging, time-consuming and variation in methods and survey effort can make it difficult to compare results [10]. In contrast, the remote sensing approach provides information at the landscape and continental scales and is consistently acquired over time and space [9]. While many studies used space-borne sensors to map mangroves at the global [11,12], continental [13,14], and local scales [15], few have used these sensors to monitor mangrove phenology. [16] used MODIS (Moderate Resolution Imaging Spectroradiometer) data between 2000–2014 to detect mangrove phenology using four different spectral indices in the Yucatan peninsula in Mexico. Similarly, [17] compared the phenology of mangroves to that of the surrounding forests using MODIS imagery. While the temporal resolution of the MODIS sensors is very high (1–2 days), the spatial resolution is coarse (250–500 m). Landsat satellites offer a better spatial resolution at the cost of a lower temporal resolution. Despite this tradeoff, the Landsat archive is key to using remote sensing to monitor mangrove phenology as it provides more than 30 years of imagery at a spatial resolution of 30 m x 30 m and a temporal resolution of 8–16 days [18].

To date, most studies on plant phenology have used fully parametric models, mainly in the form of double logistic or sinusoidal functions [19–21]. These functions may perform well in deciduous or temperate forests, where there is a single, well defined period of leaf production, and a single, well defined period of leaf senescence [22,23], but these methods may not be well suited for mangroves and other evergreen forests. When detecting phenology, one of the main limitations of parametric models (e.g., logistic functions) is that they fail to detect asymmetric trends in leaf growth or senescence [22]. Considering that the growing season of some evergreen forests consists of two periods of leaf growth and death, fully parametric (or model-driven) models have the potential to oversimplify the phenology of these ecosystems. Other, more complex models have also been used to examine plant phenology, mainly in the form of artificial neural networks, however, these methods are known to have mostly been used in croplands [24]. Semi-parametric (or data-driven) models, on the other hand, may be better suited for this task as they do not necessarily assume that there will be a single peak or trough in leaf growth or death. Rather, semi-parametric models use the data to determine the shape of the phenology.

Studies have documented the dual phenology of mangroves [6] and other evergreen forests [25] in the field, but this dual phenology has not been recorded using satellite imagery. Dual phenology refers to two periods of leaf growth; unlike deciduous forests which grow their leaves during the spring, some evergreen forests have two periods of leaf growth every year. In mangrove forests these events have never been documented using satellite imagery, probably due to the use of fully parametric models to detect phenology, or because in situ data collection focuses mainly on litterfall rather than leaf production. The novelty of this study is that we use a semi-parametric method to model mangrove phenology and in doing that we present, for the first time, these two distinct periods of leaf growth described in the literature.

Generalized Additive Models (GAMs) are commonly used in ecology and climate sciences, to examine non-linear relationships between response and independent variables. Here we present a novel, data-driven method to extracting mangrove phenology from a series of Landsat images. We use discrete observations of mangrove forests (i.e., satellite images) and Generalized Additive Models (GAMs) to create a continuous curve of phenology over time, without assuming a certain shape, amplitude, or frequency. Our aims are to (1) use a semi-parametric approach (GAMs) to examine if seasonal changes in biophysical variables are related to seasonal changes in the spectral reflectance of mangrove forests; (2) compare the satellite-derived phenology with a set of field observations and measurements; (3) compare the satellite-derived phenology to peer-reviewed literature describing the phenology of mangrove forests, and (4) determine how the Enhanced Vegetation Index (EVI) responds to leaf gain, leaf fall or net leaf production in mangrove ecosystems across Australia.

This manuscript is organized in the following way: firstly, we describe the site and methods used to collect the field data. Then we describe how we use the literature to create a proxy for mangrove phenology. Afterwards, we describe the use of GAMs and satellite images to detect the apparent phenology of mangroves across northern Australia. Having done this, we present the models of apparent phenology and compare them with the data collected in the field, and the proxies from the literature. Finally, we discuss the results, limitations, and future work.

2. Materials and Methods

We selected six study sites across Australia to evaluate mangrove phenology from satellite imagery using GAMs. One site corresponds to field observations collected in the Gladstone region (Queensland) in the late 1990s, and the remaining sites ($n = 5$) correspond to qualitative data extracted from peer-reviewed publications (Figure 1). In this section, we first describe the field site followed by the peer-reviewed studies, the image acquisition process, and the time series analysis using GAMs. Finally, we describe the phenology model validation.

2.1. Field Site Description

The Gladstone region in central Queensland is home to over 100,000 ha of intertidal wetlands, out of which 30% are mangrove forests [26]. The annual mean temperature ranges between 18.4 °C and 27.5 °C, the mean annual rainfall is 874 mm, and the semi-diurnal tides often range from 1.5–3.5 m but reach up to 6 m. In this region, mangroves of the *Rhizophora*, *Avicenna*, and *Ceriops* genera are among the most common [5,27]. The field data were collected between July 1996 and August 1998 in two plots located in Fisherman's Landing and one plot on Curtis Island (Figure 2). The sites were dominated by *Rhizophora stylosa* trees and located within the Landsat World Reference System 2 (WRS-2) path 91 and rows 76 and 77.

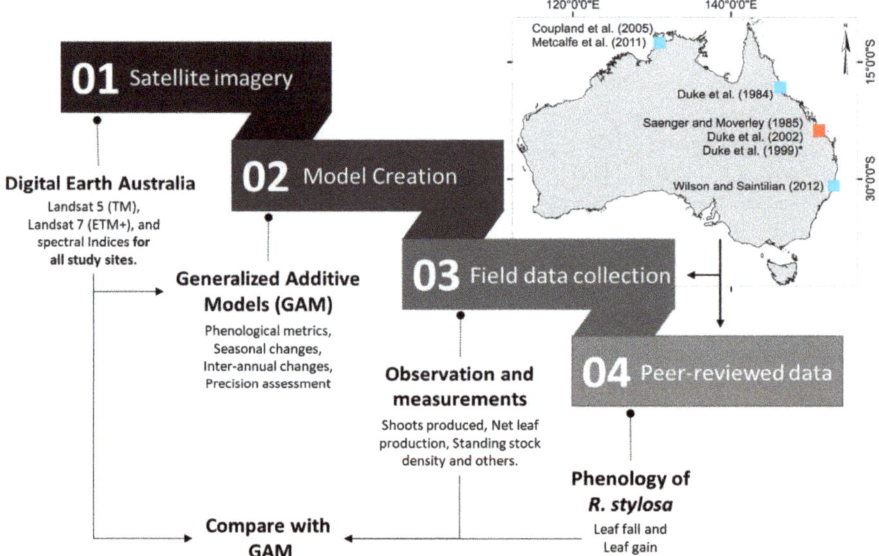

Figure 1. Workflow and location of the study sites used to validate the phenology model. * shows the location of the field site [26] and two other published studies used.

2.2. Field Observations and Measurements

The data were collected by [26] in the following way: in each plot, the authors selected mature *R. stylosa* trees between 4–9 m tall and tagged 21 leafy shoots in the upper two meters of the canopy totaling around 378 shoots. The authors conducted monthly inspections and recorded the shoot growth, number of leaves, reproductive parts and number of branch shoots. To measure the amount of litterfall, the authors suspended litter traps (1 m^2 in area) under the selected trees. The traps were suspended above the high tide mark and litter was collected, sorted and weighed on a monthly basis. During the data collection period, the average tide height was 2.49 m according to the historical record for the site [28]. Importantly, [26] never intended to validate satellite imagery with their data, therefore (1) no spectral data were collected, making these data completely independent from the satellite-derived phenology, and (2) we do not anticipate a high correlation between satellite-derived phenology and the field data. The data consist of the mean monthly values of six phenological variables, however, we selected the three that were more relevant for our study (see ([26], pp 105–108) for details). The selected biophysical variables are detailed below:

Leaves lost [leaves $\times$ m^{-2} $\times$ day^{-1}]: The number of leaves that fell into the litter traps.

Leaves gained [stipules $\times$ m^{-2} $\times$ day^{-1}]: The number of stipules that fell into the litter traps. This variable serves as a proxy for the number of leaves produced in a tree.

Net leaf production [leaves gained-leaves lost]: The difference between leaves gained and fallen leaves. This measure is an indication the net balance of leaves in the canopy with more or less leaves as leaves appear or fall, leaving the canopy in either debit (=stressed), credit (=growth) or neutral condition.

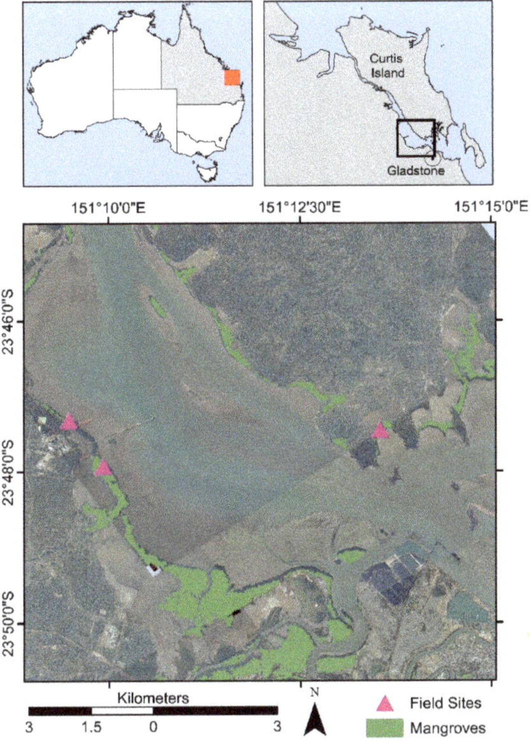

Figure 2. Location of the field sites and mangrove patches in the Gladstone region, Queensland. Aerial images of the study site for 1996, provided by the State of Queensland (QAP5402131/47).

2.3. Published Literature on the Phenology of R. stylosa

To compare the apparent phenology (i.e., from the GAMs) to other sources of information, we gathered a set of peer-reviewed papers that included *R. stylosa* as target species. The reasons for selecting this species were twofold: (1) it is common throughout northern Australia; (2) a number of studies have described its phenology over a wide geographic area across the Indo West Pacific region. We looked for papers that had a graphical interpretation of leaf fall and/or leaf gain over time and we found six examples (Table 1). We used the published graphical interpretations of leaf fall and leaf gain to determine, in a qualitative way, the times of the year where most leaves grew or fell. All published graphs show "Time" on the horizontal axis and a measure of leaf fall or gain on the vertical axis. In each study, we divided the vertical axis into five equidistant categories (i.e., very low, low, medium, high, very high) and recoded the category for each month (not shown). Finally, we calculated the net leaf production from each study by subtracting the leaf fall from leaf gain values and then compared the three variables with the apparent phenology of each site (Table 1).

From Table 1 one can see that the studies by [29,30] predate the time where Landsat imagery was collected. Between 1974 and 1989, cyclones Dawn (March 1976), Keith (January 1977), Gordon (January 1979), and Kerry (March 1979) affected Hinchinbrook Island, Gladstone, or Proserpine in Queensland. Neither [29] nor [30] mention the effects of cyclones, drought on their respective study sites, either because the field campaigns happened before the cyclones, or there was no significant damage to the trees. While there is no certainty about the effects of extreme weather events for those studies, here we assume that the phenology observed by the authors did not change until satellite imagery was acquired.

Table 1. Peer-reviewed studies used for the qualitative comparison: target species, study duration, location and description of satellite images used.

Ref.	Target Species	Leaf Fall or Leaf Gain	Study Duration	Observation Frequency	Location	Satellite Images Used	Satellite Images Date Range	WRS2 Path/Row
[29]	C. tagal var. tagal B. gymnorhiza R. apiculata R. stylosa R. X lamarckii	LF, LG	1975–1978	Monthly	Hinchinbrook Island, QLD	70	July/1987 to December/1991	95/73
[30]	A. annulata A. corniculaturn A. marina C. tagal E. agallocha L. racemosa O. octodonta R. stylosa X australasicus	LF, LG	1979–1982	Monthly	Gladstone and Proserpine, QLD	80	August/1987 to December/1991	91/76, and 91/77
[31] *	R. stylosa	LF, LG	1996–1998	Monthly	Gladstone, QLD	106	January/1995 to December/1999	91/76, and 91/77
[32]	A. marina C. australis R stylosa S. alba	LF, LG	1999–2001	Monthly	Darwin Harbor, NT	113	January/1998 to December/2002	106/69
[8]	A. marina B. exaristata C. schultzii C. australis E. ovalis L. racemosa R. stylosa S. alba	LF	1997–2000	Monthly	Darwin Harbor, NT	195	February/1996 to December/2001	106/69
[33]	R. stylosa	LF, LG	2002–2004	73 days	South West Rocks Creek, Richmond River, Brunswick River, NSW	104	January/1998 to December/2002	88/81

* Same location as Duke et al. (1999).

2.4. Landsat Image Acquisition and Processing

Digital Earth Australia holds a copy of the Landsat archive (1987–present) for the whole of Australia [18,34]. Digital Earth Australia provided all L1T images ($n = 668$) of our sites between 1987 and 2006 for the Landsat 5 (TM) and Landsat 7 (ETM+) sensors, and performed (1) geometric, (2) atmospheric and (3) Nadir-adjusted Bidirectional reflectance distribution function Reflectance (NBAR) corrections following [35]. All corrections were performed within Digital Earth Australia, using the python programming language. Digital Earth Australia uses the 'Pixel Quality Assessment' algorithm [18] to remove pixels with clouds and cloud shadows, as well as all missing pixels resulting from the Landsat 7 Scan Line Corrector failure; these pixels were removed from the datasets and not used (i.e., there was no gap-filling for ETM+ data). The main reason for not filling the missing pixels is that we wanted to use the GAM with only true values, rather than including additional uncertainty to the GAMs. Having done this, we calculated the Enhanced Vegetation Index (EVI) [36] for each pixel in each image. We chose EVI because studies show that this spectral index does not saturate with high vegetation densities [36], it is better suited than other indices for discriminating vegetation fraction in mangrove ecosystems [37], and it is commonly used for phenology investigations [16,17].

Our study leverages the high temporal density of the Landsat archive, and the overlapping footprints of two or more Landsat scenes, hence increasing the number of usable pixels in a given area. On average, the difference between the dates when field data was collected and the closest satellite image acquired is 5 days, as shown in Table 2. To compare the GAMs with peer-reviewed literature, we selected a period equal to the time of the data collection plus and minus one year, thereby ensuring that the models had enough input data. In the cases where studies were dated before 1987, we used the first three years of available imagery of the area to create the GAM (Table 1). We estimated the location of the studies from the site descriptions in each publication and created a region of interest of approximately 17 ha of mangrove forests surrounding the study area. Afterwards, we selected only the pixels that corresponded to mangrove forests using the "Mangrove Canopy Cover" product developed by Geoscience Australia ([13]) and applied the GAM to every pixel within our region of interest. This approach ensured that we captured the phenology of the mangrove community instead of a small plot.

Table 2. Comparison between field data collection dates and satellite image dates.

Date of Field Data Collection	Date of Closest Satellite Image	Difference (Days)
6-June-1996	18-May-1996	9 days
24-Jully-1996	31-July-1996	7 days
24-August-1996	16-August-1996	8 days
23-September-1996	17-September-1996	6 days
15-October-1996	19-October-1996	4 days
20-November-1996	04-November-1996	16 days
17-December-1996	22-December-1996	5 days
16-January-1997	23-January-1997	7 days
14-Mararch-1997	12-March-1997	2 days
2-May-1997	29-April-1997	3 days
18-June-1997	16-June-1997	2 days
23-July-1997	18-July-1997	5 days
21-August-1997	19-August-1997	2 days
7-October-1997	06-October-1997	1 days
11-November-1997	17-November-1997	4 days
10-December-1997	09-December-1997	1 days
4-February-1998	26-January-1998	9 days
7-May-1998	02-May-1998	5 days
18-August-1998	22-August-1998	4 days

2.5. Time Series Analysis Using Generalized Additive Models

With all images pre-processed, georeferenced and sorted by time of acquisition, we proceeded to create a model of phenology for every available pixel in our study sites using GAMs. Contrary to linear additive models, GAMs are statistical models in which the relationship of predictor and response variables is captured by smooth functions instead of coefficients [38]. Equations (1) and (2) show the respective linear and generalized additive relationships between one response variable (Y) and two predictor variables (X_i) for i observations [39]:

$$Y = \beta_0 + \beta_1 X_{i1} + \beta_2 X_{i2} + \varepsilon \tag{1}$$

$$Y = \beta_0 + f_1(X_{i1}) + f_2(X_{i2}) + \varepsilon \tag{2}$$

Noticeably, there is no change in the form of the model, however, there is no assumption that the relationship between predictor and response variables is linear. In Equation (1) an additive linear relationship between Y and X_i is captured by the slope terms β_1 and β_2, while in Equation (2) the additive relationship is captured by the "smooth" functions $f_1(\cdot)$ and $f_2(\cdot)$. The shape of the "smooth" functions $(f_n(\cdot)$, also known as "splines" or "smooth splines", is determined during the computation in an iterative way and can take many forms (e.g., polynomial, linear, quadratic) [39,40]. Each 'smooth' function $(f_n(\cdot))$, or spline, is comprised of several basis functions (b_n), their coefficients (β_n), and where K determines the maximum complexity of each smooth function:

$$f(x) = \sum_{k=1}^{K} \beta_n b_n(x) \tag{3}$$

As explained by [41], GAMs use several types of splines to determine the relationship between each predictor variable (X_i) and the response (Y) is evaluated at every data point in an iterative way, for every predictor variable (see [42] a detailed description). In this case, and following [43], the base function is a Fourier basis with 2N parameters $\delta = [a_1, b_1, \ldots, a_N, b_N]^T$, that allow the construction of a matrix of seasonality vectors for each past and future value of t. Importantly, we are neither fitting a Fourier series to the data, nor assuming that the relationship of EVI and time is symmetric. Here, the apparent phenology of each pixel results from the Fourier basis expansion evaluated at each data point, and adding the weighted basis functions. In summary, the detecting apparent phenology is a curve-fitting exercise rather than a time series decomposition one.

Another characteristic of GAMs is that measurements do not need to be evenly spaced in time [43]. This works well in our case for two reasons: (1) pixels with clouds, shadows and other errors are flagged as invalid observations leading to time series with random gaps in both length and timing; and (2) areas where the footprints of two or more scenes overlap will have more observations than areas with no overlap.

To detect mangrove phenology from a satellite-derived data series, we used the Python programming language and the "Prophet" package (version 0.3.post2) developed by Facebook [43]. Facebook designed this package to analyze user engagement with the social network at different time scales, and to investigate how periodic events such as holidays affect that engagement. Similar to mangrove phenology, user engagement on the social media platform is affected by regular and irregular events such as weekends (i.e., regular events) and public holidays, which change every year [43]. In a similar fashion, mangroves are affected by regular changes in temperature and rainfall (i.e., seasons) and irregular events such as cyclones or drought. While the time scales may differ, the concept of tracing an event (e.g., phenology or user engagement) over time remains the same. We selected this package due to its ease of use and re-purposed it to extract seasonal variations in greenness and phenological metrics from satellite images of mangrove forests.

2.6. Phenological Metrics

With the data ingested, Prophet separates the seasonal components of the time series from the trend and the residual components. The seasonal component of the GAMs is used as an approximation of phenological cycles and several techniques can be adapted to extract the start, end, and duration of the mangrove "green-up" season. In this study, we adopted similar definitions to those by [44] to identify the Start of Season, End of Season and Peak Growing Season, however, as our study does not involve a sinusoidal curve the definitions vary slightly. For simplicity, we define the start of season and end of season as the lowest points, and peak growing season as the highest points of the de-trended time series (Figure 3). We also define the length of the growing season as the time between the start of season and end of season. Because we use start of season, end of season and peak growing season as phenological metrics of the landscape, they do not represent individual species.

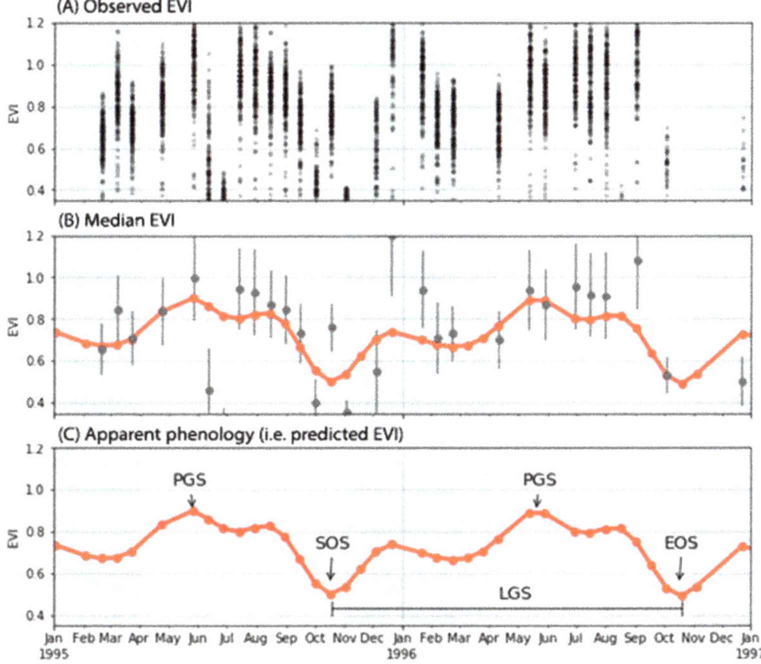

Figure 3. Panel (**A**) shows every available Enhanced Vegetation Index (EVI) observation for every pixel in the 17-ha region of interest from February 1995 to December 1996 for the Gladstone region. Panel (**B**) shows the median and standard deviation of the observed EVI values in grey dots and lines respectively, and the apparent phenology (i.e., GAM) in red. Panel (**C**) shows the apparent phenology, the definitions of start and end of season (SOS, EOS), peak growing season (PGS) and length of the growing season (LGS). Shaded areas represent the wet season months.

We extracted the start of season and peak growing season for each pixel in our field study site in the following way: from the seasonal component, we selected the predicted index values from the GAMs that were lower or higher than the 5 or 95 percentile as the potential start of season or peak growing season dates, respectively (Figure 3C). Then we selected the median of the image acquisition dates as the start of season, peak growing season and end of season dates. In case the selected date was not a date in which an image was acquired, we searched for the image with the closest date and used that date instead.

Because we wanted to determine if the GAMs were correlated with biophysical processes described in the literature (i.e., leaf gain, leaf loss, net leaf production), we decided to shift the models (i.e., displace the models along the time axis) by one, two and three months. We then examined if the EVI response was immediate or delayed. An immediate response of EVI to a biophysical process would imply that remote sensing techniques could be used for real-time monitoring. In contrast, a delayed response would help us understand which processes drive the changes in EVI. After comparing the biophysical processes to the EVI, we examined their relationship using linear regressions.

2.7. Validation of the GAMs

We assessed the precision of our model by running linear and non-linear regressions between the apparent phenology and (1) observed EVI values from satellite imagery; (2) in situ data from [26]; and (3) leaf fall, leaf gain and net leaf production values from published literature. We did this using the Scikit-Learn package for python [45], specifically, we used simple linear regressions and support vector regression with linear, polynomial and radial basis function kernels. We chose support vector regression because it is robust against outliers, it is easily implemented, and it allowed us to compare linear and nonlinear relationships between apparent phenology and the data measured in the field. In addition, we performed 5-fold cross-validation using the performance metrics functionality provided by "Prophet" (see the Supplementary Information section, Figure S2.

3. Results

3.1. Apparent Phenology

We found that some Australian mangroves display a bimodal seasonality with two periods of high EVI values and two periods of low EVI values, as shown in Figures 3 and 4. In the Gladstone region, the highest EVI values are recorded between May and August ("Peak growing season" in Figure 3), which are immediately followed by the lowest EVI values between September and November ("Start of season/End of season" in Figure 3). During the wet season, EVI values exhibit a second, less pronounced peak between December and January followed by a subtle drop between February and April. This bimodal seasonality refers to two different peaks in leaf production [6] and is also seen through time, with EVI values in mid-year being higher than those at the beginning or end of the year (wet season).

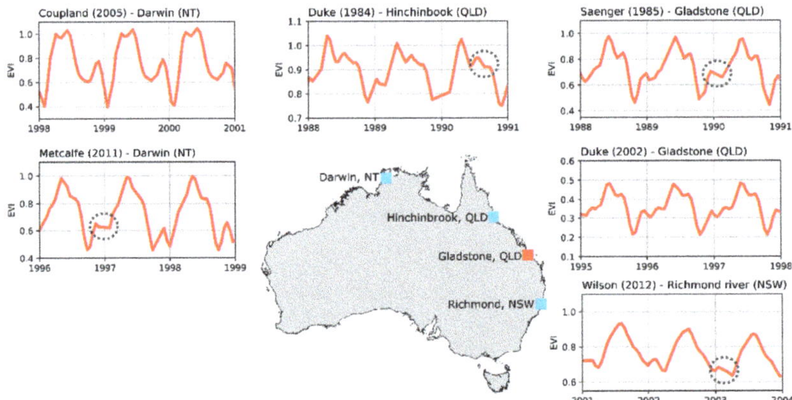

Figure 4. Apparent phenology for each study site. Grey dashed circles show examples of year-to-year variations in the apparent phenology. Blue squares represent locations where only published literature was used, while the red square represents the location of the field data site and where published literature was used.

Figure 4 shows the average phenology of all the pixels in each study site and highlights the fact that mangrove phenology varies with location and through time. For example, both Gladstone sites display similar phenology models despite being years apart. When compared to the Hinchinbrook site, however, the models are somewhat different, especially when looking immediately before and after the highest EVI values (i.e., peak growing season). On a greater scale, the phenology models across states differ greatly from one another. The site located in New South Wales has a distinctly smooth phenology curve while the Queensland sites show jagged features and the Northern Territory is in between.

Temporally, GAMs reveal subtle year-to-year differences in the phenology model that cannot be seen with fully parametric models as the latter over-simplify the phenology from satellite images. Grey circles in Figure 4 focus on certain features in the phenology models that change from year to year. Since we created the GAMs on a pixel-by-pixel basis, we can examine each pixel individually and determine the causes of such variations.

3.2. *Apparent Phenology and Field Data*

In Figure 5, we show the apparent phenology and the in situ data from [26]. We can see that each field variable has a marked seasonal pattern, where the values of the variable increase and decrease at certain times of the year (see below). Similarly, the apparent phenology shows a seasonal pattern with higher values between May and September and lower values between October and April. Both the monthly mean and the apparent phenology seem to describe some variables better than others as explained below (see also Section 4).

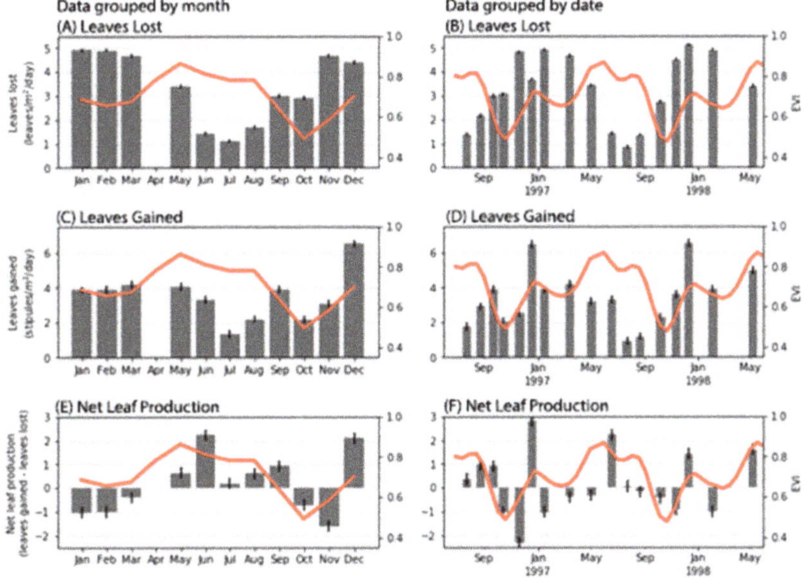

Figure 5. Apparent phenology vs. in situ data from [26]. The red line represents the apparent phenology for the Gladstone area (1995–1999). Grey bars and black lines represent the values for each variable and standard error, respectively. On the left panel, the data are grouped by month and on the right panel, the data are presented in chronological order. No in situ data were recorded for April during the experiment. Panels (**A**, **C** and **E**) display the monthly leaves lost, leaves gained, and net leaf production respectively. Panels (**B**, **D** and **F**) display the leaves lost, leaves gained, and net leaf production in chronological order.

3.2.1. Apparent Phenology and Leaves Lost

Visually, the apparent phenology appears to have an inverse relationship with the number of leaves lost. Between November and March, when the number of leaves lost is high (≥ 3 leaves/m^2/day), EVI values are often low. In contrast, EVI values are often high between May and October when fewer leaves are lost. This relationship is evident in Figure 5A,B.

3.2.2. Apparent Phenology and Leaves Gained

From Figure 5 C,D we see that the apparent phenology has a closer relationship with the number of leaves gained than with the number of leaves lost. Visually, this relationship is very strong, especially in the second half of the year. Between October and December, the number of leaves gained rises to its maximum value; this number then drops and remains stable until May. Similarly, in October, EVI rises steadily from its lowest value until December where it remains stable until March before rising to its maximum values between May and June before dropping again and restarting the cycle. In Figure 5D the apparent phenology shows peaks that coincide with periods of a high number of leaves produced (e.g., January 1997, December 1997, and May 1998). The same can be said about the troughs in the apparent phenology, which coincide with lower values of leaves produced (e.g., October 1996 and 1997).

3.2.3. Apparent Phenology and Net Leaf Production

Net leaf production presented by [26] shows two distinct peaks (i.e., June and December) and two troughs (i.e., January and November) that coincide with the peaks and troughs of the apparent phenology (Figure 5E). When the data are aggregated by month (Figure 5E), the relationship between EVI and the net leaf production is clear. Similarly, when the data are presented in chronological order, the months where net leaf production is highest (or lowest) coincide with months of high (or low) EVI values (Figure 5F). In some cases, high and low EVI values precede the highest and lowest values of net leaf production by about a month, however, this is not consistent over time.

3.2.4. Validation: Apparent Phenology vs. In Situ Variables

We validated the apparent phenology against in situ data by running a linear regression between the apparent phenology and every in situ variable from [26]. When the data are grouped by date (i.e., chronological order), the highest correlations with the apparent phenology come from the leaf production rate ($R^2 = 0.20$), total leaf area ($R^2 = 0.16$) and net leaf production ($R^2 = 0.11$). When the data are aggregated by month, however, the correlation of the variables with EVI increases in most cases (e.g., leaf production rate ($R^2 = 0.27$), standing stock ($R^2 = 0.14$)).

We also validated our model using non-linear regressions between the apparent phenology and each variable. In Table 3 we show the results from the support vector regression using RBF, linear, and polynomial kernels. For brevity, we only show the results for the regression between net leaf production and apparent phenology using the polynomial kernel, because those results show the best results. Again, monthly net leaf production has a slightly higher correlation and explained variance than chronological net leaf production.

Table 3. Explained variance, R^2, and Mean Absolute Error resulting from the support vector regression between Net leaf production and apparent phenology.

	Explained Variance	Mean Absolute Error	R^2
Data grouped by month	0.44	0.08	0.35
Data in chronological order	0.42	0.09	0.32

3.3. Apparent Phenology and Published Literature

Similar to our field data site, the apparent phenology shows a bimodal phenology curve across all sites described in the selected literature (Table 1, Figure 6). In general, the phenology models have either (1) an inverse relationship or (2) a time lag with respect to the intensity of leaf fall reported in the literature. Most studies report higher leaf fall rates between October and March and lower leaf fall rates between April and September (Figure 6), which denotes an inverse relationship with EVI. By shifting the models by three months, the visual relationship between leaf fall and EVI becomes stronger, especially for the data presented by [31,32] and [8].

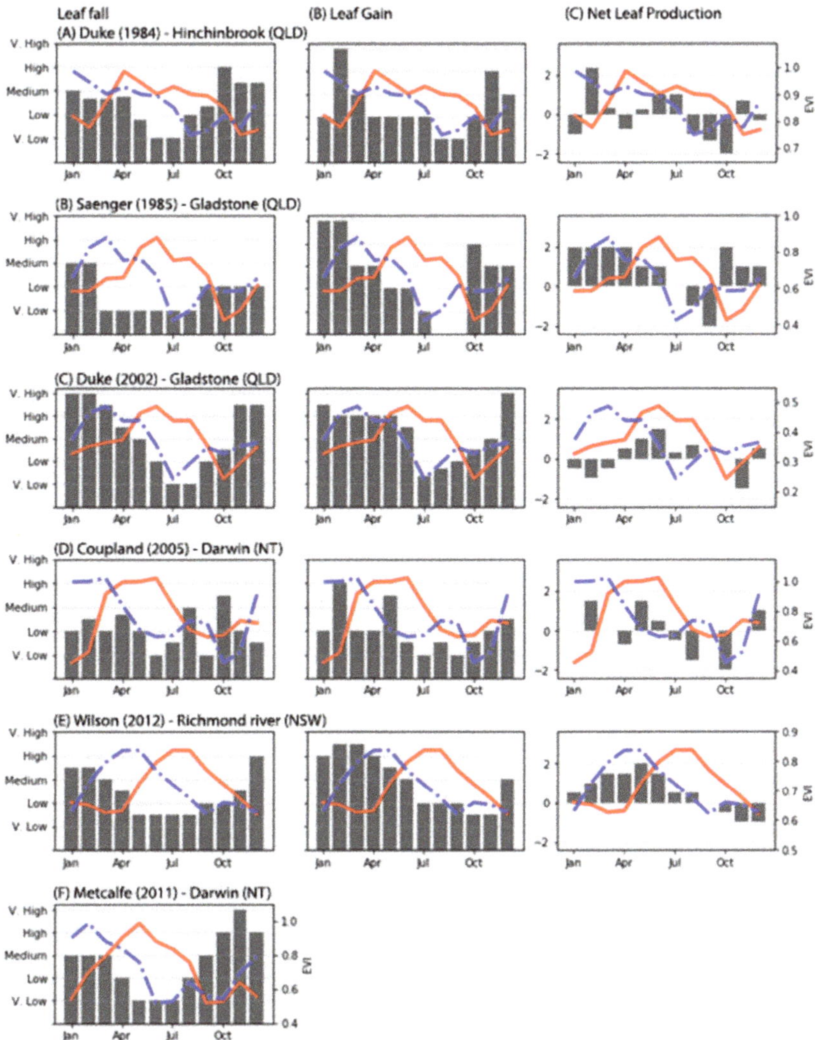

Figure 6. Panels (**A–F**) display the a qualitative measure of Leaf fall, Leaf gain and Net Leaf Production for each study site on the left, right and center respectively. Each panel represents a different study site. The red line represents the monthly value of the apparent phenology from the GAMs and the blue dotted line represents the apparent phenology shifted by three months.

Models of EVI seem to be better predictors of leaf gain intensity when compared to leaf fall intensity. On most sites, leaf gain intensity is highest between November and April and lowest between May and October. Higher values of leaf gain intensity relate well with high values of EVI. However, their timing does not match exactly. Visually, the shifted apparent phenology shows a much closer relationship with leaf gain intensity across all sites than the models with no time shift (Figure 6). Regarding net leaf production, sites in Gladstone (QLD) and Darwin (NT) show that the peaks and troughs of the apparent phenology coincide with the highest and lowest values of leaf production (Figure 6C,D). In contrast, the shifted models shown in Figure 6A,B,E have a better visual relationship with net leaf production.

3.3.1. Validation: Apparent Phenology vs. Published Data

With the exception of [29], all sites have higher correlation values with leaf gain or net leaf production when the apparent phenology is shifted by two or three months. For example, the apparent phenology correlates better with leaf gain values shifted by two months in the case of [31,33] and tree months in the case of [30,32]. The high R^2 values in Table 4 demonstrate that, in mangrove forests, the EVI response to leaf gain intensity and net leaf production is not immediate but delayed by two to three months.

Table 4. Correlation coefficients of the apparent phenology versus net leaf production, leaf fall and leaf gain for each site. Highest R^2 values per site are shown in bold.

Site	Shift (Months)	Leaf Fall		Leaf Gain		Net Leaf Production	
		R^2	*p*-Value	R^2	*p*-Value	R^2	*p*-Value
Duke_1984	−3	0.01	0.75	0.24	0.11	0.15	0.22
	−2	0.24	0.11	0.17	0.18	0.00	0.93
	−1	0.33	0.05	0.24	0.10	0.00	0.91
	0	0.33	0.05	**0.47**	**0.01**	0.05	0.49
Saegner_1985	−3	0.11	0.29	0.54	0.01	**0.55**	**0.01**
	−2	0.16	0.19	0.43	0.02	0.36	0.04
	−1	0.21	0.13	0.35	0.04	0.23	0.12
	0	0.41	0.02	0.35	0.04	0.14	0.23
Duke_2002	−3	0.38	0.03	0.53	0.01	0.01	0.81
	−2	0.74	0.00	**0.90**	**0.00**	0.04	0.53
	−1	0.71	0.00	0.36	0.04	0.43	0.02
	0	0.25	0.10	0.02	0.70	0.51	0.01
Coupland_2005	−3	0.04	0.52	**0.39**	**0.03**	0.15	0.22
	−2	0.07	0.42	0.05	0.49	0.00	0.98
	−1	0.08	0.38	0.08	0.38	0.00	0.90
	0	0.06	0.46	0.00	0.87	0.01	0.71
Wilson_2012	−3	0.02	0.63	0.42	0.02	**0.75**	**0.00**
	−2	0.06	0.45	0.71	0.00	0.53	0.01
	−1	0.33	0.05	0.66	0.00	0.16	0.20
	0	0.71	0.00	0.37	0.04	0.00	0.83
Metcalfe_2011	−3	0.06	0.46	-	-	-	-
	−2	0.55	0.01	-	-	-	-
	−1	**0.79**	**0.00**	-	-	-	-
	0	0.61	0.00	-	-	-	-

In summary: (1) the apparent phenology resulting from the GAMs is a good predictor of leaf gain and net leaf production across our study sites; and (2) apparent phenology does not respond immediately to leaf gain and in most cases has a two- to three-month delay after mangrove forests show signs of leaf gain or increased net leaf production.

3.4. Phenological Metrics

After analyzing every pixel in the Landsat images from our field site, we found that the growing season starts and ends between Day of Year (DOY) 280 and 316, that is between September and October each year (Figure 7A). The peak growing season occurs most frequently between May and July (i.e., DOY 137-165, Figure 7B). The start of season and peak growing season usually occurs before and after the wet season respectively, however establishing a relationship between the two events is beyond the scope of this study. See the Discussion section for more on the bimodal seasonality of mangrove ecosystems.

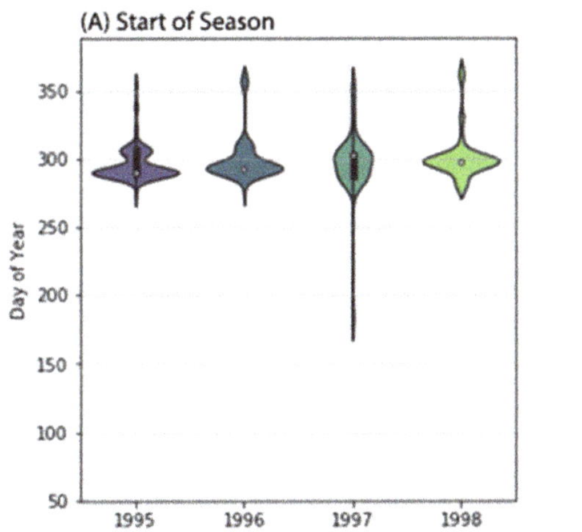
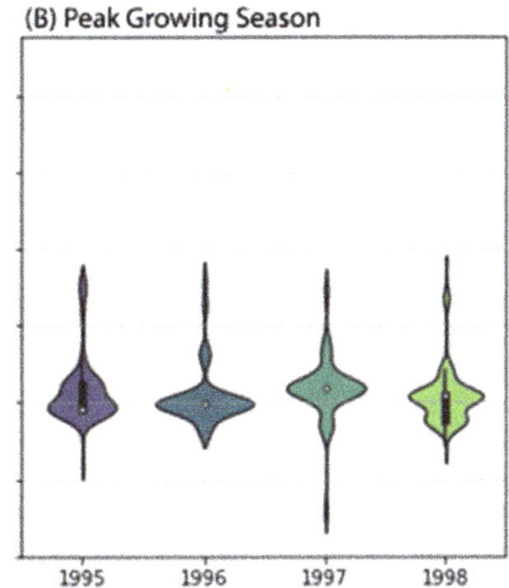

Figure 7. Violin plot of showing the Start of Season (**A**), and Peak of Growing Season (**B**) for mangroves in the Gladstone region (QLD) between 1995–1999, as determined by the apparent phenology. The height of each violin represents the range of values, and the width of each violin represents the number of values in that range.

4. Discussion

Extracting phenological metrics of mangrove forests from satellite images is an ongoing field of research. We contribute to this field by (1) presenting a novel, data-driven method to extracting phenology from satellite imagery; and (2) applying the method in evergreen forests across Australia. We also expand this field of research by presenting the dual phenology of mangrove forests, as described in the literature. By demonstrating that there is more than one period of leaf growth in these ecosystems, we also highlight the need for deeper investigation into the detection and causes of dual phenology in evergreen forests and the need for long term field studies that validate satellite observations. Importantly, with the use of satellite images, we have demonstrated that plots with similar species can have different phenologies. Phenology is, in turn, site-dependent and hence should not be described using a single logistic or sinusoidal curve.

Because phenology is site-dependent, methods like the one presented here can be used with long term imagery archives to determine the baseline of mangrove phenology in stable ecosystems, and compare it to stressed ecosystems. Early detection of ecosystem stress, indicated by changes in phenology, could lead us to detect, predict, and hopefully prevent, mangrove dieback events.

4.1. The Phenology of Rhizophora Stylosa

Authors have described, in situ, the phenological traits of *R. stylosa* around the world, however, few have attempted to compare mangrove phenology across regions. [16], for example, used satellite imagery and a sinusoidal model to describe seasonal variations of mangrove forests in the Yucatan peninsula in Mexico. They found that the spectral index values are lower during the dry season and higher during the wet season, an outcome that differs from our findings. Across all our study sites, we found lower EVI values during the wet season and higher EVI values during the dry season. These differences may be due to the geographical location and the species composition of both studies. *R. mangle, Laguncularia racemosa, Avicennia germinans* and *Conocarpus erectus* dominated their study site, while *R. stylosa, A. marina* and *C. tagal* dominated ours. Despite *R. stylosa* being the dominant species in our site, several species of mangroves may contribute to the apparent phenology of each pixel given the resolution of the Landsat images (i.e., 30 m). Determining the contribution of each species to the apparent phenology is an ongoing field of research.

Another big difference between our study and that of [16] is that the species that dominate mangrove forests in their study sites show a unimodal phenology response across the Yucatan peninsula. In contrast, we found bimodal phenology signals in our field study site, as well as in the sites described in the peer-reviewed literature (Figure 4). This bimodal phenology is not new [29], and later [6] described *R. stylosa* and other mangrove species as having two distinct periods of leaf growth each year. The novelty of our analysis has allowed us to demonstrate that these two periods of leaf growth can be detected from satellite imagery using semi-parametric methods. While not all mangrove ecosystems display dual phenology, GAMs create the capability to detect it, giving researchers a tool for inductive reasoning and opening the doors for further mangrove research to validate satellite observations.

Moreover, the method presented here accounts for differences in leaf growth intensity between the sites (e.g., Figure 6C,E). Resolving these two periods of leaf growth is important to improve current models of carbon and water fluxes and their relationship to climate forcing. Descriptions at the plot level certainly help us understand the bimodal response of EVI over time, however, the number of leaves that a tree produces might not be the only explanation for a bimodal response in EVI.

Recently, [46] suggested that the seasonal response of EVI in an evergreen forest was bimodal due to layers of the canopy responding in opposing seasonal patterns. When there is a decrease in leaf area index of the upper canopy, the lower canopy takes advantage of the extra light to increase its leaf area. Following the findings by [46,47], propose that the seasonal response of EVI comes from variations in leaf area index and photosynthetic capacity (i.e., younger, more efficient leaves), rather than from climate and weather patterns alone. Indeed, leaf ontogeny, demography and longevity could influence satellite observations, with leaves of different ages having varying amounts of chlorophyll, carotenoids, and water, resulting in slightly different spectral signatures. These assertions need to be tested in mangrove ecosystems, to determine if the bimodal response of EVI is related to the canopy structure or net leaf production.

In this study, we have demonstrated that seasonal and inter-annual changes in leaf gain and net leaf production are related to changes in EVI. [32] suggested that leaf production of *R. stylosa* in Northern Australia is most evident during the wet season (December through May), however, this species produces new leaves throughout the year. Similarly, [33] found that *R. stylosa* has the highest values of leaf production and leaf fall between December and April. The authors also indicated that this species has a net leaf gain between January and August, and a net leaf loss between September and December, which coincides with upward and downward trends in the apparent phenology for that site (Figure 6E). Despite the strength of the apparent phenology-net leaf production relationship, when using satellite imagery, other factors affect the phenology response of mangrove forests.

Environmental and biological factors such as cyclones, rainfall and tree age are known to alter the phenology and spectral response of mangrove ecosystems [16,48,49]. With the help of satellite imagery and GAMs, we can now look at inter-annual changes in mangrove phenology (e.g., Figure 4) and relate them to environmental or biological factors. Because these factors may change one by one, or several at a time, inter-annual predictions of mangrove phenology are difficult. A way to improve the prediction of seasonal and inter-annual phenology is to include past and present observations of these factors during the creation of the GAMs. GAMs create a numerical relationship between each factor and the phenology model to determine how influential is the former over the latter. This is an ongoing avenue of research.

Similarly, biological variables such as species composition, growth rate and forest maturity may also affect the spectral response of mangroves, and thus, any model derived from satellite sensors. A mature, dense forest will have a different response to a newly-planted mangrove patch or one that is recovering from a natural or manmade disaster [50]. Because relating 30 years of satellite observations to biological processes requires that in situ data is collected frequently and over long periods, the need for long-term (five or more years) monitoring sites in mangrove forests is evident. Long-term monitoring of mangrove ecosystems is important, especially when relating variations in spectral indices, to changes in tree growth and temperature such as those presented by [51,52].

4.2. GAMs vs. Parametric Methods

Our ability to detect and forecast mangrove phenology improves our understanding of the ecosystem [1], and our approach greatly differs from others, more commonly used [53]. For example, [54] derived a mathematical function that resembles the phenological phases of forests in the northeastern United States. They aimed to use satellite images to monitor vegetation dynamics at the landscape level, and one of their biggest achievements was that the method did not require any fixed constants or thresholds to be applicable. This method has been used at the local [55] and global scales [23], however the premise that a parametrized mathematical curve fits every plot of land has remained unchanged. The method derived by [54] assumes that the selected model of phenology is: (1) correct, (2) already known and (3) is invariant through space and time [39,56]. Even recent studies [17] insist on these assumptions when applying smoothers and filter to the data prior to detecting the phenology without considering that the data they discard may provide insights into the phenomenon they are trying to model (i.e., phenology). This in itself is not a limitation of the parametric models, but of the analysis workflow selected by the authors. We, on the other hand, used semiparametric GAMs as an estimation method to describe the changing relationship between mangrove phenology and EVI. GAMs let the data define the shape of the phenology curve, allowing for bimodality and skewness to be detected and modelled [56].

As shown above, the relationship between EVI and mangrove phenology changes with space and time. The main limitation of parametric approaches is that those methods are constrained to the particular models evaluated (e.g., [54,57,58]). That is to say, parametric methods assume that the shape of the phenological curve remains invariant and only variations in the frequency and amplitude of the signals are allowed. As demonstrated by [46,47] leaf demography and ontogeny can change the way we detect phenology from remotely sensed data, and as such, we need models that can tell us more than just seasonal variations in greenness. GAMs can fill this gap.

In contrast to parametric methods, GAMs use the data itself (in this case EVI values from satellite imagery) to determine the shape of the relationship. Because the relationship between the predictor and response variable is unknown beforehand, GAMs apply a series of smooth functions and use the data to determine which function is the best fit for a given dataset. This data-driven approach enabled us to demonstrate three key things: (1) the phenology response of mangroves forests dominated by *R. stylosa* is not unimodal, but often bimodal across our study sites; (2) the second leaf growth phase varies in intensity depending on the site. For example, the second leaf flush is much lower in New South Wales (Figure 6E) than in Queensland and the Northern Territory (Figure 6B,D). The reasons behind these differences are not fully understood but may be due, in part, to differences in air temperature, rainfall and water temperature; (3) the phenology response of mangrove ecosystems is site-dependent and GAMs allow us to see small seasonal and inter-annual variations that are otherwise impossible to detect using fully parametric methods. However, to provide a better, more accurate, description of mangrove phenology we need ecophysiological descriptions that span more than 18–24 months to compare to satellite-derived phenology.

4.3. Validation of the GAMs

Peer-reviewed literature of phenological models often lacks a description of the methods used to validate the models, making it difficult to compare the performance of the GAMs versus other approaches. Despite this limitation, we validated our apparent phenology using three different, independent sources of information and found that GAMs are good tools to extract the phenology response of mangrove forests from satellite images.

Linear regressions between Observed and apparent phenology values showed a good model fit, with $R^2 > 0.40$ in half of our study sites despite variations in the raw EVI values (Figure S1, see supplementary materials). Correlation values between the GAM and in situ (Table 3, and Figure S1) data were low (as expected), but there are valid reasons for this. The study by [26] focused on three sites in the Gladstone region in Queensland and aimed at examining potential bioremediation strategies in case an oil spill hit the Queensland coast. The authors never intended to use the field data to validate satellite imagery, hence the difficulties in correlating one with the other. Furthermore, the dataset consisted of one value per date per site i.e., only three data points per date, and some dates had no values (e.g., April 1997), which reduced the potential for correlation even more. Lastly, the study only gathered data over an 18-month period, limiting our information to one and a half growing seasons. Having only one full season of information limits the number of links we could create between the apparent phenology and the field data. We need longer field studies to understand fully the phenology curves extracted from satellite imagery and to examine the response of mangrove forests to changes in weather and climate patterns.

The validation of the apparent phenology with peer-reviewed literature provided us with two important pieces of information: (1) our models correlate well with the leaf gain intensity and net leaf production reported in the literature, regardless of the year in which those data were acquired; (2) apparent phenology has a two- to three-month lag with leaf gain intensity in most of our study sites. The former is important because it highlights the usefulness of GAMs. The latter tells us that, from a biophysical perspective, EVI responds to the canopy elements that absorb red light for photosynthesis and scatter near infra-red light, while field phenology traces leaf formation and drop. The delayed response of EVI is expected for two reasons: (1) the time it takes newly formed leaves to reach their maximum size, and (2) the net leaf production varies throughout the year. In Figure 8 we show an example of this. At "t1", leaves are scarce and bud breaking, net leaf production and chlorophyll content are low but positive and the satellite captures mainly the background of the mangrove tree (i.e., exposed soil and understory water). At "t2", leaves are growing and new leaves are bud breaking. At "t3", net leaf production peaks, meaning that there are many leaves growing and bud breaking, the satellite captures mainly green material, including chlorophyll, thus EVI is high as well. When "t4" arrives EVI is still high, more leaves are dropping than bud breaking, but leaves from "t3" are still growing and reaching maturity (i.e., high chlorophyll content) hence the lag between peak EVI and peak leaf gain/production. Finally, the tree loses more leaves and the background in the satellite images starts to show again ("t5") and the cycle starts again. The implications of this delayed response of EVI need to be explored because the growth/recovery rate after a natural disaster may not be as evident from satellite imagery as previously thought [50].

As demonstrated here, GAMs can be used in different locations, and with different species. Single species mangrove ecosystems are rare, and it is common to have a variety of species, especially within a 30 m pixel. Each study site here (Figure 1) was chosen due to the dominance of *R. stylosa*, however, the associated species varied from study to study. From the Northern Territory to the New South Wales coast, our study covers different regions of the Australian coastline and uses data from different points in time, thereby we have demonstrated that GAMs are easily transferable through location and time. As a result, GAMs can be used in local, continental, and global scale models of phenology spanning years or decades. We contribute to the remote sensing community by demonstrating that GAMs can be used in combination with remotely sensed data and presenting an alternative to fully parametric models.

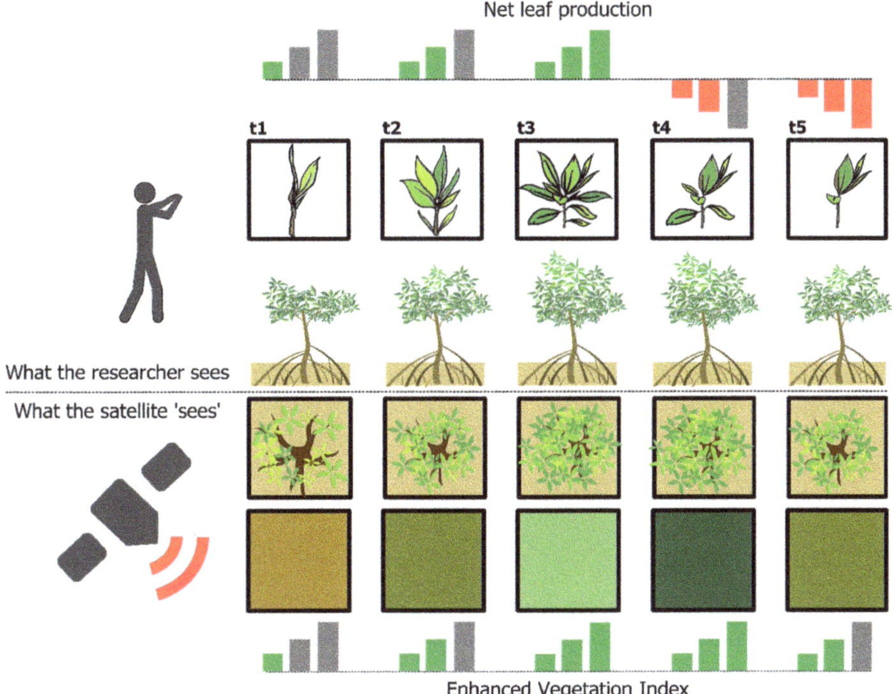

Figure 8. Time difference between peak leaf production and peak EVI during a given year for a simulated mangrove tree.

4.4. Moving Forward

Apparent phenology can be detected using a variety of spectral indices, and EVI is only one of several that has been used for coastal ecosystem investigations [14,16,59]. Future studies could use EVI in combination with other spectral indices to improve the detection of phenological events such as accurate measurement of leaf production and different stages of leaf growth. It would also be important to assess whether other spectral indices also display a time lag with relation to net leaf production and whether other phenology models show this temporal shift as well. For example, spectral indices that use the short-wave infra-red region of the spectrum could provide information on water content and indirectly inform the number of leaves in the forest. Establishing this relationship is important, especially in scenarios where mangroves are at risk of massive diebacks such as drought and heatwaves.

Besides temperature, rainfall, and other climate data, other sources of information that can potentially provide additional insights to our model: (1) Fractional Vegetation cover, and (2) radar imagery from Sentinel 1 or Advanced Land Observing Satellite (ALOS) sensors. The use of Radar datasets to monitor mangrove forests has been increasing in the past few years, mainly providing insights on mangrove zonation [60,61], canopy structure and height [62], while Fractional Vegetation Cover informs mangrove dynamics [13]. The spatial resolution of many of these sensors, including Landsat, does not allow the discrimination of species, however, estimating general trends in mangrove phenology could be more important to protect these forests rather than species-specific values.

We have also identified several ways in which the remote sensing and ecology communities can take phenology modelling to the next level. Firstly, we can evaluate accuracy by gathering and/or sharing field data with sufficient temporal resolution to compare it with satellite imagery. These data could include leaf area index, litterfall, leaf onset, biomass, and other measurements of plant phenology and growth that aid in assessing model accuracy and potential bias. For these efforts to be successful, data collection has to use identical, or at least comparable, techniques to identify and measure the variables of interest. Agreements on how to define and measure mangrove phenology, coupled with high-resolution imagery could greatly benefit this type of studies.

Secondly, GAMs and high-resolution imagery (i.e., equal or better than 1.5 × 1.5 m) can be used to model phenological changes on individual plants. By modelling phenology and the factors that affect it, users can take preventive or corrective measurements before the plant (or crop) fails or dies. More importantly, high-resolution imagery could potentially be used to create models at the same scale as the data collection plots, could be used to monitor restoration projects [63], and couple phenology to functional traits [64].

Thirdly, incorporating independent datasets to the GAMs will allow us to examine which environmental variables have the most influence on mangrove phenology at a continental scale. These datasets could include parameters like temperature, rainfall, humidity and tidal range. Besides altering spectral reflectance value values in the near and short-wave infrared bands [9], the tidal range at the time of image acquisition may play an important role in mangrove phenology. Just like temperature and rainfall, the tides vary seasonally across Australia [65] and their impact on mangrove phenology is yet to be assessed.

Lastly, we have demonstrated the usefulness of GAMs with a dense time-series of remotely sensed imagery, but the applications of this work could also be used with Moderate-Resolution Imaging Spectroradiometer (MODIS), Sentinel or other satellite sensors. Creating maps of mangroves around the world is important, but we currently have the technology to process large datasets in just hours, so why not model (and forecast) phenology under different climate change scenarios? This means detecting changes in the start of season and peak growing season dates over time and how that may correlate with changing weather and climatic patterns.

5. Conclusions

In this paper, we demonstrated that GAMs help us detect (1) the dual phenology of mangrove forests, and (2) seasonal and inter-annual changes in mangrove phenology by using 668 satellite images of different study sites across Australia. The two distinct periods of leaf growth in mangrove forests had not been detected using satellite imager until now. We compared our model to the field and published data to explore which biophysical variables help explain the seasonal changes in EVI. When compared to field data, we found that seasonal and inter-annual variations of EVI correlate well with the leaf production rate, net leaf production of mangrove forests. When compared to published data, we found that there is a time lag between leaf gain and the EVI. Overall, leaf gain and net leaf production are more closely related to higher EVI values than leaf fall. Regarding the phenological metrics, in our Gladstone site, the start of season occurs more frequently between September and October each year and the peak growing season between May and July.

Rather than imposing a parameterized mathematical curve to the data, our study leverages the ability of GAMs to let the data determine the type of relationship between a given spectral index and plant phenology. This data-driven approach helped us detect a bimodal phenology in mangrove forests dominated by *R. stylosa;* bimodal phenology has been reported in the literature but it has never been seen with remote sensing techniques. More importantly, GAMs allowed us to determine that mangrove phenology is site-dependent. Fully parametric methods, when applied to remotely sensed data, have over-simplified the phenology of mangrove ecosystems and other evergreen forests worldwide.

By understanding how phenology changes from site to site, and year to year, this study provides a tool for regional and continental-scale assessments of mangrove phenology. We expect to see an increase in the use of GAMs, especially in conjunction with the Landsat and other long-term, worldwide imagery archives.

Remote Sens. **2020**, *12*, 4008

Supplementary Materials: The following are available online at http://www.mdpi.com/2072-4292/12/24/4008/s1, Figure S1: Observed EVI vs Apparent phenology for all sites, Figure S2: Mean Absolute error of the Cross-validation predictions of EVI.

Author Contributions: Conceptualization, N.Y., K.E.J., L.L. and S.W.M.; methods, T.D.N., L.L. and N.Y.; field data collection, N.C.D.; data analysis, N.Y., T.D.N., N.C.D.; writing—original draft preparation, N.Y.; review and editing, N.Y., K.E.J., S.W.M., T.D.N., L.L., N.C.D. All authors have read and agreed to the published version of the manuscript.

Funding: This research received grants the following support Wet Tropics Management Authority Student Research Grant (NY), National Environment Science Program (NESP) Tropical Water Quality (TWQ) Hub Research Grant (NY), and a Centre for Tropical Water & Aquatic Ecosystem Research (TropWater) Student Research Grant (NY).

Acknowledgments: This project is supported by NIESGI Cia. Ltda. This research used resources from the National Computational Infrastructure (NCI) and Digital Earth Australia.

Conflicts of Interest: The authors declare no conflict of interest.

References

1. Chambers, L.E.; Altwegg, R.; Barbraud, C.; Barnard, P.; Beaumont, L.J.; Crawford, R.J.M.M.; Durant, J.M.; Hughes, L.; Keatley, M.R.; Low, M.; et al. Phenological Changes in the Southern Hemisphere. *PLoS ONE* **2013**, *8*, e75514. [CrossRef] [PubMed]

2. Garonna, I.; Jong, R.; Schaepman, M.E. Variability and evolution of global land surface phenology over the past three decades (1982–2012). *Glob. Chang. Biol.* **2016**, *22*, 1456–1468. [CrossRef]

3. Morellato, L.P.C.; Alberton, B.; Alvarado, S.T.; Borges, B.; Buisson, E.; Camargo, M.G.G.; Cancian, L.F.; Carstensen, D.W.; Escobar, D.F.E.E.; Leite, P.T.P.P.; et al. Linking plant phenology to conservation biology. *Biol. Conserv.* **2016**, *195*, 60–72. [CrossRef]

4. Menzel, A. Phenology: Its Importance To the. *Clim. Chang.* **2002**, *54*, 379–385. [CrossRef]

5. Duke, N.C.; Kleine, D.; University of, Q. *Australia's Mangroves: The Authoritative Guide to Australia's Mangrove plants*; University of Queensland: Brisbane, Australia, 2006; ISBN 0646461966.

6. Tomlinson, P.B. *The Botany of Mangroves*; Cambridge University Press: Cambridge, UK, 1986.

7. Duke, N.C. Phenological Trends with Latitude in the Mangrove Tree Avicennia Marina. *J. Ecol.* **1990**, *78*, 113–133. [CrossRef]

8. Metcalfe, K.N.; Franklin, D.C.; McGuinness, K.A. Mangrove litter fall: Extrapolation from traps to a large tropical macrotidal harbour. *Estuar. Coast. Shelf Sci.* **2011**, *95*, 245–252. [CrossRef]

9. Younes Cárdenas, N.; Joyce, K.E.; Maier, S.W. Monitoring mangrove forests: Are we taking full advantage of technology? *Int. J. Appl. Earth Obs. Geoinf.* **2017**, *63*, 1–14. [CrossRef]

10. Cresswell, I.D.; Semeniuk, V. Mangroves of the Kimberley coast: Ecological patterns in a tropical Ria coast setting. *J. R. Soc. West. Aust.* **2011**, *94*, 213–237.

11. Giri, C.; Ochieng, E.; Tieszen, L.L.; Zhu, Z.; Singh, A.; Loveland, T.; Masek, J.; Duke, N.C. Status and distribution of mangrove forests of the world using earth observation satellite data. *Glob. Ecol. Biogeogr.* **2011**, *20*, 154–159. [CrossRef]

12. Hamilton, S.E.; Casey, D. Creation of a high spatio-temporal resolution global database of continuous mangrove forest cover for the 21st century (CGMFC-21). *Glob. Ecol. Biogeogr.* **2016**, *25*, 729–738. [CrossRef]

13. Lymburner, L.; Bunting, P.; Lucas, R.; Scarth, P.; Alam, I.; Phillips, C.; Ticehurst, C.; Held, A. Mapping the multi-decadal mangrove dynamics of the Australian coastline. *Remote Sens. Environ.* **2019**, *238*, 111185. [CrossRef]

14. Rogers, K.; Lymburner, L.; Salum, R.; Brooke, B.P.; Woodroffe, C.D. Mapping of mangrove extent and zonation using high and low tide composites of Landsat data. *Hydrobiologia* **2017**, *803*, 49–68. [CrossRef]

15. Asbridge, E.; Lucas, R.; Ticehurst, C.; Bunting, P. Mangrove response to environmental change in Australia's Gulf of Carpentaria. *Ecol. Evol.* **2016**, *6*, 3523–3539. [CrossRef] [PubMed]

16. Pastor-Guzman, J.; Dash, J.; Atkinson, P.M. Remote sensing of mangrove forest phenology and its environmental drivers. *Remote Sens. Environ.* **2018**, *205*, 71–84. [CrossRef]

17. Songsom, V.; Koedsin, W.; Ritchie, J.R.; Huete, A. Mangrove Phenology and Environmental Drivers Derived from Remote Sensing in Southern Thailand. *Remote Sens.* **2019**, *11*. [CrossRef]

18. Dhu, T.; Dunn, B.; Lewis, B.; Lymburner, L.; Mueller, N.; Telfer, E.; Lewis, A.; McIntyre, A.; Minchin, S.; Phillips, C. Digital earth Australia—Unlocking new value from earth observation data. *Big Earth Data* **2017**, *1*, 64–74. [CrossRef]

19. Broich, M.; Huete, A.; Paget, M.; Ma, X.; Tulbure, M.; Coupe, N.R.; Evans, B.; Beringer, J.; Devadas, R.; Davies, K.; et al. A spatially explicit land surface phenology data product for science, monitoring and natural resources management applications. *Environ. Model. Softw.* **2015**, *64*, 191–204. [CrossRef]

20. Pastor-Guzman, J.; Atkinson, P.M.; Dash, J.; Rioja-Nieto, R. Spatiotemporal Variation in Mangrove Chlorophyll Concentration Using Landsat 8. *Remote Sens.* **2015**, *7*, 14530–14558. [CrossRef]

21. Zhang, X.; Tan, B.; Yu, Y. Interannual variations and trends in global land surface phenology derived from enhanced vegetation index during 1982–2010. *Int. J. Biometeorol.* **2014**, *58*, 547–564. [CrossRef]

22. Melaas, E.K.; Sulla-Menashe, D.; Gray, J.M.; Black, T.A.; Morin, T.H.; Richardson, A.D.; Friedl, M.A. Multisite analysis of land surface phenology in North American temperate and boreal deciduous forests from Landsat. *Remote Sens. Environ.* **2016**, *186*, 452–464. [CrossRef]

23. White, M.A.; De Beurs, K.M.; Didan, K.; Inouye, D.W.; Richardson, A.D.; Jensen, O.P.; O'keefe, J.; Zhang, G.; Nemani, R.R.; Van Leeuwen, W.J.D.; et al. Intercomparison, interpretation, and assessment of spring phenology in North America estimated from remote sensing for 1982–2006. *Glob. Chang. Biol.* **2009**, *15*, 2335–2359. [CrossRef]

24. Xin, Q.; Li, J.; Li, Z.; Li, Y.; Zhou, X. Evaluations and comparisons of rule-based and machine-learning-based methods to retrieve satellite-based vegetation phenology using MODIS and USA National Phenology Network data. *Int. J. Appl. Earth Obs. Geoinf.* **2020**, *93*, 102189. [CrossRef]

25. Liu, L.Y.; Tang, H.; Caccetta, P.; Lehmann, E.A.; Hu, Y.; Wu, X.L. Mapping afforestation and deforestation from 1974 to 2012 using Landsat time-series stacks in Yulin District, a key region of the Three-North Shelter region, China. *Environ. Monit. Assess.* **2013**, *185*, 9949–9965. [CrossRef] [PubMed]

26. Duke, N.C.; Burns, K.A.; Swannell, R.P.J. *Research into the Bioremediation of Oil Spills in Tropical Australia: With Particular Emphasis on Oiled Mangrove and Salt Marsh Habitat*; AMSA: Townsville, Australia, 1999.

27. Trewin, C. Mangrove & Saltmarsh Monitoring: Literature Review. *Report prepared by Sinclair Kn. Merz For the Gladstone Ports Corp.* **2013**, *CA120019 R*, 53, Australia.

28. TMR—Queensland Dept. of Transport and Main Roads 1998–1999—Fishermans Landing Tide Gauge Archived Interval Recordings. Available online: https://www.data.qld.gov.au/dataset/fishermans-landing-tide-gauge-archived-interval-recordings/resource/ef1d0409-06ee-498d-83a0-649f8478a786 (accessed on 15 November 2020).

29. Duke, N.C.; Bunt, J.S.; Williams, W.T. Observations on the floral and vegetative phenologies of north- eastern Australian mangroves. *Aust. J. Bot.* **1984**, *32*, 87–99. [CrossRef]

30. Saenger, P.; Moverley, J. Vegetative phenology of mangroves along the Queensland coastline. *Proc. Ecol. Soc. Aust.* **1985**, *13*, 257–265.

31. Duke, N.C. Sustained high levels of foliar herbivory of the mangrove Rhizophora stylosa by a moth larva Doratifera stenosa (Limacodidae) in north-eastern Australia. *Wetl. Ecol. Manag.* **2002**, *10*, 403–419. [CrossRef]

32. Coupland, G.T.; Paling, E.I.; McGuinness, K.A. Vegetative and reproductive phenologies of four mangrove species from northern Australia. *Aust. J. Bot.* **2005**, *53*, 109–117. [CrossRef]

33. Wilson, N.C.; Saintilan, N. Growth of the mangrove species Rhizophora stylosa Griff. at its southern latitudinal limit in eastern Australia. *Aquat. Bot.* **2012**, *101*, 8–17. [CrossRef]

34. Lewis, A.; Lymburner, L.; Purss, M.B.J.; Brooke, B.; Evans, B.; Ip, A.; Dekker, A.G.; Irons, J.R.; Minchin, S.; Mueller, N.; et al. Rapid, high-resolution detection of environmental change over continental scales from satellite data—The Earth Observation Data Cube. *Int. J. Digit. Earth* **2016**, *9*, 106–111. [CrossRef]

35. Lewis, A.; Oliver, S.; Lymburner, L.; Evans, B.; Wyborn, L.; Mueller, N.; Raevksi, G.; Hooke, J.; Woodcock, R.; Sixsmith, J.; et al. The Australian Geoscience Data Cube—Foundations and lessons learned. *Remote Sens. Environ.* **2017**, *202*, 276–292. [CrossRef]

36. Huete, A.; Didan, K.; Miura, T.; Rodriguez, E.; Gao, X.; Ferreira, L. Overview of the radiometric and biophysical performance of the MODIS vegetation indices. *Remote Sens. Environ.* **2002**, *83*, 195–213. [CrossRef]

37. Younes, N.; Joyce, K.E.; Northfield, T.D.; Maier, S.W. The effects of water depth on estimating Fractional Vegetation Cover in mangrove forests. *Int. J. Appl. Earth Obs. Geoinf.* **2019**, *83*, 101924. [CrossRef]

38. Hastie, T.; Tibshirani, R. Generalized Additive Models. *Stat. Sci.* **1986**, *1*, 297–318. [CrossRef]

39. Jones, K.; Almond, S. Moving out of the Linear Rut: The Possibilities of Generalized Additive Models. *Trans. Inst. Br. Geogr.* **1992**, *17*, 434. [CrossRef]

40. Zuur, A.F. *A Beginner's Guide to Generalized Additive Models With R*; Highland Statistics Ltd.: Newburgh, NY, USA, 2012; ISBN 9780957174122;0957174128.

41. Zuur, A.F.; Saveliev, A.A.; Ieno, E.N. *A Beginner's Guide to Generalised Additive Mixed Models with R 2014*; Highland Statistics Ltd.: Scotland, UK, 2014.

42. Pedersen, E.J.; Miller, D.L.; Simpson, G.L.; Ross, N. Hierarchical generalized additive models in ecology: An introduction with mgcv. *PeerJ* **2019**, *7*, e6876. [CrossRef]

43. Taylor, S.J.; Letham, B.; Taylor, S.J.; Letham, B. Forecasting at Scale. *Am. Stat.* **2018**, *72*, 37–45. [CrossRef]

44. Restrepo-Coupe, N.; Huete, A.; Davies, K. Satellite Phenology Validation. In *AusCover Good Practice Guidelines: A Technical Handbook Supporting Calibration and Validation Activities of Remotely Sensed Data Product*; Held, A., Phinn, S., Soto-Berelov, M., Jones, S., Eds.; TERN AusCover: Canberra, Australia, 2015; pp. 155–157. ISBN 978-0-646-94137-0.

45. Pedregosa, F.; Varoquaux, G.; Gramfort, A.; Michel, V.; Thirion, B.; Grisel, O.; Blondel, M.; Prettenhofer, P.; Weiss, R.; Dubourg, V.; et al. Scikit-learn: Machine Learning in Python. *J. Mach. Learn. Res.* **2011**, *12*, 2825–2830.

46. Smith, M.N.; Stark, S.C.; Taylor, T.C.; Ferreira, M.L.; de Oliveira, E.; Restrepo-Coupe, N.; Chen, S.; Woodcock, T.; dos Santos, D.B.; Alves, L.F.; et al. Seasonal and drought-related changes in leaf area profiles depend on height and light environment in an Amazon forest. *New Phytol.* **2019**, *222*, 1284–1297. [CrossRef]

47. Wu, J.; Albert, L.P.; Lopes, A.P.; Restrepo-Coupe, N.; Hayek, M.; Wiedemann, K.T.; Guan, K.; Stark, S.C.; Christoffersen, B.; Prohaska, N.; et al. Leaf development and demography explain photosynthetic seasonality in Amazon evergreen forests. *Science* **2016**, *351*, 972–976. [CrossRef]

48. Lovelock, C.E.; Cahoon, D.R.; Friess, D.A.; Guntenspergen, G.R.; Krauss, K.W.; Reef, R.; Rogers, K.; Saunders, M.L.; Sidik, F.; Swales, A.; et al. The vulnerability of Indo-Pacific mangrove forests to sea-level rise. *Nature* **2015**, *526*, 559–563. [CrossRef] [PubMed]

49. Kovacs, J.M.; Wang, J.; Blanco-Correa, M. Mapping Disturbances in a Mangrove Forest Using Multi-Date Landsat TM Imagery. *Environ. Manage.* **2001**, *27*, 763–776. [CrossRef] [PubMed]

50. Giri, C.; Long, J.; Tieszen, L. Mapping and Monitoring Louisiana's Mangroves in the Aftermath of the 2010 Gulf of Mexico Oil Spill. *J. Coast. Res.* **2011**, *27*, 1059–1064. [CrossRef]

51. Coldren, G.A.; Barreto, C.R.; Wykoff, D.D.; Morrissey, E.M.; Langley, J.A.; Feller, I.C.; Chapman, S.K. Chronic warming stimulates growth of marsh grasses more than mangroves in a coastal wetland ecotone. *Ecology* **2016**, *97*, 3167–3175. [CrossRef]

52. Coldren, G.A.; Langley, J.A.; Feller, I.C.; Chapman, S.K. Warming accelerates mangrove expansion and surface elevation gain in a subtropical wetland. *J. Ecol.* **2018**, *0*. [CrossRef]

53. Helman, D. Land surface phenology: What do we really 'see' from space? *Sci. Total Environ.* **2018**, *618*, 665–673. [CrossRef]

54. Zhang, X.; Friedl, M.A.; Schaaf, C.B.; Strahler, A.H.; Hodges, J.C.F.; Gao, F.; Reed, B.C.; Huete, A. Monitoring vegetation phenology using MODIS. *Remote Sens. Environ.* **2003**, *84*, 471–475. [CrossRef]

55. Fisher, J.I.; Mustard, J.F.; Vadeboncoeur, M.A. Green leaf phenology at Landsat resolution: Scaling from the field to the satellite. *Remote Sens. Environ.* **2006**, *100*, 265–279. [CrossRef]

56. Yee, T.W.; Mitchell, N.D. Generalized Additive Models in Plant Ecology. *J. Veg. Sci.* **1991**, *2*, 587–602. [CrossRef]

57. Pasquarella, V.J.; Holden, C.E.; Woodcock, C.E. Improved mapping of forest type using spectral-temporal Landsat features. *Remote Sens. Environ.* **2018**, *210*, 193–207. [CrossRef]

58. Zhu, Z.; Woodcock, C.E.; Olofsson, P. Continuous monitoring of forest disturbance using all available Landsat imagery. *Remote Sens. Environ.* **2012**, *122*, 75–91. [CrossRef]

59. Murray, N.J.; Phinn, S.R.; DeWitt, M.; Ferrari, R.; Johnston, R.; Lyons, M.B.; Clinton, N.; Thau, D.; Fuller, R.A. The global distribution and trajectory of tidal flats. *Nature* **2019**, *565*, 222–225. [CrossRef]

60. Held, A.; Ticehurst, C.; Lymburner, L.; Williams, N. High resolution mapping of tropical mangrove ecosystems using hyperspectral and radar remote sensing. *Int. J. Remote Sens.* **2003**, *24*, 2739–2759. [CrossRef]

61. Worthington, T.A.; Ermgassen, P.S.E.zu; Friess, D.A.; Krauss, K.W.; Lovelock, C.E.; Thorley, J.; Tingey, R.; Woodroffe, C.D.; Bunting, P.; Cormier, N.; et al. A global biophysical typology of mangroves and its relevance for ecosystem structure and deforestation. *Sci. Rep.* **2020**, *10*, 14652. [CrossRef] [PubMed]

62. Simard, M.; Fatoyinbo, L.; Smetanka, C.; Rivera-Monroy, V.H.; Castañeda-Moya, E.; Thomas, N.; Van der Stocken, T. Mangrove canopy height globally related to precipitation, temperature and cyclone frequency. *Nat. Geosci.* **2019**, *12*, 40–45. [CrossRef]
63. Lee, S.Y.; Hamilton, S.; Barbier, E.B.; Primavera, J.; Lewis, R.R. Better restoration policies are needed to conserve mangrove ecosystems. *Nat. Ecol. Evol.* **2019**. [CrossRef]
64. Aguirre-Gutiérrez, J.; Rifai, S.; Shenkin, A.; Oliveras, I.; Bentley, L.P.; Svátek, M.; Girardin, C.A.J.; Both, S.; Riutta, T.; Berenguer, E.; et al. Pantropical modelling of canopy functional traits using Sentinel-2 remote sensing data. *Remote Sens. Environ.* **2021**, *252*, 112122. [CrossRef]
65. Bishop-Taylor, R.; Sagar, S.; Lymburner, L.; Beaman, R.J. Between the tides: Modelling the elevation of Australia's exposed intertidal zone at continental scale. *Estuar. Coast. Shelf Sci.* **2019**. [CrossRef]

Publisher's Note: MDPI stays neutral with regard to jurisdictional claims in published maps and institutional affiliations.

Article

Integration of GF2 Optical, GF3 SAR, and UAV Data for Estimating Aboveground Biomass of China's Largest Artificially Planted Mangroves

Yuanhui Zhu [1], Kai Liu [2,3], Soe W. Myint [2,3,4], Zhenyu Du [5], Yubin Li [4], Jingjing Cao [2], Lin Liu [1,6,*] and Zhifeng Wu [7]

1   Center of GeoInformatics for Public Security, School of Geographical Sciences, Guangzhou University, Guangzhou 510006, China; zhuyhui2@gzhu.edu.cn
2   Guangdong Provincial Engineering Research Center for Public Security and Disaster, Guangdong Key Laboratory for Urbanization and Geo-simulation, School of Geography and Planning, Sun Yat-sen University, Guangzhou 510275, China; liuk6@mail.sysu.edu.cn (K.L.); caojj5@mail.sysu.edu.cn (J.C.)
3   Southern Marine Science and Engineering Guangdong Laboratory, Zhuhai 519000, China
4   School of Geographical Sciences and Urban Planning, Arizona State University, Tempe, AZ 85287, USA; Soe.Myint@asu.edu (S.W.M.); yubinli@asu.edu (Y.L.)
5   School of Geographic and Oceanographic Sciences, Nanjing University, Nanjing 210023, China; duzhy5@mail2.sysu.edu.cn
6   Department of Geography, University of Cincinnati, Cincinnati, OH 45221-0131, USA
7   Guangdong Province Engineering Technology Research for Geographical Conditions Monitoring and Comprehensive Analysis, School of Geographical Sciences, Guangzhou University, Guangzhou 510006, China; zfwu@gzhu.edu.cn
*   Correspondence: liulin1@gzhu.edu.cn or lin.liu@uc.edu; Tel.: +1-513-556-3429

Received: 15 May 2020; Accepted: 24 June 2020; Published: 25 June 2020

Abstract: Accurate methods to estimate the aboveground biomass (AGB) of mangroves are required to monitor the subtle changes over time and assess their carbon sequestration. The AGB of forests is a function of canopy-related information (canopy density, vegetation status), structures, and tree heights. However, few studies have attended to integrating these factors to build models of the AGB of mangrove plantations. The objective of this study was to develop an accurate and robust biomass estimation of mangrove plantations using Chinese satellite optical, SAR, and Unmanned Aerial Vehicle (UAV) data based digital surface models (DSM). This paper chose Qi'ao Island, which forms the largest contiguous area of mangrove plantation in China, as the study area. Several field visits collected 127 AGB samples. The models for AGB estimation were developed using the random forest algorithm and integrating images from multiple sources: optical images from Gaofen-2 (GF-2), synthetic aperture radar (SAR) images from Gaofen-3 (GF-3), and UAV-based digital surface model (DSM) data. The performance of the models was assessed using the root-mean-square error (RMSE) and relative RMSE (RMSEr), based on five-fold cross-validation and stratified random sampling approach. The results showed that images from the GF-2 optical (RMSE = 33.49 t/ha, RMSEr = 21.55%) or GF-3 SAR (RMSE = 35.32 t/ha, RMSEr = 22.72%) can be used appropriately to monitor the AGB of the mangrove plantation. The AGB models derived from a combination of the GF-2 and GF-3 datasets yielded a higher accuracy (RMSE = 29.89 t/ha, RMSEr = 19.23%) than models that used only one of them. The model that used both datasets showed a reduction of 2.32% and 3.49% in RMSEr over the GF-2 and GF-3 models, respectively. On the DSM dataset, the proposed model yielded the highest accuracy of AGB (RMSE = 25.69 t/ha, RMSEr = 16.53%). The DSM data were identified as the most important variable, due to mitigating the saturation effect observed in the optical and SAR images for a dense AGB estimation of the mangroves. The resulting map, derived from the most accurate model, was consistent with the results of field investigations and the mangrove plantation sequences. Our results indicated that the AGB can be accurately measured by integrating images from the optical, SAR, and DSM datasets to adequately represent canopy-related information, forest structures, and tree heights.

Keywords: mangrove plantation; aboveground biomass estimation; optical images; SAR; DSM

---

1. Introduction

Mangrove ecosystems are highly efficient blue carbon sinks, owing to their high productivity and low respiration rates, which allow them to store large amounts of biomass and organic carbon for a long time [1,2]. The capability for carbon sequestration of coastal ecosystems, including mangrove forests, has been reported to be 10–50 times higher than that of the terrestrial ecosystem [3]. Being among the most productive ecosystems, they can effectively mitigate climate change [4,5]. Therefore, the accurate estimation of the aboveground biomass (AGB) of mangrove plantations is essential for identifying the patterns of distribution in tropical and subtropical coastal zones to assess emissions from deforestation and carbon sinks from reforestation [6].

By the end of the 1990s, the total area occupied by mangrove forests in China was smaller than 15,000 ha, and had been reduced by 68.7% since its historical peak [7], owing to urban expansion, tidal flat reclamation, and deforestation for cultivation [8,9]. Since then, afforestation and reforestation projects have been implemented to conserve and restore mangroves. Mangrove plantation has been encouraged such that, by 2015, the area occupied by mangrove forests in China reached 22,419 ha [10]. The AGB of mangrove plantation should be accurately measured when monitoring, restoring, and managing wetland ecosystems, because it can help support global climate change mitigation programs, such as the Reducing Emissions from Deforestation and Forest Degradation in (REDD+), as well as the Payments for Ecosystem Services (PES) schemes [11,12].

However, field measurements on the biomass of mangroves are challenging, because they are distributed in intertidal zones that are difficult to access [13]. Thus, remotely sensed images have been widely used for this purpose [14–16]. Accordingly, regression models have been proposed by constructing relationships between the AGB and variables derived from various data sources. In general, optical and synthetic aperture radar (SAR) images are commonly used for AGB estimation studies [17,18]. The bands and vegetation indices (VIs) derived from the optical images vary according to water and chlorophyll content, and the structure of the leaf cavity of the vegetation that are correlated to the type of plant or its stages of growth. Thus, they can be used to monitor the biomass of forests [19–21]. Optical images are widely available—for example, Moderate Resolution Imaging Spectroradiometer (MODIS) [22], Landsat [19], IKONOS [21], SPOT [23], and WorldView-2 images [6]. However, these remote sensors are not capable of penetrating the surface of the canopies of forests to obtain their structure and the heights of trees needed for biomass estimation [24,25]. Since SAR images penetrate the canopy [26], they can be used to determine the structure of the canopy by emitting radiation to detect and measure branches and trunks [27]. Hence, the C-band and X-band of SAR images such as Rardarsat-2, ALOS PALSAR, and airborne SAR images are useful for monitoring the biomass of mangrove plantations [14,28–31]. However, SAR are images acquired by receiving transmitted signals contain speckle noise usually caused by the constructive or destructive interference of backscattered microwave signals that degrade image quality, and thus, may not provide accurate information concerning the target objects [32].

Optical and SAR images have considerable limitations in estimating forest biomass accurately [33]. Past studies have shown that optical and SAR images can be integrated to acquire the spectral and structural features of forest canopies to improve the accuracy of their predicted biomass [17,18,34,35]. This integration usually involves incorporating variables derived from the optical and SAR images or fusing them into new datasets (such as in wavelet transforms or principal component analysis) [18]. However, these variables exhibit saturation effects for highly dense mangrove forests that limit their availability to estimate only within a specific range of the AGB [33,36]. Hence, they can only be used to estimate AGB for areas with low biomass.

To overcome the above limitations, recent studies have focused on the vertical structure of mangrove forests (e.g., tree height) using Interferometric Synthetic Aperture Radar (InSAR), Light Detection and Ranging (LiDAR), and stereo photogrammetry of overlapping photographs, to help resolve the saturation problem [37–40]. Aerial photographs using the structure-from-motion (SfM) algorithm and the Unmanned Aerial Vehicle (UAV) platform are a low-cost option to measure the heights of trees [41]. The UAV-based digital surface models (DSM) can determine the relative tree height of mangrove forests, because they mainly grow over even terrain [42]. Hence, the integration of optical, SAR and UAV-based DSM data to represent the spectrum of the canopy, structure, and height of mangroves can improve the accuracy of the estimation of AGB for dense mangrove forests. Navarro et al. (2019) estimated mangrove AGB by combining UAV-based tree height, Sentinel-1, and Sentinel-2 images, and provided accurate estimates for young and sparse mangrove plantations [17]. However, the effectiveness of this approach needs to be examined further for dense and complicated mangroves.

This research aims to address the aforementioned gaps in the estimation of the AGB. We integrate images from multiple sources—GF2 optical, GF3 SAR, and fixed-wing UAV-based DSM data— to estimate the AGB of mangrove plantations. The objectives of this study are (1) to develop prediction models for the AGB using original and composite bands generated from optical, SAR, and UAV-based DSM data; (2) to evaluate the effectiveness of the AGB models and select the best one; (3) to determine the importance of the chosen parameters; and (4) to map the AGB to observe the spatial pattern of the biomass of a mangrove plantation in comparison with field surveys and its sequence of growth.

2. Materials and Methods

2.1. Study Area

This study was conducted in the mangrove nature reserve area of Guangdong Province in Dawei Bay of Qi'ao Island, in the Pearl River estuary of China (113°36′40″ E–113°39′15″ E, 22°23′40″ N– 22°27′38″ N) [43,44]. It is a typical tropical and subtropical wetland ecosystem. The area of the reserve is 5103 ha, covering about 700 ha of mangrove forests: the largest contiguous area of artificially planted mangroves in China [6,45]. Mangrove forests are characterized by high spatial variability that represents a dynamic landscape. The mangrove forests planted in the study area were composed mainly of a fast-growing species, *Sonneratia apetala* (*S. apetala*), which was introduced from Bengal. They belong to the woody mangrove species, with features of high adaptability and seed production [46]. The heights of their trees usually increase by about 1.5 m during each of the first few years [47,48]. They were the tallest tree species in the study area, ranging from 2 m to 20 m, and the diameter at breast height (DBH) of older trees can be as large as 30 cm.

S. apetala has been artificially planted on the island since 1999 to control invasive species, (i.e., *Spartina alterniflora*) and reconstruct mangrove forests. *S. apetala* generally has an afforestation specification of 1–2 m × 1.5 m with high densities. The tree ages range from 1 year to 17 years with high biomass variability and complication. Its afforestation process runs seaward from land, which implies a gradient distribution of tree age and AGB. The mangrove plantations had extended throughout the study area, and *S. apetala* had become the dominant species, covering more than 80% of the mangrove forest.

2.2. Field Investigation

We conducted field investigations to obtain biomass samples from June to July in 2016. To ensure the availability of samples to build estimation models for the AGB, we collected 127 samples along almost all the accessible tidal creeks to account for the variation in the biomass of all stages of tree growth. The samples were located in the low-, middle-, and high-intertidal zones (Figure 1d). For each sample, we measured the height and diameter at breast height (DBH) of each tree and recorded data for trees within a 10 × 10 m quadrat. Tree height was measured by a handheld laser range finder (precision of 1 m; Trueyard SP-1500H, Trueyard Optical Instruments co.), and DBH higher than 5 cm was recorded by a breast diameter ruler at 1.3 m above the ground. To match the locations of the samples and overlaid images coordinates, the four vertices and center of each quadrat were recorded by a submeter-accurate GPS. The precise locations of the plots were recorded with the assistance of the GPS and high-resolution images (UAV images with 0.12 m spatial resolution). Details of the locations of the quadrats, such as distance to the shore, were identified in and marked on the images. Using the measured tree heights and DBHs during the field survey, we calculated the AGB of each mangrove tree using allometric equations [49], and computed the sum of the AGB of all trees within a quadrat, to represent the AGB of a sample.

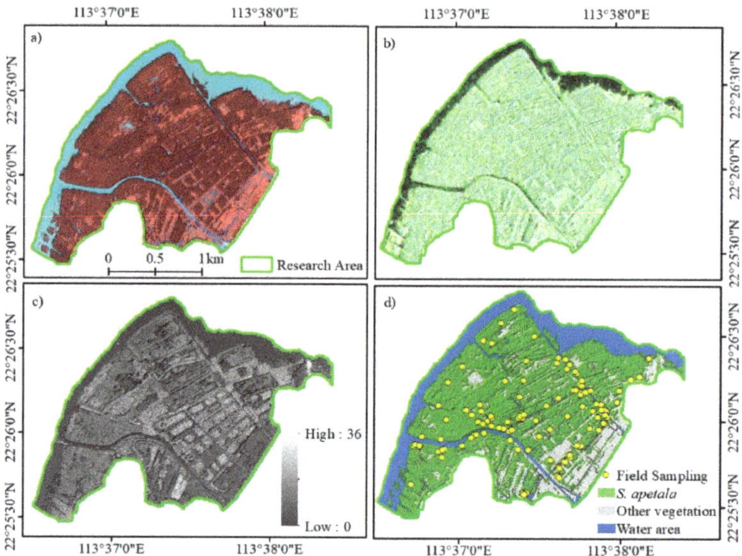

Figure 1. Images of the study area. (**a**) Gaofen-2 (GF-2) images (bands 4, 3, 2 false-color combinations); (**b**) HH, HV, and VV color composition of Gaofen-3 (GF-3) images; (**c**) digital surface model (DSM) data derived from Unmanned Aerial Vehicle (UAV) images, and (**d**) spatial distribution of *S. apetala* and field sampling on Qi'ao Island.

2.3. Remote Sensing Data and Preprocessing Procedure

2.3.1. GF2 Optical Data

Gaofen-2 (GF-2) captures high-resolution images. It was launched by China National Space Administration (CNSA), Beijing, China, in August 2014. It has been applied to land monitoring, urban planning, and resource surveys [50]. GF-2 images have a panchromatic band (1-m resolution) and four multispectral bands (4-m resolution): red (R), green (G), blue (B), and near-infrared (NIR). We obtained GF-2 multispectral images on 15 February 2017 from Land Observation Satellite Data Service Platform (Figure 1a).

The pre-processing of the GF-2 images—including radiation calibration, atmospheric correction, and geometric correction—was carried out using the ENVI 5.4.1 software package. Atmospheric correction of the images was carried out using the fast line-of-sight atmospheric analysis of the spectral hypercubes (FLAASH) model with the ENVI module. The images were also geo-rectified to a 1:10,000 topographic map using ground control points, to ensure that the position error was smaller than 0.5 pixels.

After preprocessing, the images were used to calculate four vegetation indices (VIs)—difference vegetation index (DVI), ratio vegetation index (RVI), normalized difference vegetation index (NDVI) and soil-adjusted vegetation index (SAVI)—as input variables to estimate the AGB of the mangrove forest (Table 1).

Table 1. List of vegetation indices extracted from GF-2 optical data.

Vegetation	Acronym	Formula	Reference
Difference Vegetation Index	DVI	$DVI = NIR - R$	[51]
Ratio Vegetation Index	RVI	$RVI = \frac{NIR}{R}$	[52]
Normalized Difference Vegetation Index	NDVI	$NDVI = \frac{NIR-R}{NIR+R}$	[53]
Soil-Adjusted Vegetation Index	SAVI	$SAVI = \left[\frac{NIR-R}{NIR+R+L}\right](1+L)$	[54]

2.3.2. GF3 SAR Data

The Gaofen-3 (GF-3), launched in August 2016, was developed by the China National Space Administration (CNSA), and is the first Chinese satellite that collects multi-polarized C-band SAR data. GF-3 images are the only radar images in the Chinese High-resolution Earth Observation System [55]. They have 12 imaging modes, ranging from single to dual and full polarization, with a resolution of 1 to 500 m and have a revisiting period of 3.5 days at most to the same point on Earth. Such characteristics render the GF-3 suitable for resource monitoring. We obtained fully polarimetric (FP) SAR data (HH, HV, VH, VV) from the Land Observation Satellite Data Service Platform on 5 August 2017, in the Quad-Polarization Stripmap 1 (QPS1) imaging mode at an azimuth resolution of 5.3 m, range resolution of 2.25 m, range of incidence angle of 29.63°, and in the single-look complex (SLC) format (Figure 1b and Table 2). Based on the original data, the image-related data were preprocessed to preserve phase and amplitude information in the complex images.

Table 2. Characteristics of GF-3 synthetic aperture radar (SAR) images.

Level	Imaging Mode	Format	Polarization Mode	Incidence Angle	Coordinate
Level 1A	SLC	TIFF + RPC	Full	29.63°	WGS-1984
Azimuth resolution	Range resolution	Size	Center Longitude	Center Latitude	Time
5.30 m	2.25 m	7435 × 7880	113.7°	22.4°	5 August 2017

(1) Preprocessing of GF-3 images

FP SAR images contain speckle noise and geometric distortions that can have a significant negative impact on features of the polarization of the target objects. Thus, a preprocessing sequence consisting of data import, multi-look processing, adaptive filtering, radiometric calibration, and the geometric correction was applied to the GF-3 images using the SARscape 5.4.1 module embedded into the ENVI software. The steps of the preprocessing are as follows:

a) The metadata of the SAR images was imported to obtain the slant-range resolution and angle of incidence. The ground-range resolution was then calculated using the following:

$$R_{ground} = \frac{R_{slant}}{\sin(\theta)} \tag{1}$$

where R_{ground} represents ground-range resolution (4.5480 m), R_{slant} is the slant-range resolution (2.2484 m), and θ is the incidence angle (29.6281°).

b) The multi-look method was applied to de-speckle and re-sample the full-polarization GF-3 images. The GF-3 SAR images had a 1 × 1 multilook (azimuth × range), and were resampled to a regular grid with 5 × 5 m pixels in terms of azimuth and range resolution (5.30 × 2.25 m).

c) After the multi-look processing, the images were preprocessed using adaptive filters to reduce speckle and to enhance the edges and other features. We applied the refined Lee method of adaptive filtering by setting a 5 × 5 m filter window.

d) Radiometric calibration was carried out to convert the intensity values to the calibrated backscattering coefficient σ^0 (dB) of the normalized radar using the following equation [56]:

$$\sigma^\circ = 10 \times \lg(DN^2) + K \tag{2}$$

where DN represents the pixel values of the complex images, and K is the calibration constant. This equation referred to ALOS satellite processing. The radiometric calibration is still being explored to better process GF-3 SAR images. Previous studies used equation 2 to perform radiometric calibration, and demonstrated its utility and obtained satisfactory results [57,58]. Further study for radiometric calibration of Gaofen-3 images is needed.

e) Finally, the corrected backscatter map was generated from the backscattering coefficients to reduce the negative effect of the incidence angle on the radar's backscatter. Geometric correction is then performed to match the positions of the ground control points selected in the GF2 images to corresponding points in a 1:10000 topographic map, with the positional error smaller than 0.5 pixels.

(2) Variables derived from GF-3 images

The preprocessed images were used to acquire HV/HH-, VH/HH-, HV/VV-, and VH/VV-polarized data, by calculating the ratio of the backscattering coefficients of different polarimetric channels.

The FP SAR data allow us to identify the scattering mechanisms of different types that can significantly improve the depiction of features of the target object. This was achieved by polarimetric decomposition techniques, to separate the received signals of the radar. Such analysis can help repose a simpler object susceptible to an easier physical interpretation as a combination of the scattering [59]. Coherent decomposition is used to measure the scattering matrix by the responses of coherent scatters [60]. The targets of coherent scatter are analyzed based on the Sinclair matrix (S) representing all polarimetric information. With linear horizontal (H) and vertical (V) polarizations, the Sinclair matrix can be expressed as follows:

$$S = \begin{bmatrix} s_{HH} & s_{HV} \\ s_{VH} & s_{VV} \end{bmatrix} \tag{3}$$

The Pauli and Krogager decomposition approaches were used to analyze the targets of coherent scatter based on the Sinclair matrix [61]. Both could be applied to a homogeneous distribution of mangrove species in the study area [62]. Pauli decomposition was used to extract features of the polarization of the objects by defining different polarization fundamental matrices representing various types of objects. Pauli's polarimetric parameters were then decomposed into three elementary scattering mechanisms: odd-bounce scattering (P1), even-bounce scattering (P2), and volume scattering (P3). The Krogager decomposition aims to decompose the scattering matrix of a complex symmetric radar target into the physical interpretation of three components: sphere (KS1), diplane (KD3), and helix (KH2) [63].

The radar vegetation index (SAR-RVI), derived from FP SAR images, was used as a measure of the randomness of scattering from vegetation. It models the vegetation canopy as a collection of randomly oriented dipoles [64], and has yielded a good correlation between the SAR-RVI and the AGB. It was calculated by preprocessed FP data as follows [65]:

$$\text{SAR-RVI} = \frac{8 \times \sigma_{HV}}{2 \times \sigma_{HV} + \sigma_{HH} + \sigma_{VV}} \tag{4}$$

where σ_{HH}, σ_{HV}, σ_{VH}, and σ_{VV} represent the backscattering coefficients of the polarimetric channels HH, HV, VH, and VV, respectively.

In this study, the 11 predictors derived from GF-3 SAR images—the four backscattering coefficients (HH, HV, VH, and VV) of the FP channels, three Pauli polarimetric parameters (P1, P2, and P3), three Krogager polarimetric parameters (KS1, KD3, and KH2), and the SAR-RVI—were used as input parameters to build and predict the AGB of mangrove forests. They can reflect the different properties of mangrove forests. The backscatter of HH is linked to both trunk and crown biomass,

HV and VH return crown biomass, VV is dominated by branch biomass [66,67]. For the ratio of backscattering, they can potentially reduce topographic effects and forest structural effects, thereby increasing estimation accuracy [68]. The SAR-RVI reflects the canopy vegetation characteristics [64,69]. The Pauli decomposition can be used to separate the scattering matrix into simpler scattering responses related to single bouncing (e.g., canopy surface), double bouncing (e.g., trunk) and volume scattering (e.g., crown) [70,71]. The Krogager decomposition is related to surface, two, and three-sided corner reflectors [72]. The partial variables are shown in Figure 2.

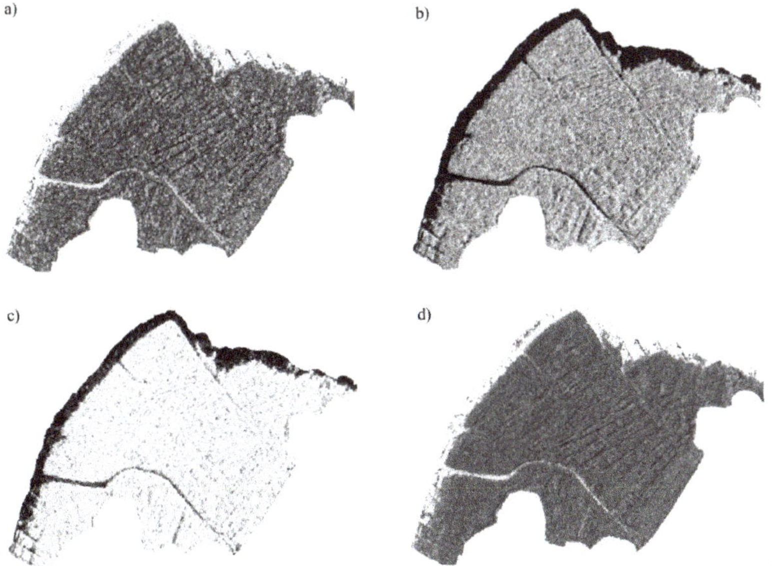

Figure 2. The partial variables derived from SAR images. (**a**) HV/HH, (**b**) P1 of Pauli decomposition, (**c**) KD3 of Krogager decomposition, and (**d**) radar vegetation index (SAR-RVI).

2.3.3. UAV-Based DSM

DSM data were derived from a fixed-wing UAV with an onboard SONY NEX-5T camera and GPS/inertial measurement unit in 2016. The configuration of the UAV was set to an altitude of 400 m, 80% frontal overlap, and 60% side overlap. A total of 349 valid photographs were captured with geolocation and altitude embedded into the EXIF data. They were processed by the Agisoft PhotoScan Professional (64 bit) software. The DSM data were generated by overlapping photographs and the SfM photogrammetry algorithm and were exported at a resolution of 0.12 m (Figure 1c). A geometric correction was then executed by a 1:10,000 topographic map and ground control points.

2.3.4. Mangrove Classification Based on GF-2 and DSM Data

In this study, we just focus on the AGB estimation of mangrove plantation species, *S. apetala*. Their spatial distribution needed to be identified to map and predict AGB. The mangrove plantation (*S. apetala*), with a homogeneous distribution, was identified prior to AGB estimation. The plantation in the study area had distinctive traits, such as the tallest trees, from those of other mangrove species. The GF-2 and DSM data were integrated to extract characteristics of the mangrove plantation accurately. We collected over 700 samples for classification by field investigation. Half of them were used to build the classification models and the other half to validate them.

The multispectral and panchromatic bands of the GF-2 data were first fused by the pan-sharpening method. After image fusion, image objects were generated using a multi-resolution segmentation algorithm in eCognition Developer 9.0 software package. For each object, the mean values of the four bands and DSMs, Vis (DVI, RVI, NDVI, SAVI), and texture features (homogeneity, contrast, entropy, mean, and correlation) were calculated and used as input features. The random forest (RF) algorithm was used to train and build the classifier, using the input features and measured training samples. Finally, the RF models were used to predict maps of mangrove plantation. Compared with the measured test samples, the overall accuracy of the RF models for mangrove species classification was 86.14%.

2.4. Modeling and Accuracy Assessment of AGB Estimation

Models for the estimation of the biomass of the mangrove forest were developed using the random forest regression algorithm (RFR), which is an ensemble machine learning technique that consists of a large number (n_{tree}) of decision trees grown by bootstraps of the original samples [73]. Each node of decision trees is separated by a random subset of input variables (m_{try}). The final results of the prediction are obtained by averaging the individual predictions of all regression trees [74]. The importance of all input variables was measured by out-of-bag (OOB) samples using the RF model and quantified by mean decrease in accuracy (MDA) [75]. Each variable's MDA was calculated by the difference in OOB error between the original dataset and the dataset with randomly permutated variables. To reduce the randomness of the RF models, the mean importance values of the input variables were measured 50 times.

To integrate data from multiple sources, all variables derived from GF-2, GF-3, and DSM were resampled at a resolution of 4 m to correspond to the GF-2 images. The variables were used as input variables, and the measured AGB samples were used as output variables to build the RF models. The spatial distribution of the AGB across the study area was predicted and mapped by the built RF models. We employed iterated five-fold cross-validation by partitioning the AGB samples into five datasets, four of which were used for training and one for validation to ensure the stability, reliability, and generalization capability of the models. All five datasets were generated using stratified random sampling, which led them to represent the entire range of biomass values. The accuracies of the built models were assessed by the root-mean-square error (RMSE) and relative RMSE (RMSEr) calculated from the observed and predicted values of the AGB.

To qualify the effect of the input variables on the accuracy of estimation of the AGB, four experiments were conducted. RF models were built to this end by combining different types of variables. In experiment 1 (Expt. 1 for short), the model used eight variables derived from optical images of the GF-2, including the four bands, DVI, RVI, NDVI, and SAVI. In experiment 2 (Expt. 2), the model employed 15 variables derived from GF-3 SAR images: the four full polarizations (HH, HV, VV, and VH); the ratio of backscattering coefficients of different polarimetric channels (HV/HH, VH/HH, HV/VV, and VH/VV); Pauli decomposition (P1, P2, and P3); Krogager decomposition (KS1, KD3, and KH2); and the SAR-RVI. In experiment 3 (Expt. 3), the model used 23 variables through a combination of GF-2 optical and GF-3 SAR data. In experiment 4 (Expt. 4), the model used 24 variables by integrating GF2 optical, GF3 SAR, and UAV-based DSM data.

2.5. Workflow for Analyses

This study focuses on developing effective models of the AGB of a mangrove plantation based on images from multiple sources, including GF-2 optical, GF-3 SAR, and UAV-based DSM images. The models were built using a machine learning approach (i.e., random forest (RF)), and the input variables were derived from multiple datasets. The corresponding accuracies were examined to study the effect of the input variables on the monitoring of the AGB. Finally, the model with the highest accuracy was used to predict and map the spatial distribution of the AGB of mangrove plantations. The workflow is provided in Figure 3.

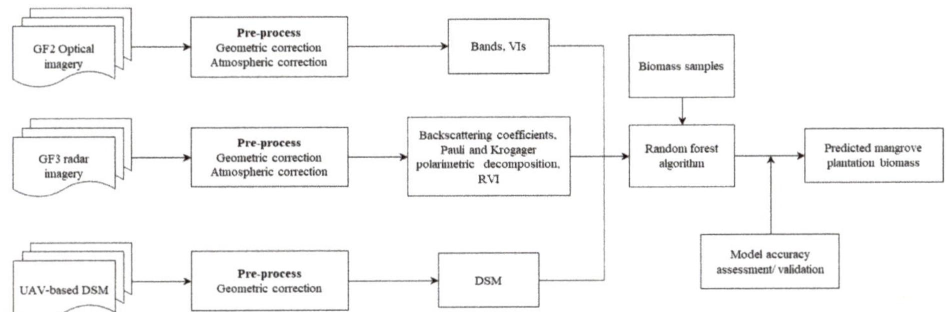

Figure 3. Workflow for the measurement of the aboveground biomass of artificially planted mangroves by integrating images from GF-2, GF-3, and UAV-based DSM datasets.

3. Results

3.1. AGB from Field Sampling

Field data of the mangrove plantation were collected a few times in the study area, and a total of 127 sampling units were obtained. The AGB was calculated by the allometric equation of the specific species. The plantation *S. apetala* had a density of 1623 trees per ha. The heights of the tree ranged from 2 to 20 m, with an average of 13.64 m. As it is a fast-growing species, the AGB of the mangrove plantation ranged from 90.65 to 237.74 t/ha, with an average of 159.70 t/ha, exhibiting a wide extent (43.88 t/ha of standard deviations), owing to their different ages. The field data revealed decreasing trends of AGB values in accordance with the sequence of growth from shore to sea.

3.2. Importance of Input Variables for AGB Estimation

The importance of the input variables was quantified by the RF algorithm to evaluate the relationship between them and the AGB (Figure 4). The results showed that the most important variable was the UAV-based DSM, implying that it is key to the AGB estimation. The next most-important variables were the VIs (RVI, NDVI, etc.) derived from GF-2 optical images, followed by those (KD3, HV, etc.) derived from the GF-3 SAR images.

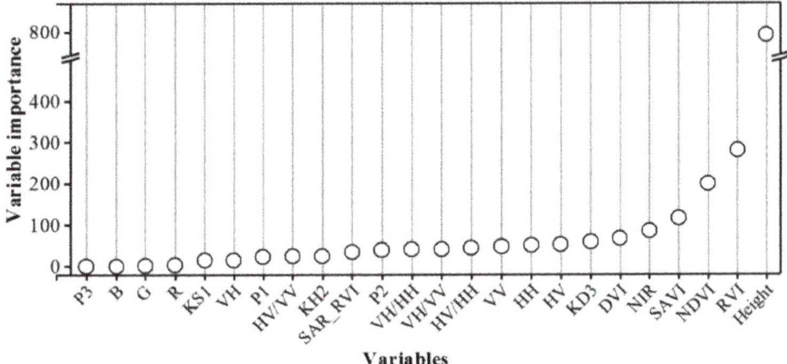

Figure 4. Importance of variables according to the random forest (RF) model.

3.3. Results and Accuracy Assessment of AGB Model

The RF models were developed using the observed AGB as output variables and the variables derived from images from multiple sources as input variables. The RMSE and RMSEr were acquired by the observed and the predicted AGB values based on five-fold cross-validation. As shown in Table 3, the mean of the predicted AGB values in the four experiments was 156 t/ha, in line with the mean observed value. The model using input variables derived from the GF-3 SAR images yielded the lowest estimation accuracy of the AGB of mangroves, followed by the model that used the GF-2 optical images. The estimation accuracy of the model obtained through the integration of GF-2 and GF-3 images was better than the model that used either GF-2 or GF-3 data, with a reduction of 2.32% and 3.49% in RMSEr, respectively. The combination of GF-2, GF-3, and DSM data produced the highest accuracy (RMSE = 25.69 t/ha, RMSEr = 16.53%) of all models. A two-sided t-test revealed significant differences ($p \geq 0.95$) in the predicted AGB values between models, using the combination of GF-2, GF-3, and DSM data, and those using only GF-2 or GF-3 images, but no significant difference ($p < 0.95$) was observed between the values obtained by models formed by a combination of GF-2, GF-3, and DSM data, and those obtained by models formed through the combination of GF-2 and GF-3 images.

Table 3. The accuracy of mangrove biomass estimation based on different input variables.

	Observed Values	GF2	GF3	GF2 and GF3	GF2, GF3, and DSM
Average (t/ha)	155.43	156.14	156.54	156.56	156.23
Standard deviations (t/ha)	37.75	18.81	14.72	15.25	19.08
Range (t/ha)	90.65–237.74	108.26–193.84	120.91–191.14	123.32–190.63	118.33–200.72
RMSE (t/ha)	/	33.49	35.32	29.89	25.69
RMSEr (%)	/	21.55	22.72	19.23	16.53

Scatterplots of the predicted versus the measured AGB values are presented to show the accuracy of the models with different input variables using the RF algorithm and five-fold cross-validation in the study area. As shown in Figure 5, the predicted AGB values of all models were above the 1:1 line at lower values, indicating that AGB values of the mangrove plantation had been overestimated, but they were the opposite at higher values. The coefficient of determination (R^2) of the model derived by integrating GF-2, GF-3, and UAV-based DSM data was 0.61, followed by the model obtained by a combination of GF-2 and GF-3 images. The other two models yielded lower R^2 values.

3.4. Mapping AGB of Mangrove Plantation

The model derived using a combination of GF-2, GF-3, and UAV-based DSM data produced the highest accuracy, and was used to map the spatial distribution of the AGB across the study area (Figure 6). The AGB map exhibited significant spatial variability, ranging from 106.163 t/ha to 266.162 t/ha, with an average of 137.89 t/ha. The AGB values of the species of mangrove in the southeast of the study area were higher than those in the west and northwest. The biomass decreased progressively from shore to sea, and trees near the shore had larger canopies and higher AGB values because they were older than those in other areas. Trees growing outside the forest edge and off the shore were younger and exhibited smaller biomass. The mangrove plantation map was consistent with the results of the field surveys, visual interpretation of remotely sensed images, and prior knowledge of Qi'ao Island.

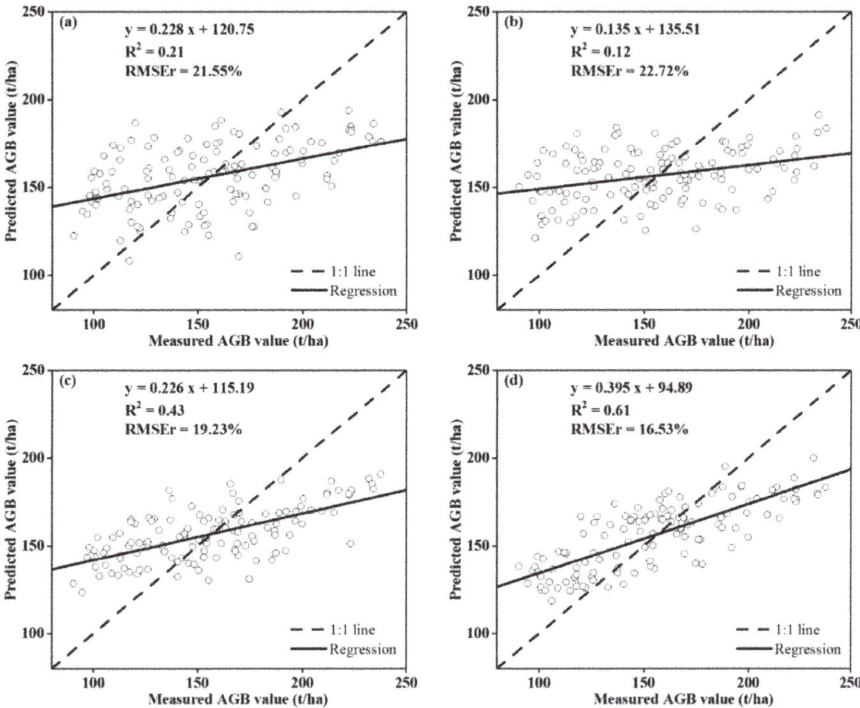

Figure 5. Scatter diagram of regression models detailing the linear regression, coefficient of determination (R^2), and relative root-mean-square error (RMSEr) between field-measured aboveground biomass (AGB) and predicted AGB from (**a**) GF-2 optical images; (**b**) GF-3 SAR images; (**c**) a combination of GF-2 and GF-3; and (**d**) a combination of GF-2, GF-3, and DSM data.

Figure 6. Spatial distribution of mangrove biomass.

4. Discussion

Estimating the AGB of forests based on satellite remote-sensing images remains challenging for tropical and subtropical mangrove forests, owing to various factors that interfere in the relationship between the AGB and the variables of images, such as the complex nature of their environments and complex forest structures [17]. The AGB of a forest has a close relationship, with its canopy-related information, structure, and height of trees, where single remote-sensing data cannot simultaneously provide this information [76]. Previous studies have investigated the combination of optical images (Landsat, SPOT, Sentinel-2, etc.) and SAR images (Sentinel-1, ALOS, etc.) to improve the accuracy of estimation of the AGB of forests [17,18]. The feasibility and applicability of data from the new GF-2 and GF-3 satellites from the Chinese civilian High-definition Earth Observation Satellite (HDEOS) program, launched in 2014 and 2016, respectively, by the China National Space Administration (CNSA), need to be tested.

4.1. Overall Performance of Random Forest Model

The objective of this study was to develop an accurate and robust biomass estimation of mangrove plantations. A random forest model was selected to establish non-linear relationships between the AGB and the input variables, because of its ease of use and prediction accuracy [43]. Previous studies already used machine-learning algorithms, such as support vector regression (SVR) or artificial neural network (ANN) for AGB estimation, which produces satisfactory results. Wang et al. (2016) investigated the applicability of RF, SVR, and ANN for remotely estimating wheat biomass, and the results indicated that the RF model produced more accurate estimates of wheat biomass than the SVR and ANN models at each stage [77].

A major advantage of RF is bootstrap sampling and variable sampling, in which the subset of all variables is randomly selected using the best split for each node of the standard regression tree. In these situations, The RF model can decrease the algorithm's risk for overfitting and multicollinearity, due to relative insensitivity to variations in input variables, thereby improving generalization and robustness to predicted data. Therefore, some variables in this study are correlated. However, as demonstrated by Cutler et al., the RF model is not sensitive to collinearity, and has the ability to model complex and nonlinear interactions among predictor variables [78]. This is helpful, as it is commonly hard to determine which variables need to be removed when two or more variables correlate with each other [79]. Therefore, the RF algorithm provides a useful exploratory and predictive tool for estimating mangrove biomass.

4.2. Contribution of Input Variables to Measuring AGB of Mangrove Plantation

This study addressed the above issues using GF-2 optical and GF-3 data. The results indicated that the potential of the optical images and C-band FP SAR images for the AGB prediction of artificially planted mangroves were similar. The SAR-based results have been slightly weaker than the results with optical data, mainly due to weaker spatial resolution of the available SAR data in this study (5×5 m pixel resolution after multi-look method processing). On the other hand, the radiometric calibration referred to ALOS satellite processing using the SARscape, which may cause the errors of biomass estimation [57,58]. The advantages of SAR images are their multi-temporal acquisitions and independence of cloud cover, making the atmospheric correction of optical images more difficult [18].

The model of the AGB developed using a combination of GF-2 and GF-3 images yielded a higher estimation accuracy than those built using GF-2 or GF-3 images alone. This finding is consistent with previous studies, which have noted that integrating optical and SAR images can improve the accuracy of estimation of the AGB of forests, mainly because factors influencing biomass estimation, such as canopy-related information (canopy density, vegetation status) and forest structure, can be reflected by them [18,80].

However, optical or SAR images often incur saturation problems in canopies owing to dense vegetation, leading to the underestimation of biomass [81]. When considering variables of the DSM based on GF-2 and GF-3 images, the accuracy of the AGB model improved by 2.7% in terms of RMSEr. DSM data were also identified as the most important variable by the RF algorithm, because they were collected on a similar date with field measurements and can reflect the relative height of trees in a mangrove plantation, which is important for biomass estimation [42]. Tree height is usually computed from corresponding digital terrain models (DTM) subtracted from digital surface models (DSM). However, the DTM for dense mangrove forests is unavailable, due to the inability to penetrate their dense and complex canopy structures. The DTM is a stable constant for mangrove forests, because they mainly grow over even terrain [42]. The DSM can thus be considered to measure the relative heights of mangrove trees instead of the canopy height model (CHM). Previous studies have shown that the DSM derived from SfM and aerial photographs can solve the saturation problem and improve the estimation accuracy of biomass [82,83].

In this study, the P3, B, G, and R variables are least important (Figure 4). These variables with relatively low importance may be caused by the insensitivity to AGB estimation or be affected by multicollinearity. For example, the near-infrared (NIR) band of optical images is more widely used to estimate vegetation biomass content, because of its spectral reflection features in green vegetation, and visible (red, green, and blue) bands are usually used to emphasize vegetation health and classification, thus, they may be insensitive to AGB estimation. However, if the features have a correlation, it can be challenging to rank the importance of the features.

4.3. Spatial Distribution Patterns of AGB of Mangrove Plantation

As an initial effort for restoration and reforestation, the mangrove species *S. apetala* was introduced to the study area in 1999. As a result of its ability to spread, the artificial planting of *S. apetala* is becoming increasingly controversial as it may invade other mangrove ecosystems. The map of the AGB of the mangrove plantation derived here can provide baseline data for subsequent analyses and applications (e.g., carbon sequestration). The AGB values were predicted and mapped by a model that used GF-2 optical, GF-3 SAR, and UAV-based DSM data. The *S. apetala* afforestation process runs seaward from land, which implies a gradient distribution of AGB. The spatial distribution of the resulting AGB corresponded to the sequence of the mangrove plantation over time. To further verify the reliability of the model and understand the spatial distribution of the AGB, areas occupied by the mangrove plantation that had grown before 2011 were extracted by WorldView-2 images [43]. They exhibited significantly higher AGB values than the other areas, due to their age and fast growth. This is shown in Figure 7. The results thus indicate that the map of the AGB of the unevenly aged mangrove plantation showed a greater heterogeneity of AGB values.

4.4. Limitation and Sources of Errors

We found a reasonable relationship among GF-2 optical, GF-3 SAR images and, field measured AGB by using the RF algorithm. The combination of multi-source datasets (GF-2, GF-3 images, and DSM) yielded a higher estimation accuracy. However, there are some limitations and sources of errors from the AGB estimation using multi-source images, caused by position errors of geometric calibration, and the time difference of images and field measurements.

The errors of geometric calibration among multi-source images were unavoidable, though we used a geometric calibration to a 1:10,000 topographic map using ground control. The mismatch of the pixels derived from multi-source may cause the uncertainty of AGB estimation. Similarly, we got the AGB samples by field investigation, and the closest remotely sensed data available. The differences in time acquisitions may bring additional sources of errors in the AGB estimation, due to the fast-growing characteristic of the *S. apetala*. The study from Ren et al. (2010) suggested that AGB accumulation rates at the *S. apetala* plantations decreased with the stand ages [48]. The AGB accumulation rates were 20.3 t/ha from 4 to 5 years stand, 5.6 t/ha from 5 to 8 years stand, and 2.85 t/ha from 8 to 10 years stand, respectively. In our study area, most of the plantations have reached over 5 years stand [84], and previous study has demonstrated mean AGB accumulation of *S. apetala* over the study area to be 4.17 t/ha per year [85]. The average of observed values of this study is 155.4 t/ha, which may cause a 2.68% error of AGB—which is within the acceptable range—due to the different date between the field data and the images. Therefore, the inevitable difference of one year between the remote sensing images and in situ measurements can be deemed acceptable.

Figure 7. The predicted map of AGB values in 2016. (**a–f**) represented the partial enlarged detail of predicted mangrove AGB overlaid with the map of mangrove plantation from before 2011.

5. Conclusions

This study explored the potential of GF-2 optical, GF-3 SAR, and UAV-based DSM data for estimating the AGB of the mangrove plantation of Qi'ao Island in China. The AGB model generated using a combination of GF-2 and GF-3 images from the Chinese civilian HDEOS program yielded a higher accuracy than those of models using only one of these datasets, with a reduction of 2.32% and 3.49%, respectively, in RMSEr. When considering variables of the DSM derived from the UAV platform, the AGB model achieved the highest accuracy with a further reduction of 2.7% decrease in RMSEr. The DSM was the most important input variable for AGB estimation as it deals with saturation problems in optical and SAR images. The resulting AGB map agreed well with field surveys and the growing sequence of mangrove plantations. The results showed that accurate AGB models and spatial distribution maps of mangrove plantation can be obtained using the RF model, and images from multiple sources (GF-2 optical, GF-2 SAR, and UAV-based DSM data). The combination of these data provided canopy-related information, forest structures, and tree heights for AGB modeling.

The study focused on the integration of GF2 optical, GF3 SAR, and UAV data for estimating aboveground biomass in China's largest artificially planted mangroves. The methodology can be used to produce accurate AGB models of mangrove forests, which can be difficult to obtain by field investigation. The AGB maps of *S. apetala* can help measure mangrove carbon sinks and provide baseline data for REDD+ programs, due to mangrove plantation. Future studies should further examine and improve AGB estimation uncertainty, such as accurate radiometric calibration and noise estimation of GF-3.

Author Contributions: Conceptualization, Y.Z., K.L. and L.L. conceived and designed the paper. Y.Z., K.L., J.C. and Z.D. conducted field study and data processing. Y.Z., K.L., L.L., S.W.M., Y.L. and Z.W. wrote and revised the manuscript. All authors have read and agreed to the published version of the manuscript.

Funding: This work is supported by Research Team Program of Natural Science Foundation of Guangdong Province, China (2014A030312010), China Postdoctoral Science Foundation (2018M633023), Postdoctoral International Training Program of Guangzhou City, Science and Technology Planning Project of Guangdong Province (2017A020217003), the Natural Science Foundation of Guangdong (2016A030313261 and 2016A030313188), and the National Science Foundation of China (Grant No. 41501368).

Conflicts of Interest: The authors declare no conflict of interest

References

1. Kosoy, N.; Corbera, E. Payments for ecosystem services as commodity fetishism. *Ecol. Econ.* **2010**, *69*, 1228–1236. [CrossRef]
2. Sani, D.A.; Hashim, M.; Hossain, M.S. Recent advancement on estimation of blue carbon biomass using satellite-based approach. *Int. J. Remote Sens.* **2019**, *40*, 7679–7715. [CrossRef]
3. Laffoley, D.; Grimsditch, G.D. *The Management of Natural Coastal Carbon Sinks*; IUCN: Gland, Switzerland, 2009.
4. Bouillon, S.; Borges, A.V.; Castañeda-Moya, E.; Diele, K.; Dittmar, T.; Duke, N.C.; Kristensen, E.; Lee, S.Y.; Marchand, C.; Middelburg, J.J. Mangrove production and carbon sinks: A revision of global budget estimates. *Glob. Biogeochem. Cycles* **2008**, *22*. [CrossRef]
5. Jia, M.; Wang, Z.; Wang, C.; Mao, D.; Zhang, Y. A new vegetation index to detect periodically submerged Mangrove forest using single-tide sentinel-2 imagery. *Remote Sens.* **2019**, *11*, 2043. [CrossRef]
6. Zhu, Y.; Liu, K.; Liu, L.; Wang, S.; Liu, H. Retrieval of Mangrove Aboveground Biomass at the Individual Species Level with WorldView-2 Images. *Remote Sens.* **2015**, *7*, 12192–12214. [CrossRef]
7. Dou, Z.; Cui, L.; Li, J.; Zhu, Y.; Gao, C.; Pan, X.; Lei, Y.; Zhang, M.; Zhao, X.; Li, W. Hyperspectral Estimation of the Chlorophyll Content in Short-Term and Long-Term Restorations of Mangrove in Quanzhou Bay Estuary, China. *Sustainability* **2018**, *10*, 1127. [CrossRef]
8. Giri, C.; Ochieng, E.; Tieszen, L.L.; Zhu, Z.; Singh, A.; Loveland, T.; Masek, J.; Duke, N. Status and distribution of mangrove forests of the world using earth observation satellite data. *Glob. Ecol. Biogeogr.* **2011**, *20*, 154–159. [CrossRef]
9. Myint, S.W.; Giri, C.P.; Wang, L.; Zhu, Z.; Gillette, S.C. Identifying Mangrove Species and Their Surrounding Land Use and Land Cover Classes Using an Object-Oriented Approach with a Lacunarity Spatial Measure. *GISci. Remote Sens.* **2008**, *45*, 188–208. [CrossRef]
10. Jia, M.M.; Wang, Z.M.; Zhang, Y.Z.; Mao, D.H.; Wang, C. Monitoring loss and recovery of mangrove forests during 42 years: The achievements of mangrove conservation in China. *Int. J. Appl. Earth Obs.* **2018**, *73*, 535–545. [CrossRef]
11. Abdul Aziz, A.; Dargusch, P.; Phinn, S.; Ward, A. Using REDD+ to balance timber production with conservation objectives in a mangrove forest in Malaysia. *Ecol. Econ.* **2015**, *120*, 108–116. [CrossRef]
12. Kankare, V.; Vastaranta, M.; Holopainen, M.; Räty, M.; Yu, X.; Hyyppä, J.; Hyyppä, H.; Alho, P.; Viitala, R. Retrieval of forest aboveground biomass and stem volume with airborne scanning LiDAR. *Remote Sens.* **2013**, *5*, 2257–2274. [CrossRef]
13. Fatoyinbo, T.E.; Simard, M.; Washington-Allen, R.A.; Shugart, H.H. Landscape-scale extent, height, biomass, and carbon estimation of Mozambique's mangrove forests with Landsat ETM+ and Shuttle Radar Topography Mission elevation data. *J. Geophys. Res. Biogeosci.* **2008**, *113*. [CrossRef]
14. Pham, L.T.; Brabyn, L. Monitoring mangrove biomass change in Vietnam using SPOT images and an object-based approach combined with machine learning algorithms. *ISPRS J. Photogramm. Remote Sens.* **2017**, *128*, 86–97. [CrossRef]
15. Sasmito, S.D.; Murdiyarso, D.; Wijaya, A.; Purbopuspito, J.; Okimoto, Y. Remote sensing technique to assess aboveground biomass dynamics of mangrove ecosystems area in Segara Anakan, Central Java, Indonesia. In Proceedings of the 34th Asian Conference on Remote Sensing 2013, ACRS 2013, Bali, Indonesia, 1 January 2013; pp. 4560–4565.

16. Li, C.; Li, Y.; Li, M. Improving forest aboveground biomass (AGB) estimation by incorporating crown density and using landsat 8 OLI images of a subtropical forest in Western Hunan in Central China. *Forests* **2019**, *10*, 104. [CrossRef]

17. Navarro, J.A.; Algeet, N.; Fernández-Landa, A.; Esteban, J.; Rodríguez-Noriega, P.; Guillén-Climent, M.L. Integration of UAV, Sentinel-1, and Sentinel-2 Data for Mangrove Plantation Aboveground Biomass Monitoring in Senegal. *Remote Sens.* **2019**, *11*, 77. [CrossRef]

18. Vafaei, S.; Soosani, J.; Adeli, K.; Fadaei, H.; Naghavi, H.; Pham, T.; Tien Bui, D. Improving Accuracy Estimation of Forest Aboveground Biomass Based on Incorporation of ALOS-2 PALSAR-2 and Sentinel-2A Imagery and Machine Learning: A Case Study of the Hyrcanian Forest Area (Iran). *Remote Sens.* **2018**, *10*, 172. [CrossRef]

19. Aslan, A.; Rahman, A.F.; Warren, M.W.; Robeson, S.M. Mapping spatial distribution and biomass of coastal wetland vegetation in Indonesian Papua by combining active and passive remotely sensed data. *Remote Sens. Environ.* **2016**, *183*, 65–81. [CrossRef]

20. Jachowski, N.R.A.; Quak, M.S.Y.; Friess, D.A.; Duangnamon, D.; Webb, E.L.; Ziegler, A.D. Mangrove biomass estimation in Southwest Thailand using machine learning. *Appl. Geogr.* **2013**, *45*, 311–321. [CrossRef]

21. Proisy, C.; Couteron, P.; Fromard, F. Predicting and mapping mangrove biomass from canopy grain analysis using Fourier-based textural ordination of IKONOS images. *Remote Sens. Environ.* **2007**, *109*, 379–392. [CrossRef]

22. Le Maire, G.; Marsden, C.; Nouvellon, Y.; Grinand, C.; Hakamada, R.; Stape, J.-L.; Laclau, J.-P. MODIS NDVI time-series allow the monitoring of Eucalyptus plantation biomass. *Remote Sens. Environ.* **2011**, *115*, 2613–2625. [CrossRef]

23. Pham, T.D.; Yoshino, K.; Bui, D.T. Biomass estimation of Sonneratia caseolaris (l.) Engler at a coastal area of Hai Phong city (Vietnam) using ALOS-2 PALSAR imagery and GIS-based multi-layer perceptron neural networks. *GISci. Remote Sens.* **2017**, *54*, 329–353. [CrossRef]

24. Chadwick, J. Integrated LiDAR and IKONOS multispectral imagery for mapping mangrove distribution and physical properties. *Int. J. Remote Sens.* **2011**, *32*, 6765–6781. [CrossRef]

25. Olagoke, A.; Proisy, C.; Féret, J.-B.; Blanchard, E.; Fromard, F.; Mehlig, U.; de Menezes, M.M.; Dos Santos, V.F.; Berger, U. Extended biomass allometric equations for large mangrove trees from terrestrial LiDAR data. *Trees* **2016**, *30*, 935–947. [CrossRef]

26. Omar, H.; Misman, M.; Kassim, A. Synergetic of PALSAR-2 and Sentinel-1A SAR polarimetry for retrieving aboveground biomass in dipterocarp forest of Malaysia. *Appl. Sci.* **2017**, *7*, 675. [CrossRef]

27. Wu, G.; Guo, Z.; Guo, J.; Zhu, Y. Estimation of mangrove wetland aboveground biomass based on remote sensing data: A review. *J. South Agric.* **2013**, *44*, 693–696.

28. Cougo, M.; Souza-Filho, P.; Silva, A.; Fernandes, M.; Santos, J.; Abreu, M.; Nascimento, W.; Simard, M. Radarsat-2 backscattering for the modeling of biophysical parameters of regenerating mangrove forests. *Remote Sens.* **2015**, *7*, 17097–17112. [CrossRef]

29. Kovacs, J.; Lu, X.; Flores-Verdugo, F.; Zhang, C.; de Santiago, F.F.; Jiao, X. Applications of ALOS PALSAR for monitoring biophysical parameters of a degraded black mangrove (Avicennia germinans) forest. *ISPRS J. Photogramm. Remote Sens.* **2013**, *82*, 102–111. [CrossRef]

30. Lucas, R.; Lule, A.V.; Rodríguez, M.T.; Kamal, M.; Thomas, N.; Asbridge, E.; Kuenzer, C. Spatial ecology of mangrove forests: A remote sensing perspective. In *Mangrove Ecosystems: A Global Biogeographic Perspective*; Springer: Berlin/Heidelberg, Germany, 2017; pp. 87–112.

31. Tsui, O.W.; Coops, N.C.; Wulder, M.A.; Marshall, P.L. Integrating airborne LiDAR and space-borne radar via multivariate kriging to estimate above-ground biomass. *Remote Sens. Environ.* **2013**, *139*, 340–352. [CrossRef]

32. Haris, M.; Ashraf, M.; Ahsan, F.; Athar, A.; Malik, M. Analysis of SAR images speckle reduction techniques. In Proceedings of the 2018 International Conference on Computing, Mathematics and Engineering Technologies (iCoMET), Sukkur, Pakistan, 3–4 March 2018; pp. 1–7.

33. Zhao, P.; Lu, D.; Wang, G.; Wu, C.; Huang, Y.; Yu, S. Examining spectral reflectance saturation in Landsat imagery and corresponding solutions to improve forest aboveground biomass estimation. *Remote Sens.* **2016**, *8*, 469. [CrossRef]

34. Pham, T.D.; Yoshino, K.; Le, N.N.; Bui, D.T. Estimating aboveground biomass of a mangrove plantation on the Northern coast of Vietnam using machine learning techniques with an integration of ALOS-2 PALSAR-2 and Sentinel-2A data. *Int. J. Remote Sens.* **2018**, *39*, 7761–7788. [CrossRef]

35. Shao, Z.; Zhang, L. Estimating forest aboveground biomass by combining optical and SAR data: A case study in Genhe, Inner Mongolia, China. *Sensors* **2016**, *16*, 834. [CrossRef]
36. Hamdan, O.; Hasmadi, I.M.; Aziz, H.K.; Norizah, K.; Zulhaidi, M.H. L-band saturation level for aboveground biomass of dipterocarp forests in peninsular Malaysia. *J. Trop. For. Sci.* **2015**, *27*, 388–399.
37. Fatoyinbo, T.E.; Simard, M. Height and biomass of mangroves in Africa from ICESat/GLAS and SRTM. *Int. J. Remote Sens.* **2013**, *34*, 668–681. [CrossRef]
38. Fedrigo, M.; Newnham, G.J.; Coops, N.C.; Culvenor, D.S.; Bolton, D.K.; Nitschke, C.R. Predicting temperate forest stand types using only structural profiles from discrete return airborne lidar. *ISPRS J. Photogramm. Remote Sens.* **2018**, *136*, 106–119. [CrossRef]
39. Fonstad, M.A.; Dietrich, J.T.; Courville, B.C.; Jensen, J.L.; Carbonneau, P.E. Topographic structure from motion: A new development in photogrammetric measurement. *Earth Surf. Process. Landf.* **2013**, *38*, 421–430. [CrossRef]
40. Solberg, S.; Hansen, E.H.; Gobakken, T.; Næssset, E.; Zahabu, E. Biomass and InSAR height relationship in a dense tropical forest. *Remote Sens. Environ.* **2017**, *192*, 166–175. [CrossRef]
41. Maimaitijiang, M.; Ghulam, A.; Sidike, P.; Hartling, S.; Maimaitiyiming, M.; Peterson, K.; Shavers, E.; Fishman, J.; Peterson, J.; Kadam, S.; et al. Unmanned Aerial System (UAS)-based phenotyping of soybean using multi-sensor data fusion and extreme learning machine. *ISPRS J. Photogramm. Remote Sens.* **2017**, *134*, 43–58. [CrossRef]
42. Otero, V.; Van De Kerchove, R.; Satyanarayana, B.; Martínez-Espinosa, C.; Fisol, M.A.B.; Ibrahim, M.R.B.; Sulong, I.; Mohd-Lokman, H.; Lucas, R.; Dahdouh-Guebas, F. Managing mangrove forests from the sky: Forest inventory using field data and Unmanned Aerial Vehicle (UAV) imagery in the Matang Mangrove Forest Reserve, peninsular Malaysia. *For. Ecol. Manag.* **2018**, *411*, 35–45. [CrossRef]
43. Zhu, Y.; Liu, K.; Liu, L.; Myint, S.; Wang, S.; Liu, H.; He, Z. Exploring the potential of worldview-2 red-edge band-based vegetation indices for estimation of mangrove leaf area index with machine learning algorithms. *Remote Sens.* **2017**, *9*, 1060. [CrossRef]
44. Cao, J.; Leng, W.; Liu, K.; Liu, L.; He, Z.; Zhu, Y. Object-based mangrove species classification using unmanned aerial vehicle hyperspectral images and digital surface models. *Remote Sens.* **2018**, *10*, 89. [CrossRef]
45. Liu, K.; Li, X.; Shi, X.; Wang, S.G. Monitoring mangrove forest changes using remote sensing and GIS data with decision-tree learning. *Wetlands* **2008**, *28*, 336–346. [CrossRef]
46. Ren, H.; Lu, H.F.; Shen, W.J.; Huang, C.; Guo, Q.F.; Li, Z.A.; Jian, S.G. Sonneratia apetala Buch.Ham in the mangrove ecosystems of China: An invasive species or restoration species? *Ecol. Eng.* **2009**, *35*, 1243–1248. [CrossRef]
47. Ren, H.; Jian, S.; Lu, H.; Zhang, Q.; Shen, W.; Han, W.; Yin, Z.; Guo, Q. Restoration of mangrove plantations and colonisation by native species in Leizhou bay, South China. *Ecol. Res.* **2008**, *23*, 401–407. [CrossRef]
48. Ren, H.; Chen, H.; Han, W. Biomass accumulation and carbon storage of four different aged Sonneratia apetala plantations in Southern China. *Plant Soil* **2010**, *327*, 279–291. [CrossRef]
49. Zan, q.j.; Wang, y.j.; Liao, b.w.; Zheng, d.z. Biomass and net productivity of Sonneratia apetala, S.caseolaris mangrove man-made forest. *J. Wuhan Bot. Res.* **2001**, *19*, 391–396.
50. Wang, H.; Wang, C.; Wu, H. Using GF-2 imagery and the conditional random field model for urban forest cover mapping. *Remote Sens. Lett.* **2016**, *7*, 378–387. [CrossRef]
51. Richardson, A.J.; Wiegand, C. Distinguishing vegetation from soil background information. *Photogramm. Eng. Remote Sens.* **1977**, *43*, 1541–1552.
52. Tucker, C.J. Red and photographic infrared linear combinations for monitoring vegetation. *Remote Sens. Environ.* **1979**, *8*, 127–150. [CrossRef]
53. Rouse, J.W., Jr.; Haas, R.; Schell, J.; Deering, D. Monitoring vegetation systems in the Great Plains with ERTS. In Proceedings of the Third Earth Resources Technology Satellite—1 Symposium: NASA SP-351, Washington, DC, USA, 10–14 December 1973; pp. 309–317.
54. Huete, A.R. A soil-adjusted vegetation index (SAVI). *Remote Sens. Environ.* **1988**, *25*, 295–309. [CrossRef]
55. Xu, L.; Zhang, H.; Wang, C.; Fu, Q. Classification of Chinese GaoFen-3 fully-polarimetric SAR images: Initial results. In Proceedings of the 2017 Progress in Electromagnetics Research Symposium—Fall (PIERS—FALL), Singapore, 19–22 November 2017; pp. 700–705.
56. Shimada, M.; Isoguchi, O.; Tadono, T.; Isono, K. PALSAR Radiometric and Geometric Calibration. *IEEE Trans. Geosci. Remote Sens.* **2009**, *47*, 3915–3932. [CrossRef]

57. Li, X.-M.; Zhang, T.; Huang, B.; Jia, T. Capabilities of Chinese Gaofen-3 synthetic aperture radar in selected topics for coastal and ocean observations. *Remote Sens.* **2018**, *10*, 1929. [CrossRef]

58. Wang, H.; Li, H.; Lin, M.; Zhu, J.; Wang, J.; Li, W.; Cui, L. Calibration of the copolarized backscattering measurements from Gaofen-3 synthetic aperture radar wave mode imagery. *IEEE J. Sel. Top. Appl. Earth Obs. Remote Sens.* **2019**, *12*, 1748–1762. [CrossRef]

59. Qi, Z.; Yeh, A.G.-O.; Li, X.; Lin, Z. A novel algorithm for land use and land cover classification using RADARSAT-2 polarimetric SAR data. *Remote Sens. Environ.* **2012**, *118*, 21–39. [CrossRef]

60. Uhlmann, S.; Kiranyaz, S. Integrating Color Features in Polarimetric SAR Image Classification. *IEEE Trans. Geosci. Remote Sens.* **2014**, *52*, 2197–2216. [CrossRef]

61. Huynen, J.R. Measurement of the target scattering matrix. *Proc. IEEE* **1965**, *53*, 936–946. [CrossRef]

62. Yamaguchi, Y.; Moriyama, T.; Ishido, M.; Yamada, H. Four-component scattering model for polarimetric SAR image decomposition. *IEEE Trans. Geosci. Remote Sens.* **2005**, *43*, 1699–1706. [CrossRef]

63. Krogager, E. New decomposition of the radar target scattering matrix. *Electron. Lett.* **1990**, *26*, 1525–1527. [CrossRef]

64. Kim, Y.; Zyl, J.J.V. A Time-Series Approach to Estimate Soil Moisture Using Polarimetric Radar Data. *IEEE Trans. Geosci. Remote Sens.* **2009**, *47*, 2519–2527. [CrossRef]

65. Wiseman, G.; McNairn, H.; Homayouni, S.; Shang, J. RADARSAT-2 polarimetric SAR response to crop biomass for agricultural production monitoring. *IEEE J. Sel. Top. Appl. Earth Obs. Remote Sens.* **2014**, *7*, 4461–4471. [CrossRef]

66. Beaudoin, A.; Le Toan, T.; Goze, S.; Nezry, E.; Lopes, A.; Mougin, E.; Hsu, C.; Han, H.; Kong, J.; Shin, R. Retrieval of forest biomass from SAR data. *Int. J. Remote Sens.* **1994**, *15*, 2777–2796. [CrossRef]

67. Ranson, K.; Sun, G. Mapping biomass of a northern forest using multifrequency SAR data. *IEEE Trans. Geosci. Remote Sens.* **1994**, *32*, 388–396. [CrossRef]

68. Sarker, M.L.R.; Nichol, J.; Iz, H.B.; Ahmad, B.B.; Rahman, A.A. Forest biomass estimation using texture measurements of high-resolution dual-polarization C-band SAR data. *IEEE Trans. Geosci. Remote Sens.* **2012**, *51*, 3371–3384. [CrossRef]

69. Gao, H.; Wang, C.; Wang, G.; Zhu, J.; Tang, Y.; Shen, P.; Zhu, Z. A crop classification method integrating GF-3 PolSAR and Sentinel-2A optical data in the Dongting Lake Basin. *Sensors* **2018**, *18*, 3139. [CrossRef] [PubMed]

70. Tanase, M.A.; Panciera, R.; Lowell, K.; Tian, S.; Hacker, J.M.; Walker, J.P. Airborne multi-temporal L-band polarimetric SAR data for biomass estimation in semi-arid forests. *Remote Sens. Environ.* **2014**, *145*, 93–104. [CrossRef]

71. Du, P.; Samat, A.; Gamba, P.; Xie, X. Polarimetric SAR image classification by boosted multiple-kernel extreme learning machines with polarimetric and spatial features. *Int. J. Remote Sens.* **2014**, *35*, 7978–7990. [CrossRef]

72. Srikanth, P.; Ramana, K.; Deepika, U.; Chakravarthi, P.K.; Sai, M.S. Comparison of various polarimetric decomposition techniques for crop classification. *J. Indian Soc. Remote Sens.* **2016**, *44*, 635–642. [CrossRef]

73. Breiman, L. Random forests. *Mach. Learn.* **2001**, *45*, 5–32. [CrossRef]

74. Pal, M. Random forest classifier for remote sensing classification. *Int. J. Remote Sens.* **2005**, *26*, 217–222. [CrossRef]

75. Liaw, A.; Wiener, M. Classification and regression by randomForest. *R News* **2002**, *2*, 18–22.

76. Zhu, X.; Hou, Y.; Weng, Q.; Chen, L. Integrating UAV optical imagery and LiDAR data for assessing the spatial relationship between mangrove and inundation across a subtropical estuarine wetland. *ISPRS J. Photogramm. Remote Sens.* **2019**, *149*, 146–156. [CrossRef]

77. Wang, L.a.; Zhou, X.; Zhu, X.; Dong, Z.; Guo, W. Estimation of biomass in wheat using random forest regression algorithm and remote sensing data. *Crop J.* **2016**, *4*, 212–219. [CrossRef]

78. Cutler, D.R.; Edwards Jr, T.C.; Beard, K.H.; Cutler, A.; Hess, K.T.; Gibson, J.; Lawler, J.J. Random forests for classification in ecology. *Ecology* **2007**, *88*, 2783–2792. [CrossRef] [PubMed]

79. Fukuda, S.; Yasunaga, E.; Nagle, M.; Yuge, K.; Sardsud, V.; Spreer, W.; Müller, J. Modelling the relationship between peel colour and the quality of fresh mango fruit using Random Forests. *J. Food Eng.* **2014**, *131*, 7–17. [CrossRef]

80. Cutler, M.E.J.; Boyd, D.S.; Foody, G.M.; Vetrivel, A. Estimating tropical forest biomass with a combination of SAR image texture and Landsat TM data: An assessment of predictions between regions. *ISPRS J. Photogramm. Remote Sens.* **2012**, *70*, 66–77. [CrossRef]

Remote Sens. **2020**, *12*, 2039

81. Mutanga, O.; Adam, E.; Cho, M.A. High density biomass estimation for wetland vegetation using worldview-2 imagery and random forest regression algorithm. *Int. J. Appl. Earth Obs.* **2012**, *18*, 399–406. [CrossRef]
82. Luo, S.; Wang, C.; Xi, X.; Pan, F.; Peng, D.; Zou, J.; Nie, S.; Qin, H. Fusion of airborne LiDAR data and hyperspectral imagery for aboveground and belowground forest biomass estimation. *Ecol. Indic.* **2017**, *73*, 378–387. [CrossRef]
83. Pena, J.M.; de Castro, A.I.; Torres-Sanchez, J.; Andujar, D.; San Martin, C.; Dorado, J.; Fernandez-Quintanilla, C.; Lopez-Granados, F. Estimating tree height and biomass of a poplar plantation with image-based UAV technology. *AIMS Agric. Food* **2018**, *3*, 313–326. [CrossRef]
84. Liu, K.; Liu, L.; Liu, H.; Li, X.; Wang, S. Exploring the effects of biophysical parameters on the spatial pattern of rare cold damage to mangrove forests. *Remote Sens. Environ.* **2014**, *150*, 20–33. [CrossRef]
85. Zhu, Y.; Liu, K.; Liu, L.; Myint, S.W.; Wang, S.; Cao, J.; Wu, Z. Estimating and Mapping Mangrove Biomass Dynamic Change Using WorldView-2 Images and Digital Surface Models. *IEEE J. Sel. Top. Appl. Earth Obs. Remote Sens.* **2020**, *13*, 2123–2134. [CrossRef]

 *remote sensing*

Article

Estimating Mangrove Above-Ground Biomass Using Extreme Gradient Boosting Decision Trees Algorithm with Fused Sentinel-2 and ALOS-2 PALSAR-2 Data in Can Gio Biosphere Reserve, Vietnam

Tien Dat Pham [1], Nga Nhu Le [2,*], Nam Thang Ha [3,4], Luong Viet Nguyen [5], Junshi Xia [1], Naoto Yokoya [1], Tu Trong To [5], Hong Xuan Trinh [5], Lap Quoc Kieu [6] and Wataru Takeuchi [7]

[1] Geoinformatics Unit, RIKEN Center for Advanced Intelligence Project (AIP), Mitsui Building, 15th floor, 1-4-1 Nihonbashi, Chuo-ku, Tokyo 103-0027, Japan; tiendat.pham@riken.jp (T.D.P.); junshi.xia@riken.jp (J.X.); naoto.yokoya@riken.jp (N.Y.)

[2] Department of Marine Mechanics and Environment, Institute of Mechanics, Vietnam Academy of Science and Technology (VAST), 264 Doi Can street, Ba Dinh district, Hanoi 100000, Vietnam

[3] Faculty of Fisheries, University of Agriculture and Forestry, Hue University, Hue 530000, Vietnam; hanamthang@huaf.edu.vn

[4] Environmental Research Institute, School of Science, University of Waikato, Hamilton 3260, New Zealand

[5] Remote Sensing Application Department, Space Technology Institute, Vietnam Academy of Science and Technology (VAST), 18 Hoang Quoc Viet street, Cau Giay district, Hanoi 100000, Vietnam; nvluong@sti.vast.vn (L.V.N.); tttu@sti.vast.vn (T.T.T.); hxtrinh@sti.vast.vn (H.X.T.)

[6] Thai Nguyen University of Sciences, Tan Thinh Ward, Thai Nguyen City Thai Nguyen University of Sciences, Tan Thinh Ward, Thai Nguyen City 250000, Vietnam; lapkq@tnus.edu.vn

[7] Institute of Industrial Science, the University of Tokyo, 4-6-1 Komaba, Meguro-ku, Tokyo 153-8505, Japan; wataru@iis.u-tokyo.ac.jp

* Correspondence: lnnga@imech.vast.vn

Received: 22 January 2020; Accepted: 25 February 2020; Published: 29 February 2020

Abstract: This study investigates the effectiveness of gradient boosting decision trees techniques in estimating mangrove above-ground biomass (AGB) at the Can Gio biosphere reserve (Vietnam). For this purpose, we employed a novel gradient-boosting regression technique called the extreme gradient boosting regression (XGBR) algorithm implemented and verified a mangrove AGB model using data from a field survey of 121 sampling plots conducted during the dry season. The dataset fuses the data of the Sentinel-2 multispectral instrument (MSI) and the dual polarimetric (HH, HV) data of ALOS-2 PALSAR-2. The performance standards of the proposed model (root-mean-square error (RMSE) and coefficient of determination (R^2)) were compared with those of other machine learning techniques, namely gradient boosting regression (GBR), support vector regression (SVR), Gaussian process regression (GPR), and random forests regression (RFR). The XGBR model obtained a promising result with $R^2 = 0.805$, RMSE = 28.13 Mg ha^{-1}, and the model yielded the highest predictive performance among the five machine learning models. In the XGBR model, the estimated mangrove AGB ranged from 11 to 293 Mg ha^{-1} (average = 106.93 Mg ha^{-1}). This work demonstrates that XGBR with the combined Sentinel-2 and ALOS-2 PALSAR-2 data can accurately estimate the mangrove AGB in the Can Gio biosphere reserve. The general applicability of the XGBR model combined with multiple sourced optical and SAR data should be further tested and compared in a large-scale study of forest AGBs in different geographical and climatic ecosystems.

Keywords: Sentinel-2; ALOS-2 PALSAR-2; mangrove; above-ground biomass; extreme gradient boosting; Can Gio biosphere reserve; Vietnam

---

1. Introduction

Mangrove forests are among the most important components of natural ecosystems. They perform a wide range of crucial functions, such as mitigating the effects of tropical typhoons and tsunami, reducing coastal erosion, and storing huge amounts of blue carbon [1,2]. Despite their functions and benefits, mangrove forests have been reduced and degraded worldwide, more seriously in South East Asia, where the decimation rate reached its highest level in the last 50 years [3,4]. The driving factors of mangrove deforestation and degradation are conversion to shrimp aquaculture, agriculture (particularly rice and oil palm in West Africa and Southeast Asia), urban development, poor governance, and overexploitation [3,5]. Unfortunately, the loss of mangrove carbon on large spatial scales is little understood. Without this knowledge, we cannot mitigate the global loss of mangrove habitats [6].

Land-cover change is thought to alter the above-ground biomass (AGB) in the tropical areas [7–9]. By mapping the spatial distribution of mangrove AGB and the carbon stocks associated with external factors, we could detect the changes in mangrove ecosystems, better understand the drivers of these changes, and reduce the uncertainty in estimating the loss of mangrove ecosystem services. A precise estimation of mangrove AGB is required for sustainably preserving and protecting mangrove ecosystems from loss and degradation under climate change and accelerated global warming. However, the complex structure of mangrove ecosystems hindered quantitative estimates of mangrove AGB. Especially, the biosphere reserves of mangroves are characterized by multiple species, very high diversity, and large spatial distributions. During the last 30 years, AGB retrieval of mangroves has been investigated worldwide [10–14]. Mangrove AGB can be accurately estimated from field-based measurements or forest inventory data. However, these approaches are disadvantaged by high cost and site-selection biases [15]. Cost-effective and accurate retrieval techniques for mangrove AGB in tropical and semi-tropical areas would provide baseline data for the monitoring, reporting, and verification schemes adopted in climate-change mitigation strategies, such as Blue Carbon projects and the United Nations' Reducing Emissions from Deforestation and Forest Degradation (REDD+) program in the tropics [16].

In recent years, mangrove AGBs have been increasingly mapped using earth observation (EO) data collected by optical sensors [17–19], synthetic aperture radar (SAR) data [13,20,21], airborne LiDAR [22,23], and LiDAR data acquired form unmanned aerial vehicles (UAV) [24,25]. A few attempts combined the data of multispectral and SAR sensors for mangrove AGB retrieval in tropical regions. Fused data are particularly useful in biosphere reserves comprising multiple mangrove species and rich biodiversity. In such systems, the spatial distribution of the mangrove AGB is difficult to estimate with sufficient accuracy. By accurately estimating the mangrove AGB in biosphere reserves, we could effectively monitor their mangrove ecosystems and implement sustainable mangrove conservation and management.

Models for estimating AGB range from simple to multi-linear regression approaches [13,21,24] to sophisticated machine learning (ML) methods [17,18,26]. For mapping and estimating forest AGBs, non-parametric approaches using various ML algorithms have proven more effective than parametric methods using linear models. Meanwhile, numerous EO datasets have been compiled from optical, SAR, and LiDAR data. These data are commonly retrieved from non-parametric regression techniques such as the random forest regression (RFR) algorithm [17,25,27], artificial neuron networks (ANN) [26], and support vector regression (SVR) [28,29]. Recently, gradient boosting decision trees (GBDT) effectively solved regression problems such as evaporation prediction [30] and oil price estimation [31]. The extreme gradient boosting regression (XGBR) algorithm is a particularly potent tool in environmental problems in environmental problems such as urban heat islands [32], algal blooming [33], and energy-supply security issues [34]. However, to our knowledge, the usefulness of the XGBR algorithm in forest AGB estimation, particularly in tropical mangrove habitats, has not been quantified. Especially, the current literature seems to lack a quantitative comparison of state-of-the-art ML techniques for estimating AGBs in different forest ecosystems.

To overcome these challenges, we estimated the mangrove AGB in the Can Gio biosphere reserve (South Vietnam) using an ML model and the fused data of the Sentinel-2 (S2) MSI and ALOS-2 PALSAR-2 sensors. We selected Sentinel-2 MSI because the multispectral bands of S-2 reflect the forest stand structures such as stem volume, whereas the longer wavelengths of the dual polarimetric (HH, HV) mode of the ALOS-2 PALSAR-2 sensor can penetrate mangrove forest canopies. The fused S2 MSI and ALOS-2 PALSAR-2 data were processed by a nonlinear regression model in the XGBR algorithm, providing the first estimation of mangrove AGB in the Can Gio biosphere reserve (CGBRS). Additionally, the performance of the XGBR model was compared with those of other GBDT techniques and several well-known ML algorithms (SVR, GPR, and RFR) on mangrove AGB estimation in the same study area. Incorporating the S-2 MSI and ALOS-2 PALSAR-2 data into the proposed model was found to improve the mangrove AGB estimation in a Vietnamese biosphere reserve and is potentially applicable to mangrove conservation in other biosphere reserves.

2. Materials and Methods

2.1. Study Area

The present study was conducted in Can Gio, a coastal district located approximately 50 km south of Ho Chi Minh City (formerly Sai Gon) along the Southern coast of Vietnam. The geographical coordinates are 10°22′–10°40′ latitude and 106°46′–107°01′ longitude. The climate is tropical monsoon and has two typical seasons. The dry season begins in April and ends in November of the following year, whereas the rainy season occurs between May and October. The average temperature is approximately 26 °C, the annual rainfall is roughly 1300–1400 mm, and the relative humidity is approximately 80% [35]. This district is well-known for its mangrove reforestation and rehabilitation programs, not only in Vietnam but also throughout Southeast Asia [36]. The wetland ecosystem of Can Gio is diverse and includes the mangrove areas distributed in zone IV, which contains the largest mangrove forest among the four mangroves zones (See Figure 1) in Vietnam [37].

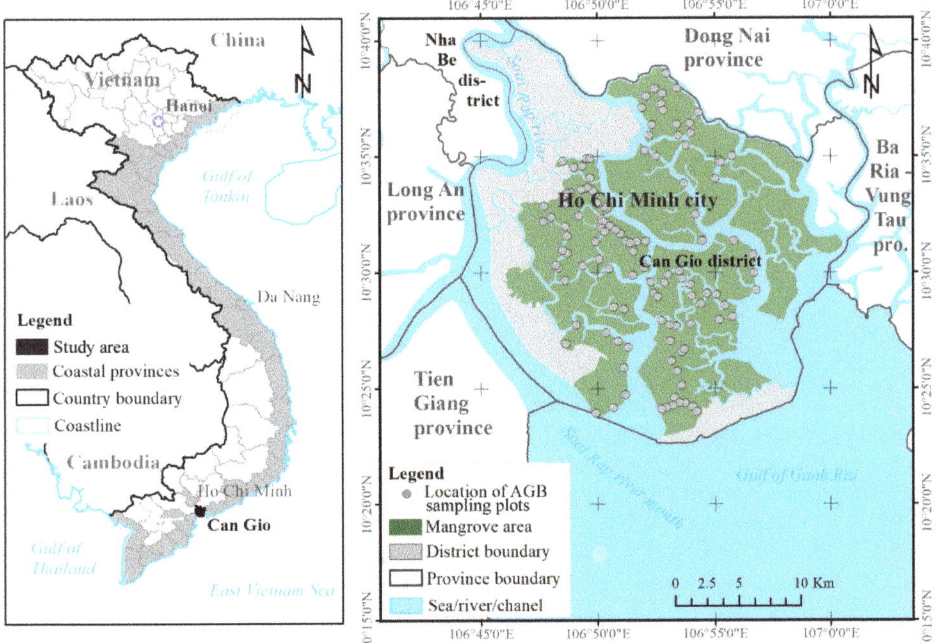

Figure 1. Location map of study areas.

The Can Gio mangrove forests were declared as a biosphere reserve by the United Nations Educational, Scientific, and Cultural Organization (UNESCO) in 2000 [38]. The dominant species are *Rhizophora apiculate, Sonneratia alba, Avicennia alba, Rhizophora mucronata,* and others. Approximately 33 species belonging to 15 families have been identified in the CGBRS [36].

2.2. Field Survey Data Collection

With permission from the local authorities, the 2018 field survey of the CGBSR was conducted during the dry season, when the coastal tides impacting the mangrove forest were lowest. A total of 121 plots were sampled by the stratified random sampling approach. Each plot sampling was initially assisted by a local counterpart to guarantee the whole range of AGB values over the reserve. During the surveying, the experimenters measured the diameter at breast height (DBH), tree height (H), and tree density. All living mangrove forest stands with DBH > 5 cm in a strata plot size of 25 × 20 m (0.05 ha) were measured. The location (accuracy ± 2 m) of each sampling plot was measured by the Garmin eTrex global positioning system (GPS) (Figure 2).

(a) (b)

Figure 2. Aboveground biomass measurements in the study area. (a & b) Biophysical parameters measurement (Photographs were taken by L.V. Nguyen during the 2018 dry season).

The mangrove AGB of each species was estimated by a specific allometric equation (see Table 1).

Table 1. Allometric equations for estimating the mangrove species in the study site.

Species	Allometric Equation	Reference
Rhizophora apiculata	$AGB = 0.235 \times DBH^{2.42}$ ($R^2 = 0.98$)	[39]
Avicennia alba	$AGB = 0.140 \times DBH^{2.40}$ ($R^2 = 0.97$)	[40]
Bruguiera gymnorrhiza	$AGB = 0.186 \times DBH^{2.31}$ ($R^2 = 0.99$)	[41]
Bruguiera parviflora	$AGB = 0.168 \times DBH^{2.42}$ ($R^2 = 0.99$)	[41]
Sonneratia caseolaris	$AGB = 0.199 \times \varphi \times 0.90 * DBH^{2.22}$ ($R^2 = 0.99$)	[40]
Lumnitzera racemosa	$AGB = 0.740 \times DBH^{2.32}$ ($R^2 = 0.99$)	[42]
Ceriops zippeliana	$AGB = 0.208 \times DBH^{2.36}$ ($R^2 = 0.96$)	[43]
Xylocarpus granatum	$AGB = 0.082 \times DBH^{2.59}$ ($R^2 = 0.99$)	[41]

Note: AGB is the above-ground biomass (kg) of a mangrove tree, DBH is the diameter (cm) at breast height (1.3 m), φ is the wood density (tons dry matter per m^3 fresh volume).

2.3. Remote Sensing Data Acquisition and Image Processing

2.3.1. Data Acquisition

The mangrove AGB in the CGBRS was estimated by fusing the ALOS-2 PALSAR-2 L-band dual polarimetric data level 2.1 obtained in high-sensitivity mode with Sentinel-2 (S-2) MSI images. Table 2 presents the S-2 and the ALOS-2 PALSAR-2 data at the study site, acquired on 23 and 24 March during the 2018 dry seasons, respectively.

Table 2. Acquired earth observation data for this study.

Earth Observation Sensor	Scene ID	Acquisition Data	Processing Level	Spectral Band/Polarizations
ALOS-2 PALSAR-2	ALOS2206940200	23 March 2018	2.1	L band (HH, HV)
	ALOS2206940190			
Sentinel-2 MSI	S2A_MSI_T48PXS	24 March 2018	1C	11 Multispectral bands
	S2A_MSI_T48PYS			

To pre-process the satellite remotely sensed data, we resampled both multispectral bands of Sentinel-2 and the dual polarization model of ALOS-2 PALSAR-2 data at a ground sampling distance (GSD) of 10 m. The satellite images were processed as described in Section 2.3.2. To validate the model's performance and optimize the hyperparameters for AGB retrieval in the CGBRS, the model was combined with the measured field data. Figure 3 is a flowchart of the satellite-image processing and the generation of mangrove AGB estimation models using the ML techniques in the current study.

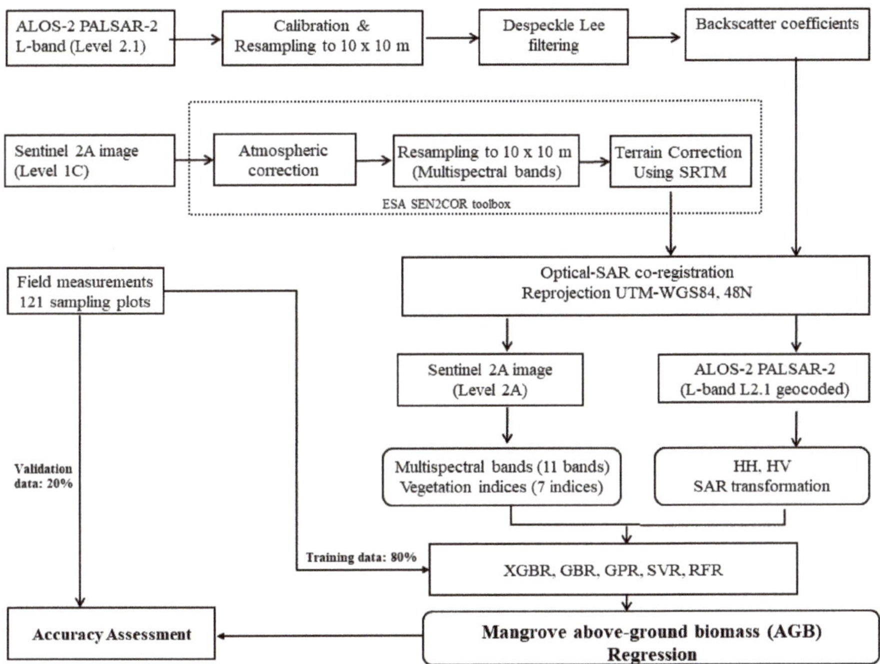

Figure 3. Flowchart for satellite-image processing and the generation of AGB models based on ML techniques.

2.3.2. Satellite Image Processing

Two scenes of the ALOS-2 PALSAR-2 Level 2.1 data acquired on 23 March 2018 during the dry season were download from https://auig2.jaxa.jp/ips/home, the website of the Aerospace Exploration Agency (JAXA). The DN (Digital Number) of the ALOS-2 PALSAR-2 imagery was converted to normalized radar sigma-zero using Equation (1):

$$\sigma^0 \text{ [dB]} = 10.\ \log 10\ (DN)^2 + CF \tag{1}$$

where σ^0 is backscatter coefficients, and CF is the calibration factor. For HH and HV polarizations, CF = −83 dB [44]. Equation (1) converts the DN of each pixel to sigma naught (σ^0) in decibel (dB).

Two scenes of the Sentinel-2 (S-2) Level-1C sensors acquired on 24 March 2018 during the dry season were retrieved from Copernicus Open Access Hub of the European Space Agency (ESA). The radiometric and geometric corrections of the S-2 data were made to the UTM/WGS84, Zone 48 North projection at top-of-atmosphere (TOA) reflectance [45]. The S-2 MSI Level-1C data were processed to Level-2A at the bottom-of-atmospheric (BOA) reflectance using the Sen2Cor algorithm of ESA (http://step.esa.int/main/third-party-plugins-2/sen2cor/). The S-2 and ALOS-2 PALSAR-2 images were processed by the SNAP toolbox, and the modeling process was performed in Python 3.7 environment using the Scikit-learn library [46].

2.3.3. Transformation of Multispectral and SAR Data

As a commonly employed method in previous mangrove AGB retrievals [13,47,48], image transformation was applied to the multispectral and SAR data of the present study. The image transformation of SAR data involves a combination of multi-polarizations such as HV/HH, HH/HV, and HH-HV, as suggested in [26]. Meanwhile, multispectral data are transformed using the vegetation indices, as each index is sensitive to mangrove structure and biomass. Table 3 shows the seven vegetation indices chosen for mangrove AGB retrieval at the CGBRS after referring to related studies [49–51]. The 23 predictor variables included five variables of ALOS-2 PALSAR-2 data (HV, HH, HV/HH, HH/HV, and HH-HV), 11 multispectral bands of S-2, and seven vegetation indices. Using the predictor variables, we computed the explanatory variables in the prediction model of mangrove AGB retrieval (Table 3). Figure 4 illustrates the image composites of different sensors and vegetation indices, along with the SAR transformation, in the study area.

Table 3. List of vegetation indices used in the current study.

Vegetation Index	Acronyms	Formula	References
Ratio Vegetation Index	RVI	$\frac{Band8}{Band4}$	[28]
Normalized Difference Vegetation Index	NDVI	$\frac{Band8-Band4}{Band8+Band4}$	[29]
Soil Adjusted Vegetation Index	SAVI	$(1+L)\left(\frac{Band8-Band4}{Band8+2.4Band4+L}\right) L = 0.5$ in most conditions	[31]
Normalized Difference Index using bands 4 and 5 of Sentinel-2	NDI45	$\frac{Band5-Band4}{Band5+Band4}$	[32]
Difference Vegetation Index	DVI	Band 8–Band 4	[33]
Green Difference Vegetation Index	GNDVI	$\frac{Band8-Band3}{Band8+Band3}$	[34]
Inverted Red-Edge Chlorophyll Index	IRECl	$\frac{Band7-Band4}{Band5/Band6}$	[35]

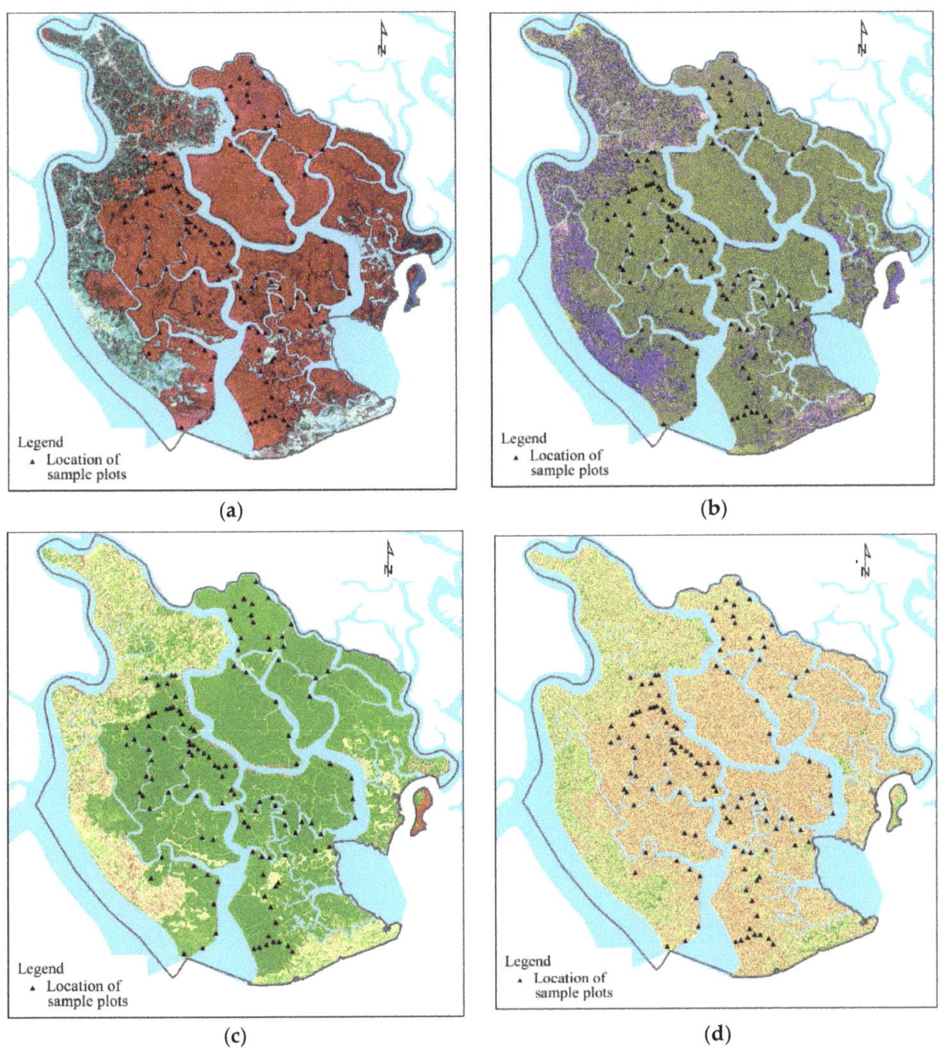

Figure 4. Illustrations of input variables in the study area. (**a**) Pseudo color composite of Sentinel-2 (RGB: Bands 8-4-3), (**b**) Pseudo color composite of ALOS-2 PALSAR-2 (RGB: HH-HV-HH/HV), (**c**) NDVI, (**d**) SAR transformation (HH-HV).

2.4. Selection of Machine Learning Model

To identify the best model for AGB retrieval in CGBSR, we compared the performances of several ML techniques (XGBR, GBR, GPR, RFR, and SVR). The SVR model best predicted the mangrove AGB in a coastal area of North Vietnam [9], whereas the RFR model delivered the best monitoring results of mangrove biomass changes in South Vietnam [10]. Therefore, SVR and RFR were selected for the present study. The other ML algorithms were chosen because they are commonly used for solving regression problems in various fields [40–42].

2.4.1. Gradient Boosting Decision Trees Algorithms

a. Gradient Boosting Regression (GBR)

GBR is an ensemble-based decision tree method that boosts the performance of weak learners to those of stronger ones. Each regression tree of the GBR learns the residual of each tree conclusion. The main purpose is to reduce the previous residuals and thereby decrease the model residual along the gradient direction. The results of all regression trees are integrated to give the final result [52,53]. The GBR model can handle mixed data types and is robust to outliers [54]. As GBR has not been widely applied to mangrove biomass estimation, it was considered for testing in the present study.

The parameters to be determined are the learning rate, number of trees, minimum number of samples required at a leaf node, maximum depth, and the number of features for the best split. The hyperparameters of the GBR model were optimized by five-fold cross-validation (CV) techniques.

b. Extreme Gradient Boosting Regression (XGBR)

The Extreme Gradient Boosting (XGB) algorithm, proposed by Chen and Guestrin [55], is a novel GBR technique that develops strong learners by an additive training process. To resolve the drawbacks of weakly supervised learning, the additive learning is divided into two phases: A learning phase fitted to the entire input data, followed by adjustment to the residuals. The fitting process is repeated many times until the stopping criteria are achieved. This algorithm is based on "boosting decision trees", which handle both classification and regression tasks in weakly supervised machine learning by the additive training strategies. The XGBR technique alleviates the undesired over-fitting problem.

The XGBR algorithm optimizes the loss function not by the first-order derivative (as in GBR) but by an efficient second-order expression. To avoid the over-fitting problem, the objective function treats the model complexity as a regularization term, and the regular term is added to the cost functions [55]. The XGBR model is quite generalizable and avoids both over-fitting and under-fitting. It also supports parallel computing to reduce computational time.

The parameters of XGBR are those of the GBR algorithm, and an additional parameter gamma (γ) representing the minimum loss of further partitioning a leaf node of the tree. The larger the γ, the more conservative is the algorithm. The XGBR model was also optimized by five-fold CV in the Python environment.

2.4.2. Support Vector Regression (SVR)

SVM is a supervised learning technique based on the statistical learning theory developed by Vapnik [56]. This method is widely used for classification and regression tasks in computer vision, pattern recognition, and environmental problems. SVR is an SVM method that solves specific regression problems. A nonlinear kernel function in SVR transforms the dataset into a higher dimensional feature space, where the data can be treated by simple linear regression. In this study, the selected kernel function was the radial basis function (RBF), the most widely adopted kernel for optimizing forest AGBs in prior studies [29,50].

The SVR model is generally configured by three hyperparameters: Epsilon (ε), the regulation parameter (C), and the kernel width (γ) of the RBF. In the present study, these parameters were optimized through five-fold CV.

2.4.3. Random Forests (RF)

RF [57] is the most common bagging model applied to both classification and regression problems. For training, RFR creates multiple uncorrelated trees from a randomly selected subset of 2/3 of the total samples (in-bag). The remaining 1/3 of the total samples (out-of-bag, OOB) are used for estimating the OOB error and validating the method. A tree is grown from in-bag samples with m features for optimizing the split at each node. In the absence of pruning, the tree reaches its largest possible extent.

The RFR model produces (1) an OOB error and (2) the relative importance of each variable. From these outputs, it assesses the prediction accuracy and the contribution of each variable.

RFR is a high-performance non-parametric method that processes nonlinear data without overestimation during the training and testing phases. Accordingly, it has been widely employed in remote sensing [58,59]. The RFR requires the number of trees and the number of features m for the split. In this study, both RFR parameters were optimized by five-fold CV in the Python environment.

2.4.4. Gaussian Processes (GP)

Based on the non-parametric Bayesian theory, GPs are applicable to both classification and nonlinear regression problems. The GPR model learns the fit function from a small dataset using various kernels, finding the probability distribution that best describes the data. The input data are assumed to follow a multivariate Gaussian distribution, and the noise is independent of the data measurements [60]. The mean vector and covariance matrix are estimated from the training data by mean and covariance functions, respectively, creating a detailed posterior distribution from which the confidence interval and uncertainty of the prediction results can be interpreted. The mean value of a GP represents the best estimation from the model, and the variance (σ^2) helps to measure the confidence level. GPs are well-known as good predictors of biophysical parameters [61].

2.5. Model Evaluation

2.5.1. Input Data for Model Running

To create the input data for training models, the 121 sampling plots were divided into training set (80%) and testing dataset (20%) using the well-known Scikit-learn [46] library in Python programming environment. Because the measured plot size (500 m^2) greatly exceeded the image pixel size (10 m), all satellite data were smoothed through a median filter with a window size of 5×5 pixels in the SciPy library [62].

2.5.2. Hyperparameters Tuning in XGBR, GBR, RFR, SVR, and GPR

Hyperparameter tuning is often required when optimizing machine learning techniques. In this work, the parameters of each ML model were optimized by grid searching and five-fold CV. The results are listed in Table 4.

Table 4. Optimized hyperparameters of the ML applied in this study.

Algorithm	Learning_Rate/Epsilon	Min_Samples_Leaf Min_Child_Weight	Gamma	Max_Depth/Max Features	n_Estimators or C Value
RFR	NA	2	NA	5, 15	50
SVR	0.01	NA	1000	NA	1000
GBR	0.2	5	NA	7, 3	100
XGBR	0.2	3	1	3	100

In the GPR, we combined the RBF with a length scale of 100 and WhiteKernel with a noise level of 1.0. The hyperparameters and kernels were maintained during the training and testing phases.

2.5.3. Feature Importance

The variables in RFR and gradient boosting machine algorithms, such as XGBR and GBR are often ranked by the variable-importance approach [55,63,64]. Relative variable importance is computed as follows. The first step searches for a candidate subset of variables (in this case, by the grid search approach). Initially, the grid search includes all variables derived from the S-2, VIs, and ALOS-2 PALSAR-2 datasets. The datasets are input to the XGBR model, which ranks the variables in descending order of their importance based on the root mean squared error (RMSE) and the coefficient of determination (R^2). Next, a certain number of the least important variables are removed, and the surviving variables form a variable subset. In this paper, the search/selection iterations were terminated when the R^2 of the prediction model of the subset did not improve the performance in the test set. The final step validates the selected variable subset and determines the relative variable importance (in this case, by the five-fold CV approach).

The modeling and generated variable importance of the XGBR model were implemented in the Python environment.

2.5.4. Model Evaluation

The model performances of the various ML techniques were evaluated and compared by the RMSE (Equation (2)) and R^2 (Equation (3)), which are widely employed in estimates of forest AGB biomass. Both standards evaluate the errors in a regression model from the differences between the measured data (the mangrove forest measurements) and the estimated AGB data [50]. A well-performing model will achieve a high R^2 and a low [24,47].

$$\text{RMSE} = \sqrt{\sum_{1}^{n} \frac{(ye_i - ym_i)2}{n}} \tag{2}$$

$$R^2 = \frac{\sum_{i=1}^{n} (ye_i - \overline{ye})(ym_i - \overline{ym})}{\sqrt{\sum_{i=1}^{n} (ye_i - \overline{ye})^2 (ym_i - \overline{ym})^2}} \tag{3}$$

In the above expressions, ye_i is the mangrove AGB predicted by the ML model, ym_i is the measured mangrove AGB, n is the total number of sampling plots, and $\overline{ye}$ and $\overline{ym}$ are the mean values of the predicted and measured mangrove AGBs, respectively.

3. Results

3.1. Mangrove Tree Characteristics in CGBRS

Table 5 gives the characteristics of the mangrove trees in the 121 sampling plots. The AGBs ranged from 7.26 to 305.41 Mg ha^{-1}, with a mean of 97.54 Mg ha^{-1}. The mangrove heights varied from 6.47 to 17.35 m, and their DBHs ranged from 6.69 to 22.19 cm. The mangrove tree densities ranged from 170 to 1680 trees ha^{-1} (Table 5).

Table 5. Characteristics of the mangrove trees in CGBRS.

Attribute	Min	Max	Mean	Standard Deviation (SD)
DBH (cm)	6.69	22.19	13.24	3.5
H (m)	6.47	17.35	11.87	2.5
Tree density (tree ha^{-1})	170	1680	694	26.45
AGB (Mg ha^{-1})	7.26	305.41	97.54	5.88

3.2. Modeling Results, Assessment, and Comparison

Table 6 and Figure 5 compare the performances of the five regression methods with all input variables derived from S-2 MSI, VIs, and ALOS-2 PALSAR-2 images for mangrove AGB estimation in the study area. The XGBR model incorporating the S-2 (11 MS bands), ALOS- 2 PALSAR-2 (5 bands), and VIs (7 bands) data achieved the highest performance (Table 6), with an R^2 of 0.805 and an RMSE of 28.13 Mg ha^{-1} in the testing dataset (23 predictor variables based on the fused S-2, the VIs and the ALOS-2 PALSAR-2 data), implying a good fit between the model estimates and field-based measurements. The next-highest performers were the GBR and RFR models. In contrast, the SVR and GPR models were unsuitable for retrieving the mangrove AGB at the study site (Table 6).

Table 6. Performance comparison of ML techniques on mangrove AGB estimation.

No	Machine Learning Model	R^2 Training (80%)	R^2 Testing (20%)	RMSE (Mg ha^{-1})
1	Extreme Boosting regression (XGBR)	0.992	0.805	28.13
2	Gradient Boosting regression (GBR)	0.998	0.632	39.54
3	Random Forests regression (RFR)	0.721	0.468	48.44
4	Support Vector regression (SVR)	0.480	0.421	48.49
5	Gaussian Processes regression (GPR)	0.509	0.378	50.23

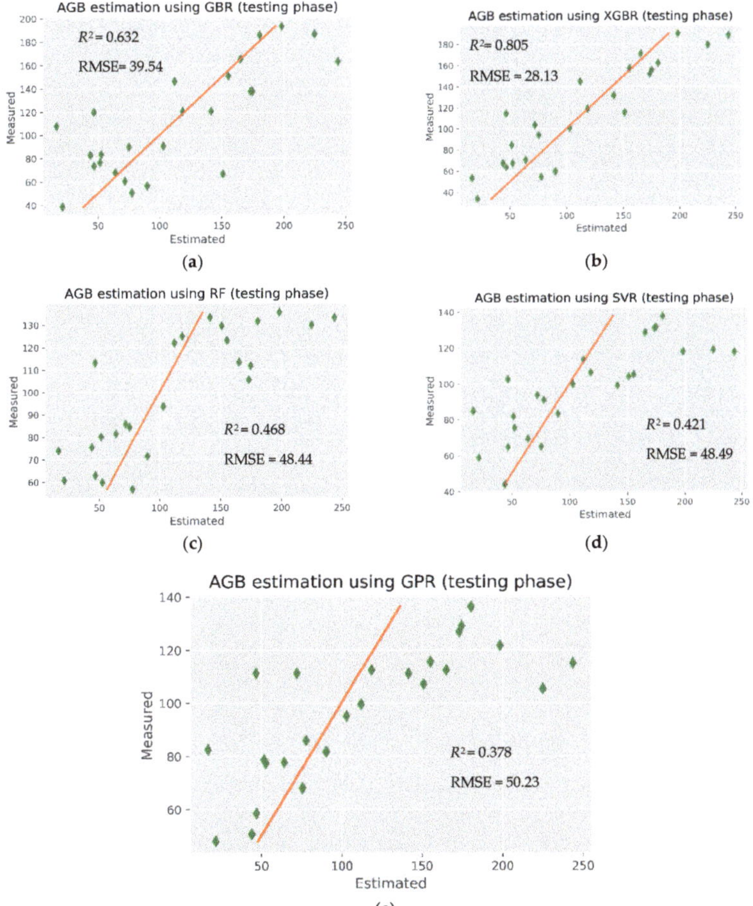

Figure 5. Scatter plots of the estimated (X axis) versus the measured (Y axis) mangrove AGB in the five ML models, integrating the data of S-2, ALOS-2 PALSAR-2, and VIs in the testing phase. (**a**) GBR, (**b**) XGBR, (**c**) RFR, (**d**) SVR, (**e**) GPR.

Table 7 lists the performances of the XGBR method in five scenarios (SCs) of mangrove AGB prediction, using different combinations of the S-2, ALOS-2 PALSAR-2, and VIs data.

Table 7. Performance of the XGBR model using different numbers of variables. (Bold values highlight the best-performing model).

Scenario (SC)	Number of Variables	R^2 Testing Set	RMSE (Mg ha^{-1})
SC1	11 variables from MS bands of S2 data	0.600	36.54
SC2	5 variables from ALOS-2 PALSAR-2 data	0.492	39.48
SC3	18 variables from MS bands and VIs from S2	0.739	34.86
SC4	23 variables (11 MS bands + 7 vegetation indices + 5 bands from ALOS-2 PALSAR-2)	0.805	28.13
SC5	16 variables (11 MS bands + 5 bands from ALOS-2 PALSAR-2)	0.656	43.25

As clarified in Table 7, the XGBR model yielded a promising result in SC3 using the combined S-2 and VIs, but the model achieved a poor result in SC2 using the ALOS-2 PALSAR-2 alone. The performance in SC1 using the S-2 dataset alone was moderate. We concluded that fusing all data in SC4 boosted the prediction performance of XGBR for estimating the mangrove AGB in the study area. The visual results of the testing phase (Figure 5) reconfirm the high performance of mangrove AGB estimation by XGBR with the 23 variables of the fused data. Particularly, the green scatter points cluster around the blue line and the RMSE is small.

3.3. Variable Importance

Among the multispectral bands of S-2 MSI, the Red (665 nm), Vegetation Red Edge (704 nm), and the narrow NIR (864 nm) spectra were most sensitive to the mangrove AGB of the present study, followed by the SWIR spectrum (MS band 11 at 1610 nm). Interestingly, among the seven VIs indices, the Inverted Red-Edge Chlorophyll Index (IRECl) and the Normalized Difference Index (NDI45) (bands 4 and 5 of S-2) were likely sensitive to the mangrove AGB in the study area. The band ratios derived from the incorporated HH and HH polarizations in the ALOS-2 PALSAR-2 data were also important for retrieving mangrove AGB in the biosphere reserve (see Figure 6). The backscatter coefficients of the crossed-polarimetric HV in ALOS-2 PALSAR-2 are likely more important than those of the HH for estimating the mangrove AGB in the study region (Figure 6).

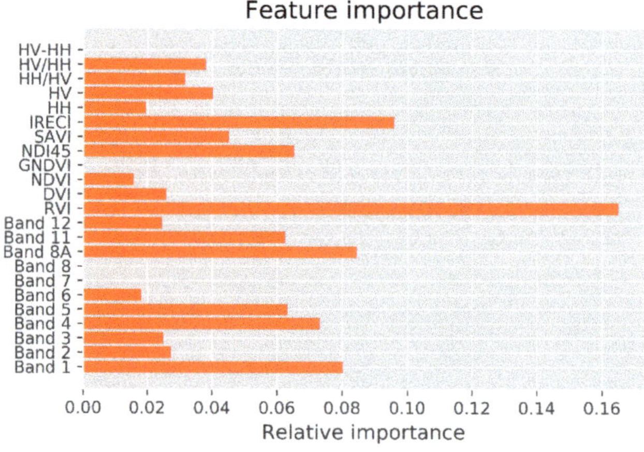

Figure 6. Variable importance comparison of S-2, VIs, and ALOS-2 PALSAR-2 data in this study.

3.4. Generation and Analysis of the AGB Map

The prediction performance of the XGBR model in mangrove AGB retrieval was improved by integrating the Sentinel-2 multispectral bands, vegetation indices, and ALOS-2 PALSAR-2 datasets. Thus, the XGBR model was selected for retrieving mangrove AGB in a biosphere reserve. The final results were computed to a raster in GeoTiff format for visualizing in QGIS. The AGB map was interpreted by seven classes (Figure 7), obtaining mangrove AGBs from 11 to 293 Mg ha^{-1} (average = 106.93 Mg ha^{-1}). As can be seen from Figure 7, the biomass is highest in the core zone of the biosphere reserve and lower in the transition and buffer zones. These results are consistent with prior mangrove AGB estimates [17] and [65], in which the high biomass was mainly distributed in the core zone of the biosphere reserve, and the lower biomass was observed in the remaining zones.

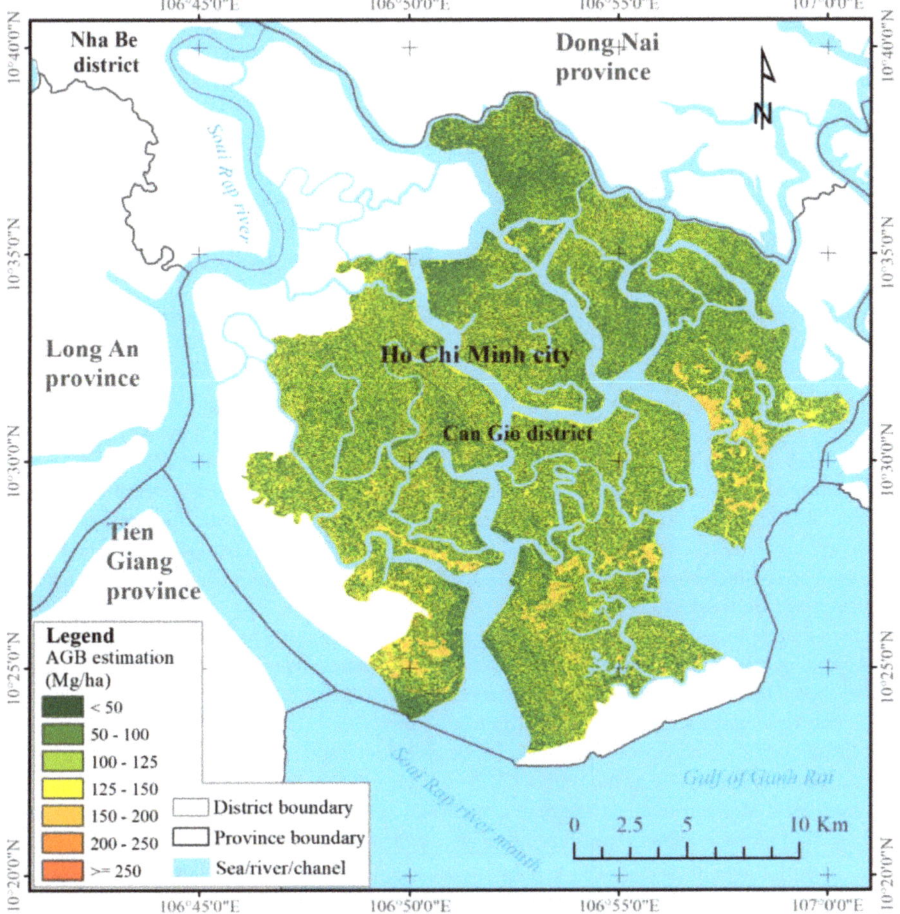

Figure 7. Estimated spatial distribution of mangrove AGB in the study area.

4. Discussion

The modeling results of mangrove AGB retrieval in the CGBSR obtained by the five ML models (XGBR, GBR, GPR, SVR, and RFR) are given in Table 6. Clearly, the XGBR model yielded the highest performance, with an R^2 and RMSE of 0.805 and 28.13 Mg ha^{-1}, respectively. The worst performing model was GPR, with an R^2 and RMSE of 0.378 and 50.23 Mg ha^{-1}, respectively. Both the XGBR model

($R^2 = 0.805$) and GBR model ($R^2 = 0.632$) were good predictors of mangrove AGB, indicating that the GBDT regression models were applicable to the study area, where the mangrove biomass is higher than in other mangrove regions of Vietnam. As shown in Table 7, the combined S-2 and ALOS-2 PALSAR data significantly improved the performance of estimating the mangrove AGB in the study area. These results are consistent with a recent previous study [50]. Overall, the XGBR model outperformed the existing algorithms in retrieving the mangrove AGB in a Vietnamese biosphere reserve.

Previous studies reported that long-wavelength PolSAR data, such as the L and the P bands, are well correlated with mangrove forest structures. Among these data, crossed-polarized HV appears to be most correlated with biophysical attributes [13,66,67]. The variable-importance analysis revealed that crossed-polarization HV is more sensitive to mangrove AGB in the study area than HH polarization (Figure 6), consistent with previous results [26,29]. However, mangrove forests in a biosphere reserve exhibit unique stand structures and species compositions that may saturate multispectral and SAR sensors. Data saturation of multispectral sensors such as Landsat TM, ETM+ or OLI, and the S-2 sensor degrades the prediction accuracy of mangrove AGBs in dense forest canopies. The saturation range of multispectral data reaches 100–150 Mg ha^{-1} in complex tropical forests, much higher than in mixed and pine forest ecosystems (with a saturation range of >150 to <160 Mg ha^{-1}) [68,69]. In several recent investigations, the saturation levels of the mangrove AGBs retrieved from SAR data ranged from above 100 Mg ha^{-1} [20] to below 150 Mg ha^{-1} [21,26]. This large range probably manifests from the root systems of different mangrove species in intertidal tropical and sub-tropical regions [13]. The sigma backscatter coefficients of the dual polarimetric data of ALOS-2 PALSAR-2 increased when the mangrove AGB fell below 100 Mg ha^{-1} and then saturated at a higher AGB because the high mangrove cover density extinguished the radar signals [70,71].

Biosphere reserves often consist of various mangrove species. The species types (i.e., *R. appiculata*, *B. gymnorrhiza*, and *S. caseolaris*) are densely grown and characterized by high DBH and tall height. Some species, such as *A. germinans* and *C. decandra*, form small but high-density mangrove patches in which high and low biomasses are easily underestimated and overestimated, respectively, by machine learning algorithms. In the current study, the XGBR model possibly over-estimated the low mangrove AGBs (below 50 Mg ha^{-1}) and under-estimated the high values (over 250 Mg ha^{-1}). Despite these limitations, the combined ALOS-2 PALSAR-2 and S-2 data sensitively detected mangrove AGBs exceeding 200 Mg ha^{-1} in the CGBRS (See Figure 5). Our findings agree with the conclusions of prior research on biosphere reserves [17,65]. Given the species complexity in mangrove biosphere reserves, we recommend the inclusion of species classification or richness indices for improved mangrove AGB estimation in future work [19,21].

In the variable-importance results, the mangrove AGB in the study area was largely retrieved from the Red band and the Vegetation Red Edge band. A similar result was reported elsewhere [18,72]. The vegetation red edge, narrow NIR, and SWIR reflectance are likely to be more strongly correlated with forest biomass and carbon stock volume than visible reflectance [17]. Accordingly, the new vegetation index ND145, which is computed from the Sentinel-2 data bands, is a probable sensitive indicator of mangrove AGB. Band 8A in the narrow NIR and band 11 in the SWIR (1613 nm) also played a crucial role in the AGB retrieval. Interestingly, the IRECl derived from S-2 was strongly correlated with mangrove AGB in the biosphere reserve. More in-depth studies would elucidate the effectiveness of image transformations involving new vegetation indices derived from the Narrow NIR bands, SWIR of S-2 data, and other image transformations computed from the fully polarized data (HH, HV, VH, and VV) of the Gaofeng-3 and the ALOS-2 PALSAR-2 sensors in biosphere reserves.

To accurately estimate mangrove AGBs, researchers attempted multi-linear regression, which performed poorly with R^2 ranging from 0.43–0.65 [13,21,73], and various ML algorithms such as GPR, MLPNN, SVR, and RFR [17,18,29]. ML approaches have proven more successful in mangrove AGB than multi-linear regression and other parametric methods [18,47], but the R^2 has rarely exceeded 0.70. Therefore, novel approaches for mangrove AGB estimation are urgently needed. In this research, the performance of the XGBR model was boosted by incorporating data from the ALOS-2 PALSAR-2,

S-2 sensors. The result (R^2 = 0.805 for the AGB of a mangrove biosphere reserve in the tropics) demonstrates the promise of this approach. Despite the good fit between the XGBR-predicted and measured-mean mangrove AGBs, the range of the predicted mangrove AGBs did not reach the extrema of the actual distribution range, which was maximized at 305.41 Mg ha^{-1} and minimized at 26 Mg ha^{-1} (Table 5). The predicted results may have been degraded by the saturation levels of the S2 MSI sensor and the dual polarimetric L-band ALOS-2 PALSAR-2 when retrieving mangrove AGB in intertidal areas. Although the AGB was well predicted by the XGBR model, the R^2 values in the training and testing phases were significantly different (Table 6). This difference is likely attributable to the mixed mangrove species planted in the CGBRS and the number of plots. To archive a more accurate forest AGB map, we should exploit the advantages of various novel GBDT algorithms with multi-sensor data integration [74]. In more intensive works, novel boosting decision tree techniques should exploit the full capability of multi-source EO data in different mangrove communities occupying tropical intertidal areas at different geographical locations, particularly those of biosphere reserves. Such developments are needed for rapid mangrove AGB monitoring in the future.

5. Conclusions

We report the first attempt to incorporate Sentinel-2 and ALOS-2 PALSAR-2 data into the extreme gradient boosting regression (XGBR) model and thereby estimate the mangrove AGB in Vietnam's Can Gio biosphere reserve. The XGBR model outperformed four other machine learning models in mangrove AGB retrieval in the study area. When provided with the Sentinel-2 and ALOS-2 PALSAR-2 data, XGBR estimated the mangrove AGB with satisfactory accuracy (R^2 = 0.805, RMSE = 28.13 Mg ha^{-1}). Interestingly, we found that new vegetation indices derived from the Sentinel-2 data, such as the Normalized Difference Index (NDI45) and the Inverted Red-Edge Chlorophyll Index (IRECl), sensitively detected mangrove AGB in the biosphere reserve. In future investigations, the proposed approach should be tested in other tropical forest ecosystems.

Author Contributions: Conceptualization, T.D.P., L.V.N., N.N.L.; methodology, T.D.P.; validation, T.D.P., N.N.L., N.T.H.; data analysis, N.N.L., T.D.P., N.T.H.; field investigation, L.V.N., L.Q.K., T.T.T., H.X.T.; writing—original draft preparation, T.D.P., N.N.L., N.T.H.; writing—review and editing, T.D.P., N.N.L., J.X., N.T.H., N.Y.; visualization, T.D.P., L.V.N.; supervision, N.Y., W.T., All authors have read and approved the final version of this paper. All authors have read and agreed to the published version of the manuscript.

Funding: This research received no external funding.

Acknowledgments: The authors would like to thank the Japan Aerospace Exploration Agency (JAXA) for providing the ALOS-2 PALSAR-2 data for this research under the 2nd Earth Observation Research Announcement Collaborative Research Agreement between the JAXA and RIKEN AIP. The authors are grateful to mission No. VAST 01.07/20-21 from the Vietnam Academy of Science and Technology (VAST) for data support of this research.

Conflicts of Interest: The authors declare no conflict of interest.

Abbreviations

List of abbreviations in this study

No	Abbreviation	Full Name
1	AGB	Above-Ground Biomass
2	ALOS	The Advanced Land Observing Satellite
3	ANN	Artificial Neuron Networks
4	PALSAR	Phased Array type L-band Synthetic Aperture Radar
5	TOA	Top Of Atmosphere
6	BOA	Bottom Of Atmospheric

7	CGBRS	Can Gio Biosphere Reserve in South Vietnam
8	CV	Cross-validation
9	DBH	Diameter at breast height
10	EO	Earth Observation
11	ESA	European Space Agency
12	GBDT	Gradient Boosting Decision Trees
13	GBR	Gradient Boosting Regression
14	GeoTiff	Tagged Image File Format for GIS applications
15	GP	Gaussian Processes
16	GPR	Gaussian Process Regression
17	GPS	Global Positioning System
18	JAXA	Japan Aerospace Exploration Agency
19	LiDAR	Light Detection and Ranging
20	ML	Machine Learning
21	MRV	Monitoring, Reporting, and Verification
22	MSI	Multispectral Instrument
23	NA	Not Available
24	QGIS	Quantum Geographic Information System
25	RBF	Radial Basis Function
26	REDD+	Reducing Emissions from Deforestation and Forest Degradation
27	RFR	Random Forest Regression
28	RMSE	Root Mean Square Error
29	S2	Sentinel-2
30	SAR	Synthetic Aperture Radar
31	SC	Scenarios
32	SNAP	Sentinel Application Platform
33	SVM	Support Vector Machine
34	SVR	Support Vector Regression
35	SWIR	Short-Wave InfraRed
36	VIs	Vegetation indices
37	XGB	Extreme Gradient Boosting
38	XGBR	Extreme Gradient Boosting Regression

References

1. Alongi, D.M. Carbon sequestration in mangrove forests. *Carbon Manag.* **2012**, *3*, 313–322. [CrossRef]
2. Brander, L.M.; Wagtendonk, A.J.; Hussain, S.S.; McVittie, A.; Verburg, P.H.; de Groot, R.S.; van der Ploeg, S. Ecosystem service values for mangroves in Southeast Asia: A meta-analysis and value transfer application. *Ecosyst. Serv.* **2012**, *1*, 62–69. [CrossRef]
3. Richards, D.R.; Friess, D.A. Rates and drivers of mangrove deforestation in Southeast Asia, 2000–2012. *Proc. Natl. Acad. Sci. USA* **2016**, *113*, 344–349. [CrossRef] [PubMed]
4. Friess, D.A.; Rogers, K.; Lovelock, C.E.; Krauss, K.W.; Hamilton, S.E.; Lee, S.Y.; Lucas, R.; Primavera, J.; Rajkaran, A.; Shi, S. The State of the World's Mangrove Forests: Past, Present, and Future. *Annu. Rev. Environ. Resour.* **2019**, *44*, 89–115. [CrossRef]
5. Pham, T.D.; Yoshino, K. Impacts of mangrove management systems on mangrove changes in the Northern Coast of Vietnam. *Tropics* **2016**, *24*, 141–151. [CrossRef]
6. Friess, D.A.; Webb, E.L. Variability in mangrove change estimates and implications for the assessment of ecosystem service provision. *Glob. Ecol. Biogeogr.* **2014**, *23*, 715–725. [CrossRef]
7. Lv, Z.Y.; Liu, T.F.; Zhang, P.; Benediktsson, J.A.; Lei, T.; Zhang, X. Novel Adaptive Histogram Trend Similarity Approach for Land Cover Change Detection by Using Bitemporal Very-High-Resolution Remote Sensing Images. *IEEE Trans. Geosci. Remote Sens.* **2019**, *57*, 9554–9574. [CrossRef]
8. Zhao, T.; Bergen, K.M.; Brown, D.G.; Shugart, H.H. Scale dependence in quantification of land-cover and biomass change over Siberian boreal forest landscapes. *Landsc. Ecol.* **2009**, *24*, 1299. [CrossRef]

9.   Lv, Z.; Liu, T.; Zhang, P.; Atli Benediktsson, J.; Chen, Y. Land Cover Change Detection Based on Adaptive Contextual Information Using Bi-Temporal Remote Sensing Images. *Remote Sens.* **2018**, *10*, 901. [CrossRef]

10. Clough, B.F.; Dixon, P.; Dalhaus, O. Allometric Relationships for Estimating Biomass in Multi-stemmed Mangrove Trees. *Aust. J. Bot.* **1997**, *45*, 1023–1031. [CrossRef]

11. Komiyama, A.; Ong, J.E.; Poungparn, S. Allometry, biomass, and productivity of mangrove forests: A review. *Aquat. Bot.* **2008**, *89*, 128–137. [CrossRef]

12. Hirata, Y.; Tabuchi, R.; Patanaponpaiboon, P.; Poungparn, S.; Yoneda, R.; Fujioka, Y. Estimation of aboveground biomass in mangrove forests using high-resolution satellite data. *J. For. Res.* **2014**, *19*, 34–41. [CrossRef]

13. Hamdan, O.; Khali Aziz, H.; Mohd Hasmadi, I. L-band ALOS PALSAR for biomass estimation of Matang Mangroves, Malaysia. *Remote Sens. Environ.* **2014**, *155*, 69–78. [CrossRef]

14. Darmawan, S.; Takeuchi, W.; Vetrita, Y.; Wikantika, K.; Sari, D.K. Impact of Topography and Tidal Height on ALOS PALSAR Polarimetric Measurements to Estimate Aboveground Biomass of Mangrove Forest in Indonesia. *J. Sens.* **2015**, *2015*, 13. [CrossRef]

15. Kauffman, J.B.; Donato, D.C. *Protocols for the Measurement, Monitoring and Reporting of Structure, Biomass, and Carbon Stocks in Mangrove Forests*; CIFOR: Bogor, Indonesia, 2012.

16. Ahmed, N.; Glaser, M. Coastal aquaculture, mangrove deforestation and blue carbon emissions: Is REDD+ a solution? *Mar. Policy* **2016**, *66*, 58–66. [CrossRef]

17. Pham, L.T.H.; Brabyn, L. Monitoring mangrove biomass change in Vietnam using SPOT images and an object-based approach combined with machine learning algorithms. *ISPRS J. Photogramm. Remote Sens.* **2017**, *128*, 86–97. [CrossRef]

18. Jachowski, N.R.A.; Quak, M.S.Y.; Friess, D.A.; Duangnamon, D.; Webb, E.L.; Ziegler, A.D. Mangrove biomass estimation in Southwest Thailand using machine learning. *Appl. Geogr.* **2013**, *45*, 311–321. [CrossRef]

19. Zhu, Y.; Liu, K.; Liu, L.; Wang, S.; Liu, H. Retrieval of Mangrove Aboveground Biomass at the Individual Species Level with WorldView-2 Images. *Remote Sens.* **2015**, *7*, 12192–12214. [CrossRef]

20. Lucas, R.M.; Mitchell, A.L.; Rosenqvist, A.; Proisy, C.; Melius, A.; Ticehurst, C. The potential of L-band SAR for quantifying mangrove characteristics and change: Case studies from the tropics. *Aquat. Conserv. Mar. Freshw. Ecosyst.* **2007**, *17*, 245–264. [CrossRef]

21. Pham, T.D.; Yoshino, K. Aboveground biomass estimation of mangrove species using ALOS-2 PALSAR imagery in Hai Phong City, Vietnam. *APPRES* **2017**, *11*, 026010. [CrossRef]

22. Maeda, Y.; Fukushima, A.; Imai, Y.; Tanahashi, Y.; Nakama, E.; Ohta, S.; Kawazoe, K.; Akune, N. Estimating carbon stock changes of mangrove forests using satellite imagery and airborne lidar data in the south Sumatra state, Indonesia. *ISPRS Int. Arch. Photogramm. Remote Sens. Spat. Inf. Sci.* **2016**, 705–709. [CrossRef]

23. Fatoyinbo, T.; Feliciano, E.A.; Lagomasino, D.; Lee, S.K.; Trettin, C. Estimating mangrove aboveground biomass from airborne LiDAR data: A case study from the Zambezi River delta. *Environ. Res. Lett.* **2018**, *13*, 025012. [CrossRef]

24. Wang, D.; Wan, B.; Liu, J.; Su, Y.; Guo, Q.; Qiu, P.; Wu, X. Estimating aboveground biomass of the mangrove forests on northeast Hainan Island in China using an upscaling method from field plots, UAV-LiDAR data and Sentinel-2 imagery. *Int. J. Appl. Earth Obs. Geoinf.* **2020**, *85*, 101986. [CrossRef]

25. Wang, D.; Wan, B.; Qiu, P.; Zuo, Z.; Wang, R.; Wu, X. Mapping Height and Aboveground Biomass of Mangrove Forests on Hainan Island Using UAV-LiDAR Sampling. *Remote Sens.* **2019**, *11*, 2156. [CrossRef]

26. Pham, T.D.; Yoshino, K.; Bui, D.T. Biomass estimation of Sonneratia caseolaris (L.) Engler at a coastal area of Hai Phong city (Vietnam) using ALOS-2 PALSAR imagery and GIS-based multi-layer perceptron neural networks. *Gisci. Remote Sens.* **2017**, *54*, 329–353. [CrossRef]

27. Wu, C.; Shen, H.; Shen, A.; Deng, J.; Gan, M.; Zhu, J.; Xu, H.; Wang, K. Comparison of machine-learning methods for above-ground biomass estimation based on Landsat imagery. *APPRES* **2016**, *10*, 035010. [CrossRef]

28. López-Serrano, P.M.; López-Sánchez, C.A.; Álvarez-González, J.G.; García-Gutiérrez, J. A Comparison of Machine Learning Techniques Applied to Landsat-5 TM Spectral Data for Biomass Estimation. *Can. J. Remote Sens.* **2016**, *42*, 690–705. [CrossRef]

29. Pham, T.D.; Yoshino, K.; Le, N.; Bui, D. Estimating Aboveground Biomass of a Mangrove Plantation on the Northern coast of Vietnam using machine learning techniques with an integration of ALOS-2 PALSAR-2 and Sentinel-2A data. *Int. J. Remote Sens.* **2018**, *39*, 7761–7788. [CrossRef]

30. Huang, G.; Wu, L.; Ma, X.; Zhang, W.; Fan, J.; Yu, X.; Zeng, W.; Zhou, H. Evaluation of CatBoost method for prediction of reference evapotranspiration in humid regions. *J. Hydrol.* **2019**, *574*, 1029–1041. [CrossRef]
31. Gumus, M.; Kiran, M.S. Crude oil price forecasting using XGBoost. In Proceedings of the 2017 International Conference on Computer Science and Engineering (UBMK), Antalya, Turkey, 5–7 October 2017; pp. 1100–1103.
32. Sun, Y.; Gao, C.; Li, J.; Wang, R.; Liu, J. Evaluating urban heat island intensity and its associated determinants of towns and cities continuum in the Yangtze River Delta Urban Agglomerations. *Sustain. Cities Soc.* **2019**, *50*, 101659. [CrossRef]
33. Ghatkar, J.G.; Singh, R.K.; Shanmugam, P. Classification of algal bloom species from remote sensing data using an extreme gradient boosted decision tree model. *Int. J. Remote Sens.* **2019**, *40*, 9412–9438. [CrossRef]
34. Li, P.; Zhang, J.-S. A New Hybrid Method for China's Energy Supply Security Forecasting Based on ARIMA and XGBoost. *Energies* **2018**, *11*, 1687. [CrossRef]
35. Veettil, B.K.; Ward, R.D.; Quang, N.X.; Trang, N.T.T.; Giang, T.H. Mangroves of Vietnam: Historical development, current state of research and future threats. *Estuar. Coast. Shelf Sci.* **2019**, *218*, 212–236. [CrossRef]
36. Tuan, L.D.; Oanh, T.T.K.; Thanh, C.V.; Quy, N.D. *Can Gio Mangrove Biosphere Reserve*; Agricultural Publishing House: Ho Chi Minh City, Vietnam, 2002; p. 311.
37. Hong, P.N.; San, H.T. *Mangroves of Vietnam*; IUCN: Bangkok, Thailand, 1993; p. 173.
38. Vogt, J.; Kautz, M.; Fontalvo Herazo, M.L.; Triet, T.; Walther, D.; Saint-Paul, U.; Diele, K.; Berger, U. Do canopy disturbances drive forest plantations into more natural conditions?—A case study from Can Gio Biosphere Reserve, Viet Nam. *Glob. Planet. Chang.* **2013**, *110*, 249–258. [CrossRef]
39. Ong, J.E.; Gong, W.K.; Wong, C.H. Allometry and partitioning of the mangrove, Rhizophora apiculata. *For. Ecol. Manag.* **2004**, *188*, 395–408. [CrossRef]
40. Komiyama, A.; Poungparn, S.; Kato, S. Common allometric equations for estimating the tree weight of mangroves. *J. Trop. Ecol.* **2005**, *21*, 471–477. [CrossRef]
41. Clough, B.F.; Scott, K. Allometric relationships for estimating above-ground biomass in six mangrove species. *For. Ecol. Manag.* **1989**, *27*, 117–127. [CrossRef]
42. Kangkuso, A.; Jamili, J.; Septiana, A.; Raya, R.; Sahidin, I.; Rianse, U.; Rahim, S.; Alfirman, A.; Sharma, S.; Nadaoka, K. Allometric models and aboveground biomass of Lumnitzera racemosa Willd. forest in Rawa Aopa Watumohai National Park, Southeast Sulawesi, Indonesia. *For. Sci. Technol.* **2016**, *12*, 43–50. [CrossRef]
43. Binh, C.H.; Nam, V.N. Carbon sequestration of Ceriops zippeliana in Can Gio mangroves. In *Studies in Can Gio Mangrove Biosphere Reserve, Ho Chi Minh City, Viet Nam*; ISME: Okinawa, Japan, 2014; p. 51.
44. Shimada, M.; Isoguchi, O.; Tadono, T.; Isono, K. PALSAR Radiometric and Geometric Calibration. *IEEE Trans. Geosci. Remote Sens.* **2009**, *47*, 3915–3932. [CrossRef]
45. Drusch, M.; Del Bello, U.; Carlier, S.; Colin, O.; Fernandez, V.; Gascon, F.; Hoersch, B.; Isola, C.; Laberinti, P.; Martimort, P.; et al. Sentinel-2: ESA's Optical High-Resolution Mission for GMES Operational Services. *Remote Sens. Environ.* **2012**, *120*, 25–36. [CrossRef]
46. Pedregosa, F.; Varoquaux, G.; Gramfort, A.; Michel, V.; Thirion, B.; Grisel, O.; Blondel, M.; Prettenhofer, P.; Weiss, R.; Dubourg, V. Scikit-learn: Machine learning in Python. *J. Mach. Learn. Res.* **2011**, *12*, 2825–2830.
47. Navarro, J.A.; Algeet, N.; Fernández-Landa, A.; Esteban, J.; Rodríguez-Noriega, P.; Guillén-Climent, M.L. Integration of UAV, Sentinel-1, and Sentinel-2 Data for Mangrove Plantation Aboveground Biomass Monitoring in Senegal. *Remote Sens.* **2019**, *11*. [CrossRef]
48. Castillo, J.A.A.; Apan, A.A.; Maraseni, T.N.; Salmo, S.G. Estimation and mapping of above-ground biomass of mangrove forests and their replacement land uses in the Philippines using Sentinel imagery. *ISPRS J. Photogramm. Remote Sens.* **2017**, *134*, 70–85. [CrossRef]
49. Patil, V.; Singh, A.; Naik, N.; Unnikrishnan, S. Estimation of Mangrove Carbon Stocks by Applying Remote Sensing and GIS Techniques. *Wetlands* **2015**, *35*, 695–707. [CrossRef]
50. Vafaei, S.; Soosani, J.; Adeli, K.; Fadaei, H.; Naghavi, H.; Pham, T.D.; Tien Bui, D. Improving Accuracy Estimation of Forest Aboveground Biomass Based on Incorporation of ALOS-2 PALSAR-2 and Sentinel-2A Imagery and Machine Learning: A Case Study of the Hyrcanian Forest Area (Iran). *Remote Sens.* **2018**, *10*, 172. [CrossRef]
51. Ghosh, S.M.; Behera, M.D. Aboveground biomass estimation using multi-sensor data synergy and machine learning algorithms in a dense tropical forest. *Appl. Geogr.* **2018**, *96*, 29–40. [CrossRef]

52. Friedman, J.H. Greedy function approximation: A gradient boosting machine. *Ann. Stat.* **2001**, *29*, 1189–1232. [CrossRef]
53. Friedman, J.H. Stochastic gradient boosting. *Comput. Stat. Data Anal.* **2002**, *38*, 367–378. [CrossRef]
54. Wei, Z.; Meng, Y.; Zhang, W.; Peng, J.; Meng, L. Downscaling SMAP soil moisture estimation with gradient boosting decision tree regression over the Tibetan Plateau. *Remote Sens. Environ.* **2019**, *225*, 30–44. [CrossRef]
55. Chen, T.; Guestrin, C. Xgboost: A scalable tree boosting system. In Proceedings of the 22nd Acm Sigkdd International Conference on Knowledge Discovery and Data Mining, San Francisco, CA, USA, 13–17 August 2016; ACM: New York, NY, USA; pp. 785–794.
56. Vapnik, V. *The Nature of Statistical Learning Theory*; Springer Science & Business Media: Berlin/Heidelberg, Germany, 2013.
57. Breiman, L. Random Forests. *Mach. Learn.* **2001**, *45*, 5–32. [CrossRef]
58. Belgiu, M.; Drăguţ, L. Random forest in remote sensing: A review of applications and future directions. *ISPRS J. Photogramm. Remote Sens.* **2016**, *114*, 24–31. [CrossRef]
59. Pham, T.D.; Xia, J.; Baier, G.; Le, N.N.; Yokoya, N. Mangrove Species Mapping Using Sentinel-1 and Sentinel-2 Data in North Vietnam. In Proceedings of the IGARSS 2019—2019 IEEE International Geoscience and Remote Sensing Symposium, Yokohama, Japan, 28 July–2 August 2019; pp. 6102–6105.
60. Perez-Cruz, F.; Vaerenbergh, S.V.; Murillo-Fuentes, J.J.; Lazaro-Gredilla, M.; Santamaria, I. Gaussian Processes for Nonlinear Signal Processing: An Overview of Recent Advances. *IEEE Signal. Process. Mag.* **2013**, *30*, 40–50. [CrossRef]
61. Rasmussen, C.E.; Williams, C.K. *Gaussian Processes for Machine Learning*; MIT Press: Cambridge, MA, USA, 2006; Volume 1.
62. Jones, E.; Oliphant, T.; Peterson, P. SciPy: Open Source Scientific Tools for Python. 2001. Available online: https://www.scienceopen.com/document?vid=ab12905a-8a5b-43d8-a2bb-defc771410b9 (accessed on 4 August 2019).
63. Grömping, U. Variable Importance Assessment in Regression: Linear Regression versus Random Forest. *Am. Stat.* **2009**, *63*, 308–319. [CrossRef]
64. Li, Y.; Li, C.; Li, M.; Liu, Z. Influence of Variable Selection and Forest Type on Forest Aboveground Biomass Estimation Using Machine Learning Algorithms. *Forests* **2019**, *10*, 1073. [CrossRef]
65. Nguyen Viet, L.; To Trong, T.; Luong Anh, K.; Nguyen Thanh, H. Biomass estimation and mapping of can GIO mangrove biosphere reserve in south of viet nam using ALOS-2 PALSAR-2 data. *Appl. Ecol. Environ. Res.* **2019**, *17*, 15–31.
66. Lucas, R.; Armston, J.; Fairfax, R.; Fensham, R.; Accad, A.; Carreiras, J.; Kelley, J.; Bunting, P.; Clewley, D.; Bray, S.; et al. An Evaluation of the ALOS PALSAR L-Band Backscatter-Above Ground Biomass Relationship Queensland, Australia: Impacts of Surface Moisture Condition and Vegetation Structure. *IEEE J. Sel. Top. Appl. Earth Obs. Remote Sens.* **2010**, *3*, 576–593. [CrossRef]
67. Schlund, M.; Davidson, M. Aboveground Forest Biomass Estimation Combining L- and P-Band SAR Acquisitions. *Remote Sens.* **2018**, *10*, 1151. [CrossRef]
68. Foody, G.M.; Boyd, D.S.; Cutler, M.E.J. Predictive relations of tropical forest biomass from Landsat TM data and their transferability between regions. *Remote Sens. Environ.* **2003**, *85*, 463–474. [CrossRef]
69. Cutler, M.E.J.; Boyd, D.S.; Foody, G.M.; Vetrivel, A. Estimating tropical forest biomass with a combination of SAR image texture and Landsat TM data: An assessment of predictions between regions. *ISPRS J. Photogramm. Remote Sens.* **2012**, *70*, 66–77. [CrossRef]
70. Proisy, C.; Couteron, P.; Fromard, F. Predicting and mapping mangrove biomass from canopy grain analysis using Fourier-based textural ordination of IKONOS images. *Remote Sens. Environ.* **2007**, *109*, 379–392. [CrossRef]
71. Joshi, N.; Mitchard, E.T.A.; Brolly, M.; Schumacher, J.; Fernández-Landa, A.; Johannsen, V.K.; Marchamalo, M.; Fensholt, R. Understanding 'saturation' of radar signals over forests. *Sci. Rep.* **2017**, *7*, 3505. [CrossRef] [PubMed]
72. Chrysafis, I.; Mallinis, G.; Siachalou, S.; Patias, P. Assessing the relationships between growing stock volume and Sentinel-2 imagery in a Mediterranean forest ecosystem. *Remote Sens. Lett.* **2017**, *8*, 508–517. [CrossRef]

73. Wicaksono, P.; Danoedoro, P.; Hartono; Nehren, U. Mangrove biomass carbon stock mapping of the Karimunjawa Islands using multispectral remote sensing. *Int. J. Remote Sens.* **2016**, *37*, 26–52. [CrossRef]
74. Pham, T.D.; Yokoya, N.; Bui, D.T.; Yoshino, K.; Friess, D.A. Remote Sensing Approaches for Monitoring Mangrove Species, Structure, and Biomass: Opportunities and Challenges. *Remote Sens.* **2019**, *11*, 230. [CrossRef]

 *remote sensing*

Article

Mapping the Global Mangrove Forest Aboveground Biomass Using Multisource Remote Sensing Data

Tianyu Hu [1,2], YingYing Zhang [1], Yanjun Su [1,2], Yi Zheng [3], Guanghui Lin [3] and Qinghua Guo [1,2,*]

[1] State Key Laboratory of Vegetation and Environmental Change, Institute of Botany, Chinese Academy of Sciences, Beijing 100093, China; tianyuhu@ibcas.ac.cn (T.H.); zhangyysee@pku.edu.cn (Y.Z.); ysu@ibcas.ac.cn (Y.S.)

[2] University of Chinese Academy of Sciences, Beijing 100049, China

[3] Department of Earth System Science, Tsinghua University, Beijing 100084, China; zhengyi15@tsinghua.org.cn (Y.Z.); lingh@mail.tsinghua.edu.cn (G.L.)

* Correspondence: qguo@ibcas.ac.cn; Tel.: +86-010-6283-6157

Received: 28 April 2020; Accepted: 20 May 2020; Published: 25 May 2020

Abstract: Mangrove forest ecosystems are distributed at the land–sea interface in tropical and subtropical regions and play an important role in carbon cycles and biodiversity. Accurately mapping global mangrove aboveground biomass (AGB) will help us understand how mangrove ecosystems are affected by the impacts of climatic change and human activities. Light detection and ranging (LiDAR) techniques have been proven to accurately capture the three-dimensional structure of mangroves and LiDAR can estimate forest AGB with high accuracy. In this study, we produced a global mangrove forest AGB map for 2004 at a 250-m resolution by combining ground inventory data, spaceborne LiDAR, optical imagery, climate surfaces, and topographic data with random forest, a machine learning method. From the published literature and free-access datasets of mangrove biomass, we selected 342 surface observations to train and validate the mangrove AGB estimation model. Our global mangrove AGB map showed that average global mangrove AGB density was 115.23 Mg/ha, with a standard deviation of 48.89 Mg/ha. Total global AGB storage within mangrove forests was 1.52 Pg. Cross-validation with observed data demonstrated that our mangrove AGB estimates were reliable. The adjusted coefficient of determination (R^2) and root-mean-square error (*RMSE*) were 0.48 and 75.85 Mg/ha, respectively. Our estimated global mangrove AGB storage was similar to that predicted by previous remote sensing methods, and remote sensing approaches can overcome overestimates from climate-based models. This new biomass map provides information that can help us understand the global mangrove distribution, while also serving as a baseline to monitor trends in global mangrove biomass.

Keywords: mangrove; LiDAR; random forest; GLAS; aboveground biomass

---

1. Introduction

Mangrove forests are important intertidal ecosystems that link terrestrial and marine systems [1], protecting land from the impact of storm surges, waves, and the erosion of the shore [2–4]. Mangrove plays a major role in the carbon cycle and helps maintain biodiversity. These forests cover only 2% of the world's coastal areas, yet they provide 5% of the net primary production of global coastal ecosystems [5,6]. While mangrove forests comprise only 0.7% of the area of tropical forests [7], their total carbon density is four times that of other tropical forests in the Indo-Pacific region [8]. Mangrove forests consist of approximately seventy taxonomically diverse tree, shrub, and fern species [9–11]. Moreover, mangrove is an important habit for other organisms [12], such as birds [13] and fish [14], such as mangroves in the Caribbean that have strong effect on the community structure of fish living in the coral reef [14].

Currently, mangroves are highly threatened by both climate change and human activities. As a result of global warming, suitable habitats for mangrove in tropical and subtropical areas have expanded poleward, but sea level rise may be a major threat to the mangrove forests as a result of changes in swamp duration, frequency, or salinity [15,16]. During the past century, approximately 35% of the area with mangrove forests has disappeared [17]. There is an annual deforestation rate of 1–3% [1,17–20] as these areas are converted for use in aquaculture or agriculture [21]. The amount of and change in aboveground biomass act as indicators of other ecosystem services, such as biodiversity [22]. For example, studies indicate a degraded mangrove forest in Malaysia can lose half of its aboveground biomass (AGB) when compared to a natural mangrove forest [22]. Consequently, accurate estimates of the global distribution of mangrove aboveground biomass is beneficial for our understanding of the status of mangrove ecosystems under threat from deforestation and degradation.

Field surveys are the most basic and most accurate methods for acquiring mangrove AGB at the local scale [23–28]. However, this method is time-consuming and costly when applied to larger areas while providing only discrete measurements of AGB at specified points [29,30]. Moreover, field surveys in mangrove areas are more difficult than surveys in other terrestrial ecosystems due to the muddy conditions and the peculiar structure of mangroves [9]. There are two additional methods for estimating regional or global mangrove AGB: model-based methods and remote sensing. Model-based methods usually provide mangrove AGB estimations from local to global scales based on a relationship between environmental drivers and mangrove biomass [31–33]. However, model-based methods usually reflect potential biomass distribution, which is often inconsistent with actual distribution. Remote sensing methods provide an indirect approach for obtaining mangrove AGB measurements using regression models built by linking surface measurements with remote sensing data. Development of these remote sensing methods has greatly improved the efficiency and lowered the cost of mapping mangrove AGB at large scales [34,35].

There are three popular remote sensing techniques for estimating mangrove biomass: passive optical remote sensing, radar, and light detection and ranging (LiDAR) [36]. Passive optical remote sensing and radar are the earliest and most frequently used methods for estimating mangrove extent and biomass mapping [35,37,38], since they have the benefit of complete global coverage and the data are easily accessible. However, both passive optical remote sensing and radar suffer from a saturation effect at high biomass levels. Neither of these methods can retrieve complete vertical canopy information because optical remote sensing only acquires canopy surface information and radar has limited penetration ability [39].

An active remote sensing method, light detection and ranging, effectively penetrates the forest canopy and can be used to derive information about forest structure in three dimensions [40,41]. Because of its ability to quantify forest height, AGB, and other structural parameters in a variety of forest environments, LiDAR is a major advance in the field of forestry remote sensing [42,43]. Moreover, LiDAR does not saturate at high biomass [44,45]. Current limitations in temporal and spatial coverage restrict the application of LiDAR at continental to global scales [46,47]. Airborne and spaceborne LiDAR can acquire large scale data, but neither can provide worldwide, continuous LiDAR measurements. The high cost of flight missions limits the use of airborne LiDAR to certain regions. Spaceborne LiDAR such as the Geoscience Laser Altimeter System (GLAS) onboard the Ice, Cloud, and Land Elevation Satellite (ICESat) have collected global LiDAR measurements, but the low density and discontinuous distribution of the GLAS footprint prevents direct production of continuous global data [48,49].

Recently, studies have demonstrated that using multi-source data can overcome the deficiencies associated with GLAS data [48,49]. Passive optical images along with other continuous variables, such as climate layers and a digital terrain model, can be used to build a regression model with GLAS measurements, allowing us to extrapolate from discrete GLAS pixels into spatially continuous layers [46,47]. This method has been used to estimate forest biomass at the scale of the GLAS footprint through a direct-link method proposed by Baccini et al. [50]. A second method uses airborne LiDAR as a medium [51], thereby extrapolating from discrete AGB points into full coverage layers. However, airborne LiDAR and plots in areas of mangrove are limited, and it is not possible to combine field measurements with GLAS data. Another method suggested by Su et al. [47] provides wall-to-wall estimates of forest AGB at larger scales. First, continuous remote sensing data are used to extrapolate discrete GLAS parameters into spatially continuous layers. Second, a model is built using surface observations rather than linking plot data directly with GLAS data.

Although global mangrove biomass estimates have been generated in the past using climate-based [31,32] and remote sensing [52,53] methods, these results have had little explanatory power or suffer from signal saturation. Moreover, structural information obtained using LiDAR were not fully utilized in previous efforts to map global mangrove biomass. The objectives of this study, then, were to estimate global mangrove AGB using ground inventory data, spaceborne LiDAR, and other multi-source data and then to determine if structural information provided by GLAS can improve our understanding of the distribution of mangrove AGB. To meet these objectives, a map of global mangrove AGB map at 250 m has been created and will be disseminated via the internet. This new biomass map provides information about mangrove forests, allowing us to better monitor regional and global biomass trends into the future.

2. Materials and Methods

The global map of mangrove AGB was generated using field observation data, GLAS data, the enhanced vegetation index (EVI), topographic data, and climate data. The methodology outlined in Figure 1 allowed us to successfully estimate nation-wide forest AGB for China [47] and global forest AGB [46]. A detailed description of each dataset (Table 1) and a brief introduction to the method used to estimate mangrove forest AGB are provided below.

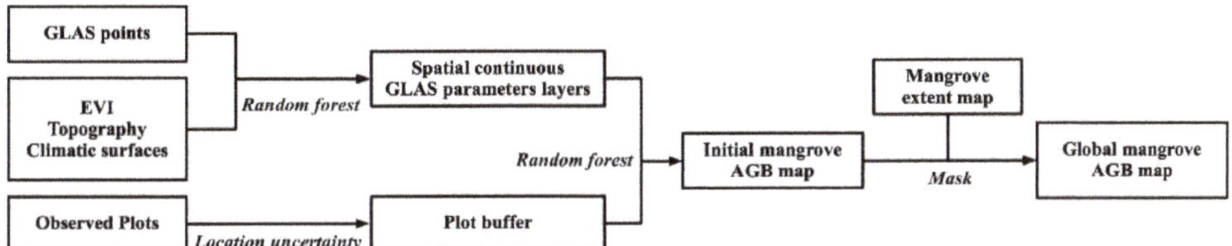

Figure 1. The workflow for producing global mangrove aboveground biomass map based on the multisource remote sensing data and ground observation data.

Table 1. The variables used in the random forest method to determine GLAS parameters and model mangrove aboveground biomass.

Variable	Dataset	Year	Resolution	Reference
Mean annual precipitation (mm)	Worldclim	1950–2000	1 km	Hijmans et al., 2005 [54]
Precipitation of driest quarter (mm)	Worldclim	1950–2000	1 km	Hijmans et al., 2005 [54]
Precipitation seasonality	Worldclim	1950–2000	1 km	Hijmans et al., 2005 [54]
Precipitation of wettest quarter (mm)	Worldclim	1950–2000	1 km	Hijmans et al., 2005 [54]
Annual mean temperature (°C)	Worldclim	1950–2000	1 km	Hijmans et al., 2005 [54]
Mean temperature of driest quarter (°C)	Worldclim	1950–2000	1 km	Hijmans et al., 2005 [54]
Mean temperature of warmest quarter (°C)	Worldclim	1950–2000	1 km	Hijmans et al., 2005 [54]

Table 1. *Cont.*

Variable	Dataset	Year	Resolution	Reference
Temperature seasonality	Worldclim	1950–2000	1 km	Hijmans et al., 2005 [54]
Elevation (m)	CSRTM	2000	30 m	Zhao et al., 2018 [55]
Slope	CSRTM	2000	30 m	Zhao et al., 2018 [55]
Enhanced vegetation index (EVI)	MOD13Q1	2004	250 m	Huete et al., 1999 [56]
Waveform extent (m)	GLAS	2004	~ a 70 m diameter spots	n/a
Leading edge extent (m)	GLAS	2004	~ a 70 m diameter spots	n/a
Trailing edge extent (m)	GLAS	2004	~ a 70 m diameter spots	n/a

2.1. Surface Measurements of Mangrove AGB

Field data are fundamental for estimating mangrove AGB from remote sensing data. In this study, we obtained 510 plot measurements from previously published articles and free-access mangrove biomass databases, such the Sustainable Wetlands Adaptation and Mitigation Program (https://data.cifor.org/dataverse/swamp) [57,58]. Since these in situ plot measurements were collected from a variety of sources using different protocols, we used three filtering criteria to ensure their quality: (1) the plot has a georeferenced location, (2) the inventory was taken after 2000, and (3) the site was not surveyed using harvesting methods. The geolocation of each individual plot was vital to this study. Using Google Earth, we manually checked each point to determine whether the plot location was in the ocean or on land. Records with the same geolocations were averaged together. In the end, 342 plot samples were retained for use in the mangrove AGB mapping procedures (Figure 2).

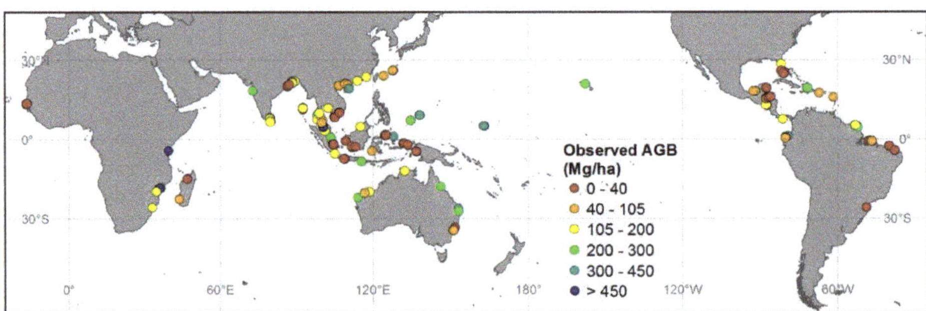

Figure 2. The collected mangrove plots distribution across the world. The color of each point indicated the value of aboveground biomass.

2.2. Spacebrone LiDAR Data

The GLAS instrument is the only waveform LiDAR instrument that has provided global coverage, and it was as an important data source for mapping global tree height and forest biomass. The GLAS instrument aboard the NASA (National Aeronautics and Space Administration) ICESat satellite was launched on 12 January 2003. After seven years in orbit and 18 laser-operation campaigns, the ICESat mission ended with the failure of the GLAS instrument. This instrument had three laser sensors, L1–L3, and each sensor used a 1064-nm laser pulse to record surface altimetry at 20 Hz. Each laser pulse had an ~65 m ellipsoidal footprint and was spaced at 170 m along a track with tens of kilometers between tracks [59]. We selected GLAS data from 2004 for use in mapping mangrove AGB since the quantity and quality of these GLAS data are better than those from later operational periods [46,47].

We downloaded three products (GLA01, GLA06, and GLA14) from the National Snow & Ice Data Center (https://nsidc.org/data). These three products were provided in HDF5 (Hierarchical Data Format) and contained full-waveform information (GLA01); geolocation and data quality information (GLA06); and surface elevation information (GLA14). Laser pulses from these products were linked together based on their unique ID and shot time.

Based on previous research [46–49], we applied four filtering criteria to quality control the GLAS data: (1) laser shots taken under cloudy conditions were removed; (2) data with saturation effects were removed; (3) the data had high signal to noise ratios (>50); and (4) data was not taken from a location significantly higher (i.e., >100 m) than the land surface elevation as indicated by the Shuttle Radar Topography Mission (SRTM) data. All GLAS data points used in this study were determined to be within Spalding et al.'s mangrove map [19]. The final GLAS dataset contained 13,686 records in areas of mangrove forests. From this dataset, three parameters were derived from the full-waveform information of each pulse (waveform extent, leading edge extent, and trailing edge extent). These GLAS parameters have been proven to be highly correlated with forest biomass, canopy height, canopy height variability, and slope of the terrain [48,60].

2.3. EVI Data

We used the MOD13Q1 Version 6 product to obtain cumulative EVI for 2004. The EVI has improved sensitivity for regions of high biomass as compared with NDVI [56]. MOD13Q1 is a composite 16-day product at a 250-m resolution. The composite algorithm chooses the best available pixel value from all acquisitions within the 16-day period, selecting pixels with low clouds, a low view angle, and the highest EVI value. Cumulative EVI can provide more accurate estimates of AGB when compared with values taken from a single time period [61,62]. Therefore, we calculated cumulative EVI from the sum of all collected MOD13Q1 data, and clipped it using a 100-km coastline buffer. These data were used as a predictor in the AGB analysis and mapping procedure.

2.4. Climate Data

In addition to using structure and spectral information from remote sensing data, we included climate data to use in model predictions of mangrove AGB (Table 1). We selected the WorldClim dataset (http://www.worldclim.org), and 50-year (1950–2000) average bioclimatic variables were calculated from monthly temperature and precipitation layers [54]. We selected eight climate variables that can be divided into two categories: precipitation and temperature (Table 1). The climate layers were obtained with a 1-km resolution and then downscaled to 250 m using a bilinear method.

2.5. Topography Data

The GLAS parameters are related to forest structure and terrain variation, so we used topography data to extrapolate from discrete GLAS data into spatial continuous layers. We selected the CSRTM digital elevation model (DEM) provided by Zhao et al. [55]. The CSRTM is a corrected product from the Shuttle Radar Topography Mission (SRTM), which reduced the vertical errors of SRTM at vegetated areas. To be consistent with other datasets, we resampled the CSRTM DEM into 250-m resolution using a bilinear method for further interpolation. The slope (denoted by tangent values of slope) was calculated from the resampled CSRTM DEM.

2.6. Mangrove AGB Estimation Methods

As mentioned, we estimated global mangrove AGB using a methodology that had been successfully implemented to estimate forest AGB at both national and global scales [46,47]. We modified the step regarding plot location uncertainty to account specifically for the distribution of mangrove. We did not use a land cover map in the random forest regression analysis as we assumed all areas were mangrove based on our data collection methods previous described. As shown in Figure 1, the estimation of mangrove AGB is generally divided into four major steps.

First, discrete GLAS points were interpolated to create continuous spatial layers using the random forest algorithm. The GLAS points were filtered using a 100-km coastline buffer and aggregated into 250-m pixels using the average value of the GLAS full waveform parameter within each pixel. These pixels were then used as training data to build the random forest model created to extrapolate the GLAS parameters along with other predictor layers (cumulative EVI, DEM, slope, climate surfaces) using the randomForest R package [63].

Second, we generated a circular buffer for each plot measurements with a 500-m radius to reduce uncertainty related to plot location. Since mangrove has a much smaller distribution than other forest types, we could not use the point-radius method suggested by Su et al. to reduce geolocation uncertainty [47]. Using their Monte-Carlo simulation method, generating plot sets with location errors of 1 or 10 km would relocate many mangrove plots into the ocean. To avoid this issue, we used the circular buffer method. Most latitudes and longitudes in our field observation data were accurate to 0.01°, corresponding to ~1km. We, therefore, adopted a 500-m radius to reduce location uncertainty.

Third, an initial global mangrove AGB map was created using the random forest method. Pixels for each explanatory layer within a plot buffer were averaged and used as explanatory variables to build a regression model from plot measurements. We randomly chose 70% of the plots (239 plots) to train the model and used the remaining 30% (103 plots) to validate the mangrove AGB estimation model. The three extrapolated GLAS parameters and the other nine parameters in Table 1 were used in the regression model to generate the outputs needed to produce this initial mangrove AGB map.

Finally, we used a mangrove extent map from Spalding et al. as a mask for our initial mangrove AGB map, eliminating areas outside of identified mangrove forests. [19]. The final global mangrove AGB map was obtained by setting AGB value in areas outside the mangrove extent to 0 Mg/ha.

2.7. Accuracy Assessment

The accuracy of the estimated AGB was assessed using the adjusted coefficient of determination (R^2) and root-mean-square error (*RMSE*). The R^2 and *RMSE* were calculated using following equations:

$$R^2 = 1 - \frac{(n-1)\sum_{i=1}^{n}(x_i - \hat{x}_i)^2}{(n-2)\sum_{i=1}^{n}(x_i - \overline{x})^2} \tag{1}$$

$$RMSE = \sqrt{\frac{\sum_{i=1}^{n}(x_i - \hat{x}_i)^2}{n-2}} \tag{2}$$

where x_i is the observed mangrove AGB, $\hat{x}_i$ is the predicted AGB based on the random forest model built with the training data, $\overline{x}$ is the average AGB of all validation plots, and n is the number of validation plots.

3. Results

3.1. The GLAS Parameters in the Mangrove Distribution Zone

The discrete GLAS parameter points were extrapolated to spatially continuous layers using the random forest method for leading edge extent, waveform extent, and trailing edge extent (Figure 3). Overall, the random forest models explained 40.32%, 59.12%, and 41.39% of the variance in leading edge extent, waveform extent and trailing edge extent, respectively. The root-mean-square residuals for leading edge extent, waveform extent, and trailing edge extent were 4.30, 6.98, and 2.35 m, respectively. According to the extrapolated results, the mean value of leading edge extent, waveform extent, and trailing edge extent for the mangroves were 11.34 ± 5.61 m, 19.06 ± 7.09 m, 4.19 ± 1.34 m, respectively. These three GLAS parameters showed similar spatial patterns of mangrove distribution. The highest values of all three parameters appeared in the Indonesian archipelago, Central America, and the Gulf of Guinea.

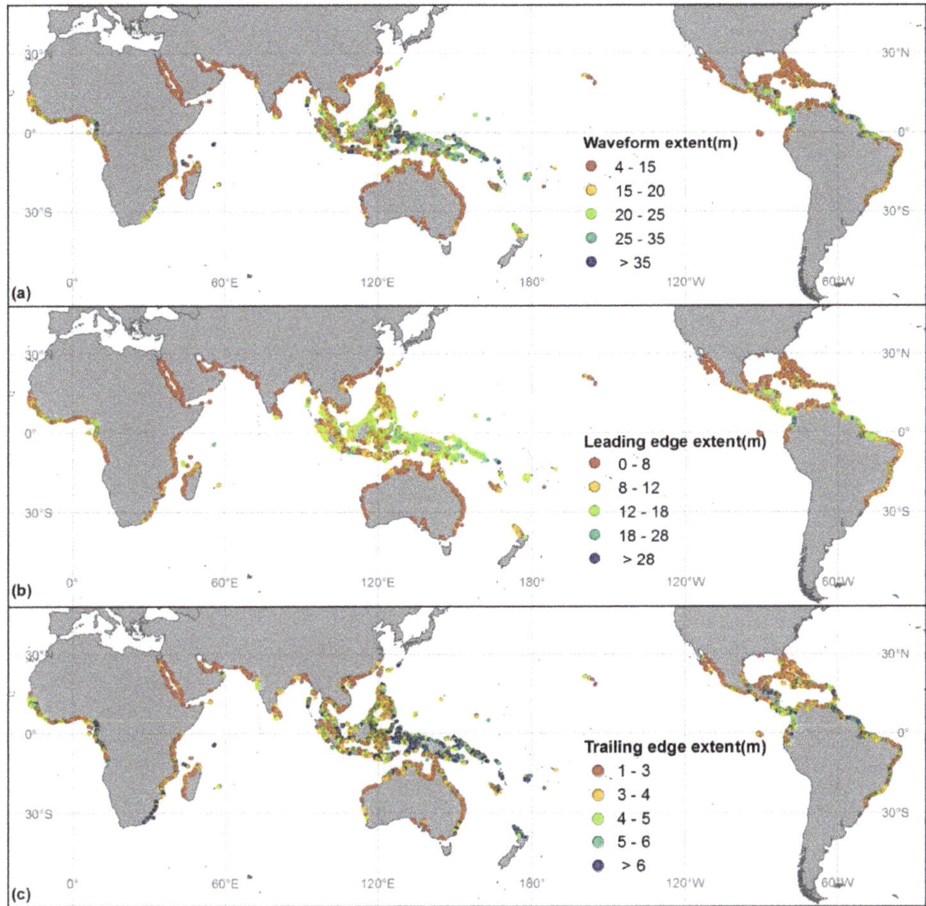

Figure 3. The spatial-continuous map of three GLAS parameters in the mangrove distribution zone, (**a**) waveform extent, (**b**) leading edge extent, and (**c**) trailing edge extent. Note that the spatially continuous map was drawn using points since the mangrove distribution zone is narrow and cannot be represented well using a raster map at the global scale.

3.2. The Global Mangrove Forest AGB Map

We used a random forest regression model with the three extrapolated GLAS parameters and other predictor variables to estimate global mangrove AGB. The random forest model explained 52.34% of the variance in AGB. The final AGB distribution pattern is similar to that of the GLAS parameters (Figure 4). The mean AGB density of global mangrove was 115.23 Mg/ha with a standard deviation of 48.89 Mg/ha. This map of AGB for mangrove forests will be shared on the GUO-Lab website (http://www.3decology.org).

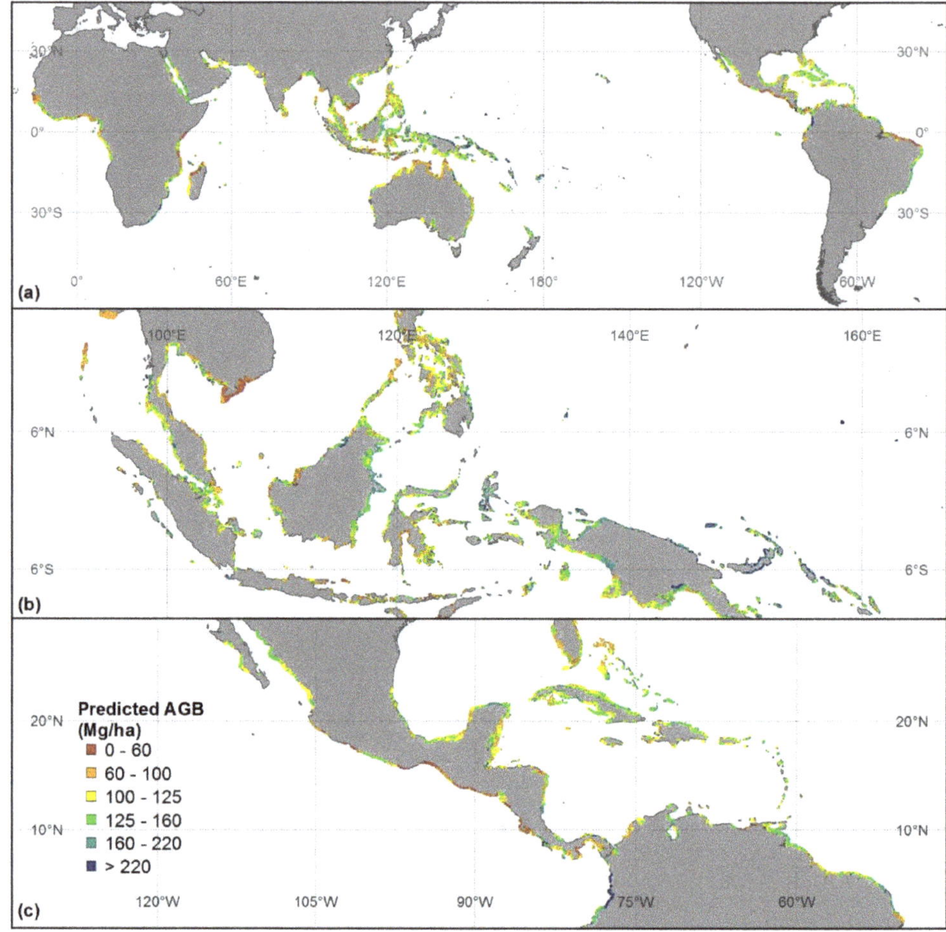

Figure 4. Predicted mangrove forest AGB distribution (**a**) throughout the world, (**b**) enlarged over Southeastern Asia, and (**c**) enlarged over Central America.

3.3. Continental and National Level Mangrove Forest AGB Density

Total global AGB for mangroves was 1.52 Pg (Table 2), but the contribution by region was not uniform. Southeastern Asia provided 34.98% of the AGB (0.53 Pg) while having both the largest area (4,044,906.25 ha) and high AGB density (131.36 ± 45.94 Mg/ha). South America encompassed the second largest area (2,062,231.25 ha) and high AGB density (111.33 ± 58.70 Mg/ha), and the second highest stock of AGB (0.15 Pg). The mangrove AGB density in Central America (110.29 ± 39.48 Mg/ha) was similar to that of South America, although the area of mangrove in Central America was much smaller (1,388,962.50 ha). Mangrove AGB stocks in Southern Asia (0.13 Pg) and Western Africa (0.12 Pg) were similar, although the density of AGB was much higher in Southern Asia (132.6 ± 29.79 Mg/ha) than in Western Africa (79.27 ± 34.32 Mg/ha). The extent of mangrove in Southern Asia (949,281.25 ha), on the other hand, was lower than that in Western Africa (1,475,343.75 ha). At the national level, Indonesia had the highest stock of AGB (0.36 Pg) because of the high AGB density (140.12 ± 41.02 Mg/ha) and large area covered by mangrove (2,547,556.25 ha). Mexico had the second largest AGB (0.1 Pg) since it has large areas (891,312.50 ha) and high AGB density (113.98 ± 34.10 Mg/ha). The mangrove AGB density and stock for other important regions and countries are listed in Table 2.

Table 2. The mean AGB density and total AGB in the different regions and countries.

Region *	Mean AGB (Mg/ha)	Mangrove Area (ha)	Total AGB (Mg)	Proportion of Global AGB (%)
Southeastern Asia	131.36 ± 45.94	4,044,906.25	531,347,520.57	34.98
South America	111.33 ± 58.70	2,062,231.25	229,594,355.48	15.12
Central America	110.29 ± 39.48	1,388,962.50	153,185,736.15	10.09
Southern Asia	132.60 ± 29.79	949,281.25	125,874,227.46	8.29
Western Africa	79.27 ± 34.32	1,475,343.75	116,944,200.85	7.70
Eastern Africa	102.79 ± 53.63	821,906.25	84,482,779.41	5.56
Caribbean	123.69 ± 31.40	571,493.75	70,690,803.56	4.65
Melanesia	149.24 ± 48.63	438,768.75	65,479,984.29	4.31
Australia and New Zealand	101.22 ± 38.71	523,643.75	53,000,621.89	3.49
Middle Africa	101.79 ± 36.88	393,006.25	40,003,008.67	2.63
Northern America	103.48 ± 44.90	300,956.25	31,142,627.18	2.05
Southern Africa	197.16 ± 52.95	43,118.75	8,501,508.15	0.56
Western Asia	134.24 ± 16.11	23,162.50	3,109,444.89	0.20
Micronesia	279.89 ± 81.67	10,162.50	2,844,427.43	0.19
Northern Africa	161.37 ± 11.47	9,481.25	1,529,943.27	0.10
Eastern Asia	114.90 ± 23.41	9,156.25	1,052,045.89	0.07
Polynesia	160.25 ± 58.88	100.00	16,025.34	<0.01
Global	115.23 ± 48.89	13,065,675.00	1,518,798,427	100
Country	Mean AGB (Mg/ha)	Mangrove Area (ha)	Total AGB (Mg)	Proportion (%)
Indonesia	140.12 ± 41.02	2,547,556.25	356,964,199.62	23.50
Mexico	113.30 ± 34.10	891,312.50	100,985,695.79	6.65
Brazil	81.09 ± 44.96	1,117,700.00	90,630,600.45	5.97
Malaysia	134.00 ± 54.85	629,643.75	84,369,792.27	5.56
Bangladesh	154.17 ± 12.84	438,487.50	67,601,229.37	4.45
Colombia	166.95 ± 66.41	371,468.75	62,016,180.38	4.08
Mozambique	131.84 ± 51.61	413,456.25	54,511,434.50	3.59
Nigeria	76.54 ± 17.80	701,337.50	53,680,631.22	3.53
Cuba	126.27 ± 30.52	421,200.00	53,186,818.02	3.50
Papua New Guinea	148.94 ± 46.75	356,356.25	53,074,570.57	3.49
Global	115.23 ± 48.89	13,065,675.00	1,518,798,427	100

* The geographic regions used to organize the final statistics results were defined by the United Nations (https://unstats.un.org/unsd/methodology/m49/).

3.4. The Accuracy of Mangrove AGB Estimation

These estimates of mangrove AGB were validated using 103 independent validation plots (Figure 5). Predicted mangrove AGB was consistent with observerd AGB. The R^2 between predicted and observed AGB is 0.48 and the RMSE is 75.85 Mg/ha. The AGB estimation method in this study tended to marginally overestimate AGB densities at low values (<125 Mg/ha; Figure 5) and tends to underestimate forest AGB density at high values (>125 Mg/ha).

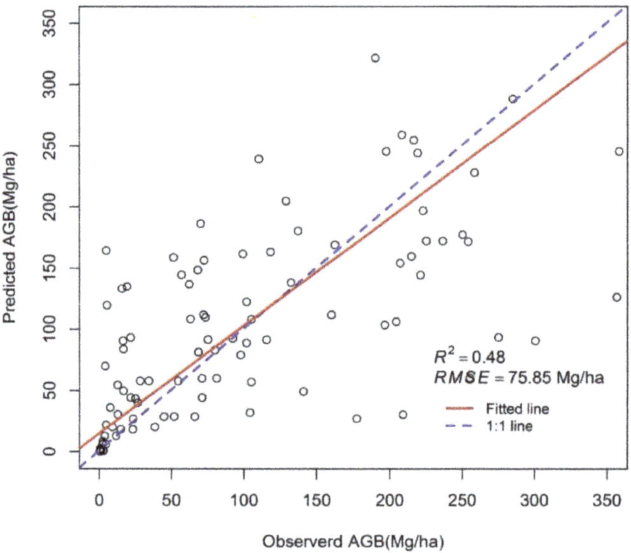

Figure 5. The validation of mangrove biomass estimated model. R^2 represents the adjusted coefficient of determination, *RMSE* represents the root-mean-square error.

4. Discussion

4.1. Comparison with Other Mangrove Models

Combining multi-source remote sensing data and surface observations helps us better understand global distribution of mangrove AGB. The model developed during this study is better able to explain the spatial variability in mangrove AGB (R^2 = 0.48) than those of other models. Rovai et al. [33] developed a set of statistical climatic-geophysical models based on the environmental signature hypothesis, which explained only 20% of the variability in mangrove AGB in the Neotropics. Twilley et al.'s [31] latitude-based model explained 7.6% of the variation in mangrove AGB at the global scale, while Hutchison et al.'s [32] climate-based model explained 26.7% of the variation. We used our plot data to test Twilley et al.'s latitude-based model and Hutchison et al.'s [32] climate-based model; the resulting explanatory power of these two models was much lower at 2.2% and 10.5%, respectively. There are three primary reasons. First, the initial mangrove AGB dataset was extremely small in both Twilley et al.'s (n = 34) and Hutchison et al.'s (n = 52) analyses. Insufficient training data cannot be used to create a robust global scale model. These models, therefore, have large uncertainty when validated against our larger, global data set (n = 342).

Second, machine learning methods are more suitable to estimating global mangrove AGB than multi-linear regression methods. Although the climate variables used in our model and Hutchison et al.'s [32] climate-based model were similar, the explanatory power of our model was greater because of the difference in regression methods. Several studies have demonstrated that random forest performs better than the linear regression method for estimating biomass [64].

Finally, structural information provided by GLAS and EVI improved the accuracy of random forest to estimate mangrove AGB biomass (Figure 6). Recent field studies have found that canopy height is strongly related to biomass for many mangrove species [65,66]. However, structure information provided by GLAS does not have the expected effects in this study when compared with other research into national and global forest AGB mapping. This may have been caused by the low-density footprint of GLAS in mangrove areas, limiting its ability to represent the structure variation in different mangrove species. Based on the statistical importance of each variable in our model (Figure 6), climate factors were more important than other variables. This is similar to Simard et al.'s results in which precipitation, temperature and cyclone frequency explain 74% of the global variation in maximum canopy height [53].

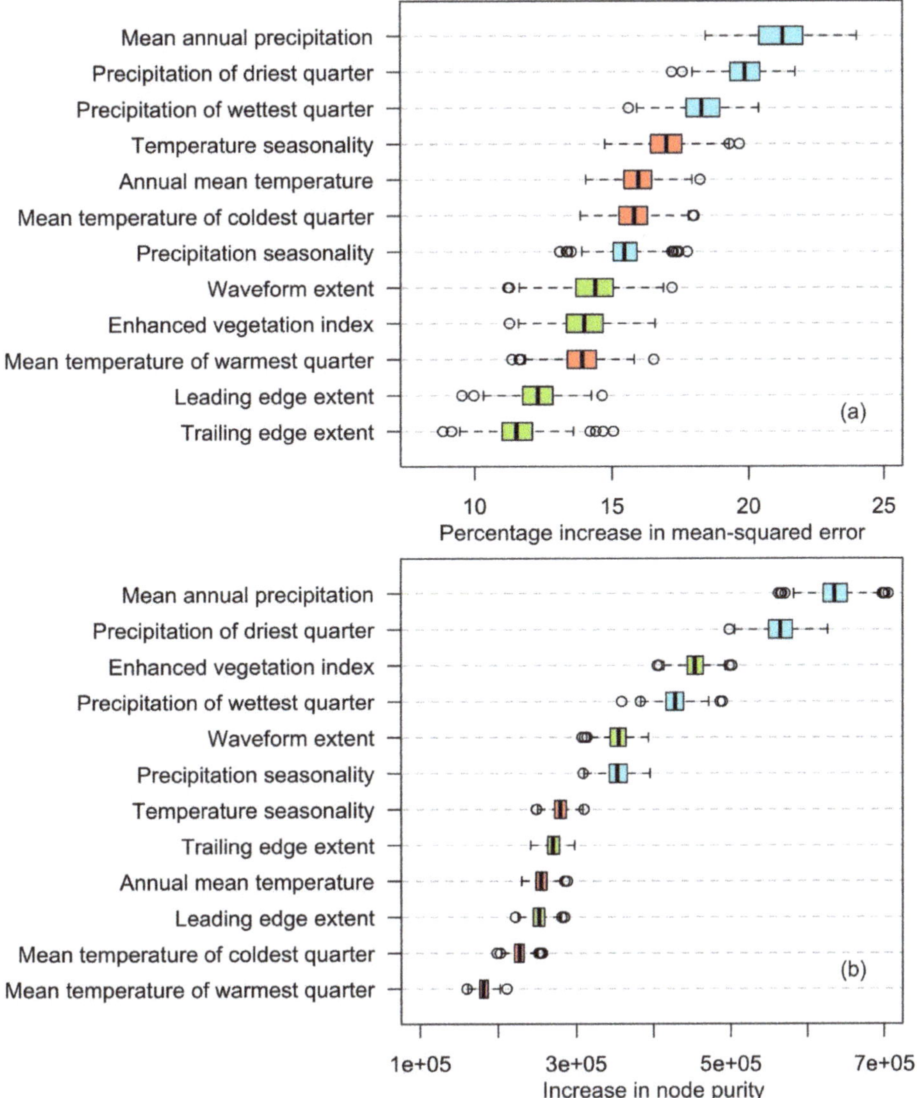

Figure 6. The mean importance of variables for AGB estimation using the randomForest model is indicated by the percentage increase of mean-squared error (**a**) and the increase in node purity (**b**) from highest to lowest. Percentage increase in mean square error is calculated by the increase in mean square error when a variable is removed in the model. The increase in node purity is calculated based on the reduction in sum of squared errors whenever a variable is chosen to split.

4.2. Comparison with Previously Published Mangrove AGB Maps

Our estimated global mangrove AGB (1.52 Pg) was similar to that of two global maps produced using other remote sensing approaches (Table 3). Tang et al. [52] reported that total global mangrove AGB was 1.908 Pg, while Simard et al. [53] estimated it to be 1.75 Pg. Although the estimated mangrove storage was similar between these three remote sensing approaches, mean AGB density (115.23 ± 48.89 Mg/ha) in our study was lower than that of Tang et al. (146.3 Mg/ha) or Simard et al. (129.1 ± 87.2 Mg/ha). These differences are mainly caused by uncertainties induced by the allometric equations. Tang et al. [52] and Simard et al. [53] predicted global mangrove biomass using SRTM's tree height and a global mangrove biomass allometry equation. The mean AGB density reported by Tang et al. was the highest of these three estimates. Compared to Saenger and Snedaker's global mangrove height-biomass relationship used by Tang et al. [52], Simard et al. [53] applied 331 in situ plots across a wide variety of mangrove forest ecotypes to fit a global equation between AGB and basal area-weighted height. Our results for global mangrove AGB storage and mean AGB density were similar to that of Simard et al. [53] because both methods utilized spaceborne LiDAR data. Traditionally, mangrove forest aboveground biomass derived using synthetic aperture radar was underestimated due to its limited ability to penetrate the mangrove canopy. Simard et al. [53] used GLAS data to correct the SRTM tree height, thereby overcoming the issue of estimating mangrove AGB from SRTM tree height.

Table 3. Comparison of total mangrove AGB and area with previously published results.

	AGB (Pg)	Mangrove Area (ha)	Year of Estimate	Mangrove Map
Hutchison et al. (2014) [32]	2.83	15,314,094	1999–2003	Spalding et al., 2010 [19]
Twilley et al. (1992) [31]	2.34	~24,000,000	1986	World Resources, 1986 [67]
Tang et al. (2018) [52]	1.908	~13,042,000	2000	Spalding et al., 2010 [19]
Simard et al. (2019) [53]	1.75 ± 0.77	~13,776,000	2000	Giri et al., 2011 [7]
This study	**1.52**	**13,065,675**	**2004**	**Spalding et al., 2010** [19]

The estimated global mangrove AGB storage in our study (1.52 Pg) was significantly lower than those from non-remote sensing approaches (Table 3). Twilley et al. [31] estimated global mangrove AGB at 2.34 Pg based on a latitude model, nearly 54% higher than our result. Hutchison et al. [32] used a climate-based model and predicted that total global mangrove AGB storage was 2.83 Pg, 86% higher than our result. The difference in the baseline mangrove extent could be a major reason for the variation in these results. Although our study and that of Hutchison et al. [32] both used the mangrove map developed by Spalding et al. [19], the final global mangrove area in our study (13,065,675.00 ha) was 15% smaller than that used by Hutchison et al. [32] (15,314,094 ha). This difference was caused, in part, by inconsistent land boundaries between our predictor variables, mangrove distribution map, and country extents. These layers have different spatial resolutions and extents, so small mangrove patches along the coast or in the islets were omitted during our analysis. These places are also areas with a large distribution of mangrove [19]. Consequently, the disparity in area led to variations in total mangrove AGB storage between the two results. Part of the variation can also be explained by the models used by Hutchison et al., which may overestimate mangrove AGB [32]. Rovai et al. [33] found that these climate- and latitude-based models overestimated mangrove AGB by 25.3% to 44.4% in the Neotropics region. In addition, the structural data provided by spaceborne LiDAR in this study can provide better information for estimating mangrove AGB at larger geographical scales, thereby reducing uncertainty in estimates of mangrove AGB storage.

Most of the 10 countries with largest total mangrove AGB stock from our study were also reported in other research, such as that of Hutchison et al. (2014) [32] and Simard et al. (2019) [53], but the order in which these countries appear on the list was different. Indonesia has the largest mangrove AGB stock, which is consist in each study, even though the mangrove AGB in Indonesia and Papua

New Guinea reported by our study was much lower than those of Hutchison et al. (2014) [32] and Simard et al. (2019) [53] (Figure 7). This phenomenon maybe caused by the model we used to predict biomass. Validation (Figure 5) showed that our model tended to underestimate mangrove AGB density at high values (>125 Mg/ha) since observations are limited in these high biomass areas. The mangrove AGB stock in Mexico, Cuba, and Colombia differed between the three studies. The difference in Mexico and Cuba was induced by a bias in predicted mean mangrove AGB density in the different studies. Adame et al. (2013) reported that the AGB in tall, medium and dwarf mangroves in the Mexican Caribbean were as much as 176.2, 114.2 and 7.1 Mg/ha, respectively [68]. The mean mangrove AGB density of Mexico in our study was 113.30 Mg/ha which is closer to that of the medium mangroves reported by Adame et al. (2013). Simard et al. (2019) [53] reported a mean AGB in Mexico of 37.9 Mg/ha, which is much lower than that of the medium mangroves. This underestimation in Simard et al. (2019) [53] may have been caused by using a global allometric equation to predict biomass. The difference in Colombia was mainly caused by inconsistencies in mangrove extent. The mangrove area in Colombia reported by Simard et al. (2019) [53] is much lower than that in our study and in Hutchison et al. (2014) [32].

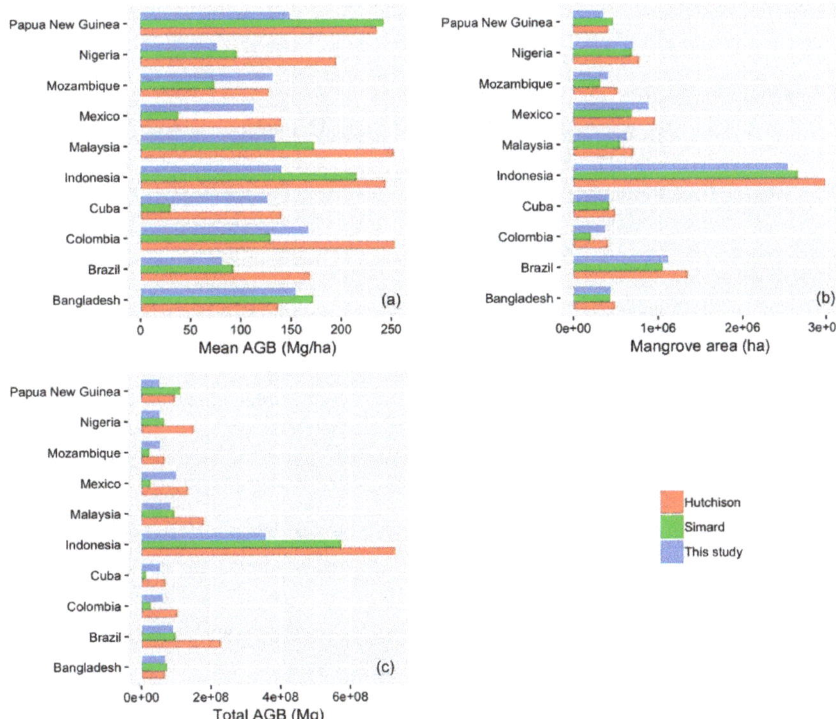

Figure 7. Comparison of (**a**) mean mangrove AGB density, (**b**) mangrove area, and (**c**) total mangrove AGB stock between our study, Hutchison et al. (2014) and Simard et al. (2019) in ten countries with the highest mangrove AGB.

4.3. Limitations and Future Studies

Although our model estimates global mangrove aboveground biomass fairly well, there are limitations to this study. Available observation data was limited when compared with other regional and global studies. Field data is fundamental for accurate estimations of global mangrove biomass. In this study, we collected 510 records from a number of sources but more than 30% of them could not be used because of uncertainty in location information. Moreover, geolocation errors in mangrove plots cannot be reduced by the point-radius model used by Su et al. [47] and Hu et al. [46], because randomly shifting mangrove plot locations had a large probability of relocating the plot into the ocean. Furthermore, the mismatch of spatial observation scales between plots and remote sensing data was a problem. Researchers have begun recently to use drone-based LiDAR to retrieve mangrove biomass [69], using it as a bridge to scale AGB from the plot level to the scale of satellite observations [70]. The increase in drone-based LiDAR data in mangrove areas will benefit global mangrove forest biomass mapping efforts in the future. Second, sparse GLAS datapoints within areas of mangrove lose some of the variability in structure during extrapolation. Even though we used GLAS data within a 100-km buffer of the coast to increase the number of GLAS datapoints, the explanatory power of the extrapolation models were nearly 10% lower than those for China and global forest mapping [46,47]. Fortunately, the Global Ecosystem Dynamics Investigation (GEDI) project [71] recently started collecting global waveform LiDAR data, which will provide higher density data with a smaller footprint than GLAS. This data will help us better understand variability in mangrove structure and biomass distribution. Third, factors such as salinity [72] and river discharge [73] that specifically control the distribution and production of mangrove forest should be added to the model in the future. With these factors, we can more accurately estimate the biomass and better understand how mangrove AGB varies under different environmental conditions.

5. Conclusions

This study produced a new global estimate of mangrove AGB for the year 2004, resulting in a 250-m resolution map that will be publicly available (http://www.3decology.org). This product was generated using methodology that was successfully implemented previously to estimate nation-wide forest AGB for China as well as global forest AGB. Three GLAS parameters and an additional nine predictor variables were used to build a random forest estimation model using plot measurements collected from published literature and free-access datasets. Based on this mangrove AGB analysis, global mangrove AGB density was estimated to be approximately 115.67 ($\pm$48.89) Mg/ha on average, with a total global AGB for mangrove forests of 1.52 Pg. Our product was compared to published global mangrove AGB products, and it has better explanatory power (R^2 = 0.48, *RMSE* = 75.85 Mg/ha) than previous climate-based models. Results showed that this estimated global mangrove AGB storage was similar to that predicted by other remote sensing methods, especially the mangrove AGB map produced by Simard et al. [53]. Future research will include better LiDAR-based measurements of mangrove biomass as well as additional factors known to affect mangrove distribution.

Author Contributions: Conceptualization, T.H., Y.Z. (YingYing Zhang), Y.S. and Q.G.; Data curation, T.H., Y.Z. (YingYing Zhang), Y.Z. (Yi Zheng) and G.L.; Formal analysis, T.H., Y.Z. (YingYing Zhang) and Y.S.; Writing—original draft, T.H. and Y.Z. (YingYing Zhang); Writing—review & editing, T.H., Y.S. and Q.G. All authors have read and agreed to the published version of the manuscript.

Funding: This study Supported by the National Key R&D Program of China (2017YFC0503905).

Acknowledgments: Thanks to the Center for International Forestry Research (CIFOR, https://www.cifor.org/) for providing mangrove field data.

Conflicts of Interest: The authors declare no conflict of interest.

References

1. Alongi, D.M. Present state and future of the world's mangrove forests. *Environ. Conserv.* **2002**, *29*, 331–349. [CrossRef]
2. Mazda, Y.; Wolanski, E.; Ridd, P. *The Role of Physical Processes in Mangrove Environments: Manual for the Preservation and Utilization of Mangrove Ecosystems*; Terrapub: Tokyo, Japan, 2007.
3. Lee, S.Y.; Primavera, J.H.; Dahdouh-Guebas, F.; McKee, K.; Bosire, J.O.; Cannicci, S.; Diele, K.; Fromard, F.; Koedam, N.; Marchand, C.; et al. Ecological role and services of tropical mangrove ecosystems: A reassessment. *Glob. Ecol. Biogeogr.* **2014**, *23*, 726–743. [CrossRef]

4. Alongi, D.M. Mangrove forests: Resilience, protection from tsunamis, and responses to global climate change. *Estuar. Coast. Shelf Sci.* **2008**, *76*, 1–13. [CrossRef]

5. Bouillon, S.; Borges, A.V.; Castañeda-Moya, E.; Diele, K.; Dittmar, T.; Duke, N.C.; Kristensen, E.; Lee, S.Y.; Marchand, C.; Middelburg, J.J.; et al. Mangrove production and carbon sinks: A revision of global budget estimates. *Glob. Biogeochem. Cycles* **2008**, *22*. [CrossRef]

6. Alongi, D.M.; Mukhopadhyay, S.K. Contribution of mangroves to coastal carbon cycling in low latitude seas. *Agric. For. Meteorol.* **2015**, *213*, 266–272. [CrossRef]

7. Giri, C.; Ochieng, E.; Tieszen, L.L.; Zhu, Z.; Singh, A.; Loveland, T.; Masek, J.; Duke, N. Status and distribution of mangrove forests of the world using earth observation satellite data: Status and distributions of global mangroves. *Glob. Ecol. Biogeogr.* **2011**, *20*, 154–159. [CrossRef]

8. Donato, D.C.; Kauffman, J.B.; Murdiyarso, D.; Kurnianto, S.; Stidham, M.; Kanninen, M. Mangroves among the most carbon-rich forests in the tropics. *Nat. Geosci.* **2011**, *4*, 293. [CrossRef]

9. Tomlinson, P. *Mangrove Botany*; Cambridge Univ. Press: Cambridge, UK, 1986.

10. Ellison, A.M.; Farnsworth, E.J.; Merkt, R.E. Origins of mangrove ecosystems and the mangrove biodiversity anomaly. *Glob. Ecol. Biogeogr.* **1999**, *8*, 95–115. [CrossRef]

11. Duke, N.; Ball, M.; Ellison, J. Factors influencing biodiversity and distributional gradients in mangroves. *Glob. Ecol. Biogeogr. Lett.* **1998**, *7*, 27–47. [CrossRef]

12. Nagelkerken, I.; Blaber, S.; Bouillon, S.; Green, P.; Haywood, M.; Kirton, L.; Meynecke, J.-O.; Pawlik, J.; Penrose, H.; Sasekumar, A. The habitat function of mangroves for terrestrial and marine fauna: A review. *Aquat. Bot.* **2008**, *89*, 155–185. [CrossRef]

13. Buelow, C.; Sheaves, M. A birds-eye view of biological connectivity in mangrove systems. *Estuar. Coast. Shelf Sci.* **2015**, *152*, 33–43. [CrossRef]

14. Mumby, P.J.; Edwards, A.J.; Ernesto Arias-González, J.; Lindeman, K.C.; Blackwell, P.G.; Gall, A.; Gorczynska, M.I.; Harborne, A.R.; Pescod, C.L.; Renken, H.; et al. Mangroves enhance the biomass of coral reef fish communities in the Caribbean. *Nature* **2004**, *427*, 533. [CrossRef] [PubMed]

15. Ward, R.D.; Friess, D.A.; Day, R.H.; MacKenzie, R.A. Impacts of climate change on mangrove ecosystems: A region by region overview. *Ecosyst. Health Sustain.* **2016**, *2*, e01211. [CrossRef]

16. Krauss, K.W.; McKee, K.L.; Lovelock, C.E.; Cahoon, D.R.; Saintilan, N.; Reef, R.; Chen, L. How mangrove forests adjust to rising sea level. *New Phytol.* **2014**, *202*, 19–34. [CrossRef] [PubMed]

17. Valiela, I.; Bowen, J.L.; York, J.K. Mangrove Forests: One of the World's Threatened Major Tropical Environments: At least 35% of the area of mangrove forests has been lost in the past two decades, losses that exceed those for tropical rain forests and coral reefs, two other well-known threatened environments. *BioScience* **2001**, *51*, 807–815. [CrossRef]

18. McLeod, E.; Chmura, G.L.; Bouillon, S.; Salm, R.; Björk, M.; Duarte, C.M.; Lovelock, C.E.; Schlesinger, W.H.; Silliman, B.R. A blueprint for blue carbon: Toward an improved understanding of the role of vegetated coastal habitats in sequestering CO_2. *Front. Ecol. Environ.* **2011**, *9*, 552–560. [CrossRef]

19. Spalding, M. *World Atlas of Mangroves*; Routledge: London, UK, 2010. [CrossRef]

20. FAO. *The World's Mangroves 1980–2005*; Food and Agriculture Organization: Rome, Italy, 2007.

21. Thomas, N.; Lucas, R.; Bunting, P.; Hardy, A.; Rosenqvist, A.; Simard, M. Distribution and drivers of global mangrove forest change, 1996–2010. *PLoS ONE* **2017**, *12*, e0179302. [CrossRef]

22. Zhila, H.; Mahmood, H.; Rozainah, M.Z. Biodiversity and biomass of a natural and degraded mangrove forest of Peninsular Malaysia. *Environ. Earth Sci.* **2014**, *71*, 4629–4635. [CrossRef]

23. Tamai, S.; Nakasuga, T.; Tabuchi, R.; Ogino, K. Standing Biomass of Mangrove Forests in Southern Thailand. *J. Jpn. For. Soc.* **1986**, *68*, 384–388. [CrossRef]

24. Slim, F.J.; Gwada, P.M.; Kodjo, M.; Hemminga, M.A. Biomass and litterfall of Ceriops tagal and Rhizophora mucronata in the mangrove forest of Gazi Bay, Kenya. *Mar. Freshw. Res.* **1996**, *47*, 999–1007. [CrossRef]

25. Ross, M.S.; Ruiz, P.L.; Telesnicki, G.J.; Meeder, J.F. Estimating above-ground biomass and production in mangrove communities of Biscayne National Park, Florida (U.S.A.). *Wetl. Ecol. Manag.* **2001**, *9*, 27–37. [CrossRef]

26. Soares, M.L.G.; Schaeffer-Novelli, Y. Above-ground biomass of mangrove species. I. Analysis of models. *Estuar. Coast. Shelf Sci.* **2005**, *65*, 1–18. [CrossRef]

27. Komiyama, A.; Ong, J.E.; Poungparn, S. Allometry, biomass, and productivity of mangrove forests: A review. *Aquat. Bot.* **2008**, *89*, 128–137. [CrossRef]

28. Chandra, I.A.; Seca, G.; Hena, M.K.A. Aboveground Biomass Production of Rhizophora apiculata Blume in Sarawak Mangrove Forest. *Am. J. Agric. Biol. Sci.* **2011**, *6*, 469–474.
29. Mitra, A.; Sengupta, K.; Banerjee, K. Standing biomass and carbon storage of above-ground structures in dominant mangrove trees in the Sundarbans. *For. Ecol. Manag.* **2011**, *261*, 1325–1335. [CrossRef]
30. Comley, B.W.T.; McGuinness, K.A. Above- and below-ground biomass, and allometry, of four common northern Australian mangroves. *Aust. J. Bot.* **2005**, *53*, 431–436. [CrossRef]
31. Twilley, R.R.; Chen, R.H.; Hargis, T. Carbon sinks in mangroves and their implications to carbon budget of tropical coastal ecosystems. *Water Air Soil Pollut.* **1992**, *64*, 265–288. [CrossRef]
32. Hutchison, J.; Manica, A.; Swetnam, R.; Balmford, A.; Spalding, M. Predicting Global Patterns in Mangrove Forest Biomass. *Conserv. Lett.* **2014**, *7*, 233–240. [CrossRef]
33. Rovai, A.S.; Riul, P.; Twilley, R.R.; Castañeda-Moya, E.; Rivera-Monroy, V.H.; Williams, A.A.; Simard, M.; Cifuentes-Jara, M.; Lewis, R.R.; Crooks, S.; et al. Scaling mangrove aboveground biomass from site-level to continental-scale: Scaling up mangrove AGB from site- to continental-level. *Glob. Ecol. Biogeogr.* **2016**, *25*, 286–298. [CrossRef]
34. Heumann, B.W. Satellite remote sensing of mangrove forests: Recent advances and future opportunities. *Prog. Phys. Geogr. Earth Environ.* **2011**, *35*, 87–108. [CrossRef]
35. Proisy, C.; Couteron, P.; Fromard, F. Predicting and mapping mangrove biomass from canopy grain analysis using Fourier-based textural ordination of IKONOS images. *Remote Sens. Environ.* **2007**, *109*, 379–392. [CrossRef]
36. Wang, L.; Jia, M.; Yin, D.; Tian, J. A review of remote sensing for mangrove forests: 1956–2018. *Remote Sens. Environ.* **2019**, *231*, 111223. [CrossRef]
37. Simard, M.; Zhang, K.; Rivera-Monroy, V.H.; Ross, M.S.; Ruiz, P.L.; Castañeda-Moya, E.; Twilley, R.R.; Rodriguez, E. Mapping Height and Biomass of Mangrove Forests in Everglades National Park with SRTM Elevation Data. *Photogramm. Eng. Remote Sens.* **2006**, *72*, 299–311. [CrossRef]
38. Pham, T.D.; Yokoya, N.; Bui, D.T.; Yoshino, K.; Friess, D.A. Remote Sensing Approaches for Monitoring Mangrove Species, Structure, and Biomass: Opportunities and Challenges. *Remote Sens.* **2019**, *11*, 230. [CrossRef]
39. Lu, D. The potential and challenge of remote sensing-based biomass estimation. *Int. J. Remote Sens.* **2006**, *27*, 1297–1328. [CrossRef]
40. Lim, K.; Treitz, P.; Wulder, M.; St-Onge, B.; Flood, M. LiDAR remote sensing of forest structure. *Prog. Phys. Geogr. Earth Environ.* **2003**, *27*, 88–106. [CrossRef]
41. Zimble, D.A.; Evans, D.L.; Carlson, G.C.; Parker, R.C.; Grado, S.C.; Gerard, P.D. Characterizing vertical forest structure using small-footprint airborne LiDAR. *Remote Sens. Environ.* **2003**, *87*, 171–182. [CrossRef]
42. Babcock, C.; Finley, A.O.; Bradford, J.B.; Kolka, R.; Birdsey, R.; Ryan, M.G. LiDAR based prediction of forest biomass using hierarchical models with spatially varying coefficients. *Remote Sens. Environ.* **2015**, *169*, 113–127. [CrossRef]
43. Dubayah, R.O.; Drake, J.B. Lidar Remote Sensing for Forestry. *J. For.* **2000**, *98*, 44–46. [CrossRef]
44. Clark, M.L.; Roberts, D.A.; Ewel, J.J.; Clark, D.B. Estimation of tropical rain forest aboveground biomass with small-footprint lidar and hyperspectral sensors. *Remote Sens. Environ.* **2011**, *115*, 2931–2942. [CrossRef]
45. Næsset, E.; Gobakken, T.; Solberg, S.; Gregoire, T.G.; Nelson, R.; Ståhl, G.; Weydahl, D. Model-assisted regional forest biomass estimation using LiDAR and InSAR as auxiliary data: A case study from a boreal forest area. *Remote Sens. Environ.* **2011**, *115*, 3599–3614. [CrossRef]
46. Hu, T.; Su, Y.; Xue, B.; Liu, J.; Zhao, X.; Fang, J.; Guo, Q. Mapping Global Forest Aboveground Biomass with Spaceborne LiDAR, Optical Imagery, and Forest Inventory Data. *Remote Sens.* **2016**, *8*, 565. [CrossRef]
47. Su, Y.; Guo, Q.; Xue, B.; Hu, T.; Alvarez, O.; Tao, S.; Fang, J. Spatial distribution of forest aboveground biomass in China: Estimation through combination of spaceborne lidar, optical imagery, and forest inventory data. *Remote Sens. Environ.* **2016**, *173*, 187–199. [CrossRef]
48. Lefsky, M.A. A global forest canopy height map from the Moderate Resolution Imaging Spectroradiometer and the Geoscience Laser Altimeter System. *Geophys. Res. Lett.* **2010**, *37*. [CrossRef]
49. Simard, M.; Pinto, N.; Fisher, J.B.; Baccini, A. Mapping forest canopy height globally with spaceborne lidar. *J. Geophys. Res. Biogeosci.* **2011**, *116*. [CrossRef]

50. Baccini, A.; Goetz, S.J.; Walker, W.S.; Laporte, N.T.; Sun, M.; Sulla-Menashe, D.; Hackler, J.; Beck, P.S.A.; Dubayah, R.; Friedl, M.A.; et al. Estimated carbon dioxide emissions from tropical deforestation improved by carbon-density maps. *Nat. Clim. Chang.* **2012**, *2*, 182. [CrossRef]

51. Boudreau, J.; Nelson, R.F.; Margolis, H.A.; Beaudoin, A.; Guindon, L.; Kimes, D.S. Regional aboveground forest biomass using airborne and spaceborne LiDAR in Québec. *Remote Sens. Environ.* **2008**, *112*, 3876–3890. [CrossRef]

52. Tang, W.; Zheng, M.; Zhao, X.; Shi, J.; Yang, J.; Trettin, C.C. Big Geospatial Data Analytics for Global Mangrove Biomass and Carbon Estimation. *Sustainability* **2018**, *10*, 472. [CrossRef]

53. Simard, M.; Fatoyinbo, L.; Smetanka, C.; Rivera-Monroy, V.H.; Castañeda-Moya, E.; Thomas, N.; Stocken, T.V.D. Mangrove canopy height globally related to precipitation, temperature and cyclone frequency. *Nat. Geosci.* **2019**, *12*, 40–45. [CrossRef]

54. Hijmans, R.J.; Cameron, S.E.; Parra, J.L.; Jones, P.G.; Jarvis, A. Very high resolution interpolated climate surfaces for global land areas. *Int. J. Climatol.* **2005**, *25*, 1965–1978. [CrossRef]

55. Zhao, X.; Su, Y.; Hu, T.; Chen, L.; Gao, S.; Wang, R.; Jin, S.; Guo, Q. A global corrected SRTM DEM product for vegetated areas. *Remote Sens. Lett.* **2018**, *9*, 393–402. [CrossRef]

56. Huete, A.; Justice, C.; Van Leeuwen, W. MODIS vegetation index (MOD13). *Algorithm Theor. Basis Doc.* **1999**, *3*, 213.

57. Murdiyarso, D.; Purbopuspito, J.; Kauffman, J.B.; Warren, M.W.; Sasmito, S.D.; Manuri, S.; Krisnawati, H.; Taberima, S.; Kurnianto, S. *SWAMP Dataset-Mangrove Biomass Vegetation-Teminabuan-2011*, V1 ed.; Center for International Forestry Research (CIFOR): Bogor, Indonesia, 2019. [CrossRef]

58. Sasmito, S.D.; Silanpää, M.; Hayes, M.A.; Bachri, S.; Saragi-Sasmito, M.F.; Sidik, F.; Hanggara, B.; Mofu, W.Y.; Rumbiak, V.I.; Hendri; et al. *SWAMP Dataset-Mangrove Necromass-West Papua-2019*, DRAFT VERSION ed.; Center for International Forestry Research (CIFOR): Bogor, Indonesia, 2019. [CrossRef]

59. Zwally, H.J.; Schutz, B.; Abdalati, W.; Abshire, J.; Bentley, C.; Brenner, A.; Bufton, J.; Dezio, J.; Hancock, D.; Harding, D.; et al. ICESat's laser measurements of polar ice, atmosphere, ocean, and land. *J. Geodyn.* **2002**, *34*, 405–445. [CrossRef]

60. Lefsky, M.A.; Keller, M.; Pang, Y.; De Camargo, P.B.; Hunter, M.O. Revised method for forest canopy height estimation from Geoscience Laser Altimeter System waveforms. *J. Appl. Remote Sens.* **2007**, *1*, 013537. [CrossRef]

61. Scholes, R.J.; Kendall, J.; Justice, C.O. The quantity of biomass burned in southern Africa. *J. Geophys. Res. Atmos.* **1996**, *101*, 23667–23676. [CrossRef]

62. Li, L.; Guo, Q.; Tao, S.; Kelly, M.; Xu, G. Lidar with multi-temporal MODIS provide a means to upscale predictions of forest biomass. *ISPRS J. Photogramm. Remote Sens.* **2015**, *102*, 198–208. [CrossRef]

63. Liaw, A.; Wiener, M. Classification and regression by randomForest. *R News* **2002**, *2*, 18–22.

64. Fassnacht, F.E.; Hartig, F.; Latifi, H.; Berger, C.; Hernández, J.; Corvalán, P.; Koch, B. Importance of sample size, data type and prediction method for remote sensing-based estimations of aboveground forest biomass. *Remote Sens. Environ.* **2014**, *154*, 102–114. [CrossRef]

65. Fromard, F.; Puig, H.; Mougin, E.; Marty, G.; Betoulle, J.L.; Cadamuro, L. Structure, above-ground biomass and dynamics of mangrove ecosystems: New data from French Guiana. *Oecologia* **1998**, *115*, 39–53. [CrossRef]

66. Smith, T.J.; Whelan, K.R.T. Development of allometric relations for three mangrove species in South Florida for use in the Greater Everglades Ecosystem restoration. *Wetl. Ecol. Manag.* **2006**, *14*, 409–419. [CrossRef]

67. World Resources Institute; International Institute for Environment and Development. World Resources 1986. In *An Assessment of the Resources Base That Supports the Global Economy*; Basic Books: New York, NY, USA, 1986.

68. Adame, M.F.; Kauffman, J.B.; Medina, I.; Gamboa, J.N.; Torres, O.; Caamal, J.P.; Reza, M.; Herrera-Silveira, J.A. Carbon Stocks of Tropical Coastal Wetlands within the Karstic Landscape of the Mexican Caribbean. *PLoS ONE* **2013**, *8*, e56569. [CrossRef] [PubMed]

69. Guo, Q.; Su, Y.; Hu, T.; Zhao, X.; Wu, F.; Li, Y.; Liu, J.; Chen, L.; Xu, G.; Lin, G.; et al. An integrated UAV-borne lidar system for 3D habitat mapping in three forest ecosystems across China. *Int. J. Remote Sens.* **2017**, *38*, 2954–2972. [CrossRef]

70. Zhu, X.; Hou, Y.; Weng, Q.; Chen, L. Integrating UAV optical imagery and LiDAR data for assessing the spatial relationship between mangrove and inundation across a subtropical estuarine wetland. *ISPRS J. Photogramm. Remote Sens.* **2019**, *149*, 146–156. [CrossRef]

71. Coyle, D.B.; Stysley, P.R.; Poulios, D.; Clarke, G.B.; Kay, R.B. Laser Transmitter Development for NASA's Global Ecosystem Dynamics Investigation (GEDI) Lidar. In Proceedings of the SPIE—The International Society for Optics and Photonics, San Diego, CA, USA, 9–13 August 2015; Volume 9612, p. 7.
72. Naidoo, G. Effects of salinity and nitrogen on growth and water relations in the mangrove, Avicennia marina (Forsk.) Vierh. *New Phytol.* **1987**, *107*, 317–325. [CrossRef]
73. Wösten, J.H.M.; de Willigen, P.; Tri, N.H.; Lien, T.V.; Smith, S.V. Nutrient dynamics in mangrove areas of the Red River Estuary in Vietnam. *Estuar. Coast. Shelf Sci.* **2003**, *57*, 65–72. [CrossRef]

MDPI

St. Alban-Anlage 66

4052 Basel

Switzerland

Tel. +41 61 683 77 34

Fax +41 61 302 89 18

www.mdpi.com

Remote Sensing Editorial Office

E-mail: remotesensing@mdpi.com

www.mdpi.com/journal/remotesensing